FRONTIERS OF FUNDAMENTAL AND COMPUTATIONAL PHYSICS

To learn more about AIP Conference Proceedings, including the Conference Proceedings Series, please visit the webpage
http://proceedings.aip.org/proceedings

FRONTIERS OF FUNDAMENTAL AND COMPUTATIONAL PHYSICS

9th International Symposium

Udine and Trieste, Italy   7 – 9 January 2008

EDITORS

B. G. Sidharth

International Institute for Applicable Mathematics and Information Sciences, Adarsh Nagar, Hyderabad, India

F. Honsell

O. Mansutti

University of Udine, Italy

K. Sreenivasan

The Abdus Salam International Centre for Theoretical Physics Trieste, Italy

A. De Angelis

University of Udine, Italy and Instituto Superior Técnico, Lisbon, Portugal

SPONSORING ORGANIZATIONS

University of Udine, Italy

The Abdus Salam International Centre for Theoretical Physics, Trieste, Italy

AMERICAN INSTITUTE OF PHYSICS

Melville, New York, 2008

AIP CONFERENCE PROCEEDINGS ■ VOLUME 1018

Editors:

B. G. Sidharth
F. Honsell

Department of Mathematics
University of Udine
Via delle Scienze 208
I-33100 Udine
Italy

O. Mansutti
A. De Angelis

Department of Physics
University of Udine
Via delle Scienze 208
I-33100 Udine
Italy

K. Sreenivasan
ICTP Trieste
Strada Costiera 11
I-340104 Trieste
Italy

L.C. Catalog Card No. 2008927782
ISBN 978-0-7354-0539-4
ISSN 0094-243X
Printed in the United States of America

CONTENTS

PART IV - EXPERIMENTS AND PHENOMENOLOGY AT THE FRONTIER OF FUNDAMENTAL PHYSICS

PART V - COMPUTATIONAL PHYSICS, COMPLEX SYSTEMS AND THE STRUCTURE OF MATTER

PART VI - PHYSICS EDUCATION

Preface

For over a decade the International Symposium Series "Frontiers of Fundamental Physics" has attracted some of the greatest Physicists in the world. The broad objective of the series could enable scholars working in different areas to meet on a single platform and exchange ideas and status reports and even dissenting views.

The areas covered have included Astronomy and Astrophysics, Particle Physics, Theoretical Physics, Gravitation and Cosmology, Computational Physics and related subjects. The Symposia have been held in India, Italy, Spain and Canada.

The eminent Physicists who have delivered Special Lectures over the years, sometimes more than once, have included Nobel Laureates Professors G.'t Hooft, S. Chu, C. Townes, K. von Klitzing, P. Gilles de Gennes, D.D. Osheroff, H. Kroto, A. Leggett, and several other prominent scholars as Y. Ne'eman.

The Ninth Symposium in this series was held in Udine and Trieste. Special Lectures were delivered by Sir Harold Kroto, Prof. Douglas D. Osheroff and Sir Anthony Leggett. These apart there have been invited talks, contributed papers and posters. Increasingly over the years, sessions involving students and Physics teaching have also been included as in this case. As in the past physicists from India and other countries including the Middle East, Europe, Russia, the US, South America and elsewhere have contributed to the success of the Symposium.

A special thank is due to Dr. Michela De Maria for her work as scientific secretary, and to Dr. Fausto Freschi for his cooperation. The team at the University of Udine and at ICTP helped in making this event possible: Dr. Giuseppe Cabras working to the web site, and Ms. Silvia Zuccaro, Ms. Marianna Kozikowski, Ms. Antonella Di Marzio, Ms. Sonia Di Giorgio, Ms. Daniela Giombi and Ms. Maria Grazia Cecutto working at the secretariat. The computing support is due to the CSIT of Udine and in particular to Dr. Claudio Castellano, Dr. Ester Orlandi, Dr. Nicola Susan. Thanks are also due to Ms. Maya Flikop, Ms. Kristen Girardi and the American Institute of Physics for making the Proceedings of the Ninth Frontiers of Fundamental Physics International Symposium a reality.

PART I

SPECIAL LECTURES

Is Quantum Mechanics the Whole Truth?*

Sir Anthony J. Leggett, 2003 Nobel Laureate in Physics

University of Illinois at Urbana-Champaign

Abstract. Quantum mechanics has been enormously successful in describing nature at the atomic level and most physicists believe it is, in principle, the "whole truth" about the world even at the everyday level. However, such a view, at first glance, leads to a severe problem. In certain circumstances, the most natural interpretation of the theory implies that no definite outcome of an experiment occurs until the act of observation. For many decades this problem was regarded as merely philosophical – it was thought it had no consequences that could be tested in experiment. However, in the last dozen years or so, the situation has changed dramatically in this respect. The problem, some popular resolutions of it, the current experimental situation and prospects for the future are discussed.

Superfluid Helium Three: The First Unconventional BCS States

Douglas D. Osheroff, 1996 Nobel Laureate in Physics

Stanford University

Abstract. The ordered phases of liquid 3He represent the first known unconventional BCS states, in which the Cooper pairs contain internal structure, most notably intrinsic spin and orbital angular momentum. They are also the only observable BCS states in which the fermions order in the absence of an underlying lattice structure, although it is believed that similar order also occurs in neutron stars. Because of this, the 3He superfluids are free to exhibit the symmetry that truly minimizes their free energy, resulting in liquid crystal-like textures over macroscopic distances. The 'spontaneously broken spin-orbit symmetry' of the Cooper pairs leads to a remarkably complex set of spin dynamical properties, which allow one to probe the nature of these condensates with remarkable ease. These ordered states and some features of the superfluid 3He behaviour are discussed.

Architecture in Nanospace*

Sir Harold W. Kroto, 1996 Nobel Laureate in Chemistry

Sussex University

Abstract. Molecules have structures that are every bit as real in the mind of the chemists who create them, as are the edifices of brick, steel and concrete designed by architects and built by engineers.

**The video of the presentation is available at the homepage of the conference ffp9.fisica.uniud.it*

CP1018, *Frontiers of Fundamental and Computational Physics*,
edited by B. G. Sidharth, F. Honsell, O. Mansutti, K. Sreenivasan, and A. De Angelis

PART II

FOUNDATIONS OF PHYSICS AND QUANTUM MECHANICS

The Thermodynamic Universe

B.G. Sidharth

International Institute for Applicable Mathematics & Information Sciences, Hyderabad (India) & Udine (Italy), B.M. Birla Science Centre, Adarsh Nagar, Hyderabad - 500 063 (India)

Abstract. Using Planck scale oscillators in the background dark energy in a model that parallels the theory of phonons, we deduce the Planck mass, the elementary particle mass scale, the mass of the Universe and a recently discovered residual energy in the cosmic background. We also deduce the Beckenstein temperature formula for black holes. Finally we show that the model explains the four minute time lag in the arrival of gamma photons from a recently observed gamma flare by the MAGIC telescope.

Keywords: Thermodynamic, Photon Mass, Gamma Rays.
PACS: 03.65.-w

INTRODUCTION

The view we are trying to put forward is that the Universe is in some sense immersed in a bath of "Dark Energy" and is "Thermodynamic" in the sense that for example the temperature of some material has a certain value.
To get a flavour, let us first derive the recently discovered [1] residual cosmic energy directly from the background Dark Energy. We may reiterate that the "mysterious" background Dark Energy is the same as the quantum Zero Point Fluctuations in the background vacuum electromagnetic field. Let us recall that the background Zero Point Field is a collection of ground state oscillators [2]. The probability amplitude is

$$\psi(x) = \left(\frac{m\omega}{\pi\hbar}\right)^{1/4} e^{-(m\omega/2\hbar)x^2}$$

for displacement by the distance x from its position of classical equilibrium. So the oscillator fluctuates over an interval

$$\Delta x \sim (\hbar/m\omega)^{1/2}$$

The background electromagnetic field is an infinite collection of independent oscillators, with amplitudes X_1, X_2 etc. The probability for the various oscillators to have amplitudes X_1, X_2 and so on is the product of individual oscillator amplitudes:

$$\psi(X_1, X_2, \cdots) = exp[-(X_1^2 + X_2^2 + \cdots)]$$

wherein there would be a suitable normalization factor. This expression gives the probability amplitude ψ for a configuration $B(x,y,z)$ of the magnetic field that is described by the Fourier coefficients $X_1, X_2, \cdots$ or directly in terms of the magnetic field configuration

CP1018, *Frontiers of Fundamental and Computational Physics*,
edited by B. G. Sidharth, F. Honsell, O. Mansutti, K. Sreenivasan, and A. De Angelis

itself by,

$$\psi(B(x,y,z)) = Pexp\left(-\int\int\frac{\mathbf{B}(\mathbf{x_1})\cdot\mathbf{B}(\mathbf{x_2})}{16\pi^3\hbar c r_{12}^2}d^3x_1d^3x_2\right).$$

P being a normalization factor. At this stage, we are thinking in terms of energy without differentiation, that is, without considering Electromagnetism or Gravitation etc as separate. Let us consider a configuration where the magnetic field is everywhere zero except in a region of dimension l, where it is of the order of $\sim \Delta B$. The probability amplitude for this configuration would be proportional to

$$\exp[-((\Delta B)^2 l^4/\hbar c)]$$

So the energy of fluctuation in a region of length l is given by finally, the density [2, 3]

$$B^2 \sim \frac{\hbar c}{l^4}$$

So the energy content in a region of volume l^3 is given by

$$\beta^2 \sim \hbar c/l \tag{1}$$

This energy is minimum when l is maximum. Let us take l to be the radius of the Universe $\sim 10^{28} cms$. The minimum energy residue of the background Dark Energy or Zero Point Field (ZPF) now comes out to be $10^{-33} eV$, exactly the observed value. This observed residual energy is a cosmic footprint of the ubiquitous Dark Energy in the Universe, a puzzling footprint that, has recently been observed [1].
If on the other hand we take for l in (1) the smallest possible length, the Planck length l_P, as we will see in the sequel, then we get the Planck mass $m_P \sim 10^{-5} gm$.
The minimum mass $\sim 10^{-33} eV$ or $10^{-65} gms$, has been shown by the author to be the mass of the photon [4].Interestingly, this also is the minimum thermodynamic mass in the Universe, as shown by Landsberg from a totally different point of view, that of thermodynamics. So (1) gives two extreme masses, the Planck mass and the photon mass. We will see how it is possible to recover the intermediate elementary particle mass from these considerations later.
As an alternative derivation, it is interesting to derive a model based on the theory of phonons which are quanta of sound waves in a macroscopic body [5]. In this theory, as is well known the phonons have a maximum frequency ω_m which is given by

$$\omega_m = c\left(\frac{6\pi^2}{v}\right)^{1/3} \tag{2}$$

In (2) c represents the velocity of sound in the specific case of photons, while $v = V/N$, where V denotes the volume and N the number of atoms. In this model we write

$$l \equiv \left(\frac{4}{3}\pi v\right)^{1/3}$$

l being the inter particle distance. Thus (2) now becomes

$$\omega_m = c/l \tag{3}$$

Let us now look upon it as a general set of Harmonic oscillators as above. Then we can see that (3) and (1) are identical. So we again recover with suitable limits the extremes of the Planck mass and the photon mass. (Other intermediate elementary particle masses follow if we take l as a typical Compton wavelength).
We now examine separately, the Planck scale.
Max Planck had noticed that, what we call the Planck scale today,

$$l_P = \left(\frac{\hbar G}{c^3}\right)^{\frac{1}{2}} \sim 10^{-33} cm \tag{4}$$

is made up of the fundamental constants of nature and so, he suspected it played the role of a fundamental length. Indeed, modern Quantum Gravity approaches have invoked (4) in their quest for a reconciliation of gravitation with other fundamental interactions. In the process, the time honoured prescription of a differentiable spacetime has to be abandoned.
There is also another scale made up of fundamental constants of nature, viz., the well known Compton scale (or classical electron radius),

$$l = e^2/m_e c^2 \sim 10^{-12} cm \tag{5}$$

where e is the electron charge and m_e the electron mass. Indeed if (5) is substituted for l in (1), we get the elementary particle mass scale.
The scale (5) has also played an important role in modern physics, though it is not considered as fundamental as the Planck scale. Nevertheless, the Compton scale (5) is close to reality in the sense of experiment, unlike (4), which is well beyond foreseeable direct experimental contact. Moreover another interesting feature of the Compton scale is that, it brings out the Quantum Mechanical spin, unlike the Planck scale.
A very important question this throws up is that of a physical rationale for a route from (4) to (5). Is there such a mechanism?

THE PLANCK AND COMPTON SCALES

Our starting point [6, 3] is the model for the underpinning at the Planck scale for the Universe. This is a collection of N Planck scale oscillators where we will specify N shortly.
Let us consider an array of N particles, spaced a distance Δx apart, which behave like oscillators that are connected by springs. We then have [7, 9, 3]

$$r = \sqrt{N\Delta x^2} \tag{6}$$

$$ka^2 \equiv k\Delta x^2 = \frac{1}{2} k_B T \tag{7}$$

where k_B is the Boltzmann constant, T the temperature, r the extent and k is the spring constant given by

$$\omega_0^2 = \frac{k}{m} \tag{8}$$

$$\omega = \left(\frac{k}{m}a^2\right)^{\frac{1}{2}} \frac{1}{r} = \omega_0 \frac{a}{r} \tag{9}$$

We now identify the particles with Planck masses and set $\Delta x \equiv a = l_P$, the Planck length. It may be immediately observed that use of (8) and (7) gives $k_B T \sim m_P c^2$, which ofcourse agrees with the temperature of a black hole of Planck mass. Indeed, Rosen [10] had shown that a Planck mass particle at the Planck scale can be considered to be a Universe in itself with a Schwarzchild radius equalling the Planck length. We also use the fact that a typical elementary particle like the pion can be considered to be the result of $n \sim 10^{40}$ Planck masses.[3]

Using this in (6), we get $r \sim l$, the pion Compton wavelength as required. Whence the pion mass is given by

$$m = m_P / \sqrt{n}$$

Further, in this latter case, using (6) and the fact that $N = n \sim 10^{40}$, and (7),i.e. $k_B T = kl^2/N$ and (8) and (9), we get for a pion, remembering that $m_P^2/n = m^2$,

$$k_B T = \frac{m^3 c^4 l^2}{\hbar^2} = mc^2,$$

which of course is the well known formula for the Hagedorn temperature for elementary particles like pions .This confirms that we can treat an elementary particle as a series of some 10^{40} Planck mass oscillators.

However it must be observed from (9) and (8), that while the Planck mass gives the highest energy state, an elementary particle like the pion is in the lowest energy state. This explains why we encounter elementary particles, rather than Planck mass particles in nature. Infact as already noted [12], a Planck mass particle decays via the Bekenstein radiation within a Planck time $\sim 10^{-42} secs$.

Using the fact that the Universe consists of $N \sim 10^{80}$ elementary particles like the pions, the question is, can we think of the Universe as a collection of nN or 10^{120} Planck mass oscillators? In fact if we use equation (6) with

$$\bar{N} \sim 10^{120},$$

we can see that the extent is given by $r \sim 10^{28} cms$ which is of the order of the diameter of the Universe itself. We shall shortly justify the value for $\bar{N}$. Next using (9) we get

$$\hbar\omega_0^{(min)} \langle \frac{l_P}{10^{28}} \rangle^{-1} \approx m_P c^2 \times 10^{60} \approx Mc^2 \tag{10}$$

which gives the correct mass M, of the Universe which in contrast to the earlier pion case, is the highest energy state while the Planck oscillators individually are this time the lowest in this description. In other words the Universe itself can be considered to be

described in terms of normal modes of Planck scale oscillators.
More generally, if an arbitrary mass M, as in (10), is given in terms of $\bar{N}$ Planck oscillators, in the above model, then we have from (10) and (6):

$$M = \sqrt{\bar{N}} m_P \text{and} R = \sqrt{\bar{N}} l_P,$$

where R is the radius of the object. Using the fact that l_P is the Schwarzchild radius of the mass m_P, this gives immediately,

$$R = 2GM/c^2$$

a well known relation. In other words, such an object, the Universe included as a special case, shows up as a Black Hole.
In fact, we do not need to specify N. We have in this case rewriting the relations (6) to (9),

$$R = \sqrt{N} l, Kl^2 = kT,$$

$$\omega_{max}^2 = \frac{K}{m} = \frac{kT}{ml^2} \tag{11}$$

In (11), R is of the order of the diameter of the Universe, K is the analogue of spring constant, T is the effective temperature while l is the analogue of the Planck length, m the analogue of the Planck mass and ω_{max} is the frequency—–the reason for the subscript *max* will be seen below. We do not yet give l and m their usual values as given in (4) for example, but rather try to deduce these values.
So using the Beckenstein temperature formula for these primordial black holes [14], that is

$$kT = \frac{\hbar c^3}{8\pi Gm}$$

in (11) we get,

$$Gm^2 \sim \hbar c \tag{12}$$

which is another form of (4). We can easily verify that (12) leads to the value $m \sim 10^{-5} gms$. In deducing (12) we have used the typical expressions for the frequency as the inverse of the time - the Compton time in this case and similarly the expression for the Compton length. However it must be reiterated that no specific values for l or m were considered in the deduction of (12).
Finally we can see from (11) that, given the value of l_P and using the value of the radius of the Universe, viz., $R \sim 10^{27}$, we can deduce that,

$$N \sim 10^{120} \tag{13}$$

The Compton scale (5) comes as a Quantum Mechanical effect, within which we have zitterbewegung effects and a breakdown of causal Physics as emphasized elsewhere [3].
We will now need a relation [20]

$$G = \Theta/t \tag{14}$$

This dependence also features in the Dirac Cosmology and a few other cosmologies. We now observe the following: It is known that for a Planck mass $m_P \sim 10^{-5} gm$, all the energy is gravitational and in fact we have, as in (12),

$$Gm_P^2 \sim e^2$$

To push these considerations further, we use the following relations for a Schwarzchild black hole [14]:

$$dM = TdS, S = \frac{kc}{4\hbar G} A,$$

where T is the Beckenstein temperature, S the entropy and A is the area of the black hole. In our case, the mass $M = \sqrt{N} m_P$ and $A = Nl_P^2$, where N is arbitrary for an arbitrary black hole. Whence,

$$T = \frac{dM}{dS} = \frac{4\hbar G}{kl_P^2 c} \frac{dM}{dN}$$

If we use the fact that l_P is the Schwarzchild radius for the Planck mass m_P and use the expression for M, the above reduces to the Beckenstein temperature formula.
Further, in this theory as is known [14],

$$\frac{dm}{dt} = -\frac{\beta}{m^2}, \qquad (15)$$

where β is given by

$$\beta = \frac{\hbar c^4}{(30.8)^3 \pi G^2}$$

This leads back to the usual black hole life time given by, for any mass m,

$$t = \frac{1}{3\beta} m^3 = 8.4 \times 10^{-24} m^3 secs \qquad (16)$$

Let us now factor in the time variation (14) of G into (15). Equation (15) now becomes [12]

$$m^2 dm = -B\mu^{-2} t^2 dt, B \equiv \frac{\hbar c^4}{\lambda^3 \pi}, \mu \equiv \frac{lc^2 \tau}{m}, \lambda^3 = (30.8)^3 \pi$$

Whence on integration we get

$$m = \frac{\hbar}{\lambda \pi^{1/3}} \left\{ \frac{1}{l^6} \right\}^{1/3} t = \frac{\hbar}{\lambda \pi^{1/3}} \frac{1}{l^2} t \qquad (17)$$

If we use the pion Compton time for t, in (17), we get for m, the pion mass. In other words, due to (14), the evanescent Planck mass decays into a stable elementary particle. Before proceeding, we observe that we have deduced that a string of N Planck oscillators, N arbitrary, form a Schwarzchild black hole of mass $\sqrt{N} m_P = M$. Using the analogue of (11) and considerations before (15), we can deduce that

$$\frac{dM}{dt} = m_P / t_P,$$

$$M = (m_P/t_P) \cdot t,$$

Whence t being the "Beckenstein decay time" this is like the equation (17). For the Planck mass, $M = m_P$, the decay time $t = t_P$. For the Universe, the above gives the life time t as $\sim 10^{17} sec$, the age of the Universe! Interestingly, for such black holes to have realistic life times $\geq 1 sec$, we deduce that the mass $\geq 10^{38} gm$.

REMARKS

Recently the MAGIC gamma ray telescope in Spain detected a big gamma ray flare in the galaxy Mkn 501. It was observed that there was a four minute lag between the arrival times of gamma photons with energy greater than 10 Tev as compared to 100 Gev photons.This was sought to be explained by Ellis et al as a Quantum Gravity effect in which there is a dispersion in the velocities of the photons with respect to frequency [38]. They obtained

$$c' = c(1 - \frac{E_0}{E'})$$

where c' is the modified velocity of the photons and E' is their estimate for the Planck energy, which turned out to be about one percent of the actual value. Our above approach predicts a similar effect but this is due to the modified energy momentum dispersion relation deduced from theory [3],

$$E^2 = E_0^2(1 - \frac{E_0^2}{E''^2})$$

This gives

$$c'^2 = c^2(1 - \frac{E_0^2}{E''^2})$$

where E'' is the Planck energy. On the other hand from the the equation above of Ellis et al. we get

$$c'^2 = c^2(1 - \frac{E_0^2}{E''^2}.10^4)$$

A comparison shows that our theory leads to the correct Planck energy unlike the Ellis et al value.

REFERENCES

1. L. Mersini-Houghton, *Mod.Phys.Lett.A.* **Vol.21**, No.1, pp.1–21 (2000).
2. C.W. Misner, K.S. Thorne and J.A. Wheeler, *Gravitation*, W.H. Freeman, San Francisco, pp.819ff (1973).
3. B.G. Sidharth *The Universe of Fluctuations*, Springer, Netherlands (2005).
4. B.G. Sidharth, *The Thermodynamic Universe*, World Scientific, Singapore (2008).
5. K. Huang, *Statistical Mechanics*, Wiley Eastern, New Delhi, pp.75ff (1975).
6. B.G. Sidharth, *Found.Phys.Lett.*, 17, (5), pp.503–506 (2004).

7. B.G. Sidharth, *Found.Phys.Lett.*, 15, (6), pp.577–583 (2002).
8. D.L. Goodstein, *States of Matter*, Dover Publications, Inc., New York, pp.462ff. (1975).
9. Y. Jack Ng., and H. Van Dam, *Mod.Phys.Lett.A.*, 9, (4), pp.335–340 (1994).
10. N. Rosen, *Int.J.Th.Phys.*, 32, (8), pp.1435–1440 (1993).
11. C. Sivaram, *Am.J.Phys.*, 51 (3), pp.277 (1983).
12. B.G. Sidharth, *Found.Phys.Lett.*, **19**, 1, pp.87ff (2006).
13. B.G. Sidharth, *A Century of Ideas*, Springer, Dordrecht (2008).
14. R. Ruffini and L.Z. Zang, *Basic Concepts in Relativistic Astrophysics*, World Scientific, Singapore, p.111ff (1983).
15. C. Cercignani, *Found.Phys.Lett.*, Vol.11, No.2, pp.189–199 (1998).
16. C. Cercignani, L. Galgani and A. Scotti, *Phys.Lett.*, 38A, pp.403 (1972).
17. F. Reif, *Fundamentals of Statistical and Thermal Physics*, McGraw-Hill Book Co., Singapore (1965).
18. P.A.M. Dirac, *The Principles of Quantum Mechanics*, Clarendon Press, Oxford, pp.4ff, pp.253ff (1958).
19. S. Weinberg, *Gravitation and Cosmology*, John Wiley & Sons, New York, p.61ff (1972).
20. B.G. Sidharth, *Int.J.Mod.Phys.A.*, 13, (15), pp.2599ff (1998).
21. B.G. Sidharth, *Int.J.Th.Phys.*, Vol.37, No.4, pp.1307–1312 (1998).
22. J.V. Narlikar, *Introduction to Cosmology*, Cambridge University Press, Cambridge, p.57 (1993).
23. B.G. Sidharth, *Found.Phys.Lett.*, 18, (4), pp.387–391 (2005).
24. A.D. Sakharov, *Soviet Physics-Doklady*, Vol.12, No.11, pp.1040–1041 (1968).
25. E. Golfain, *Chaos, Solitons and Fractals*, 22, (3), pp.513–520 (2004).
26. L. De Broglie and J.P. Vigier, *Phys.Rev.Lett.*, 28, pp.1001–1004 (1972).
27. L. De Broglie, *La mecanique ondulatoire du photon Une nouvelle theorie dellumiere* Vol.I, Paris, Hermann (1940).
28. L. De Broglie, *Les interactions entre les photons et la matiere*, Vol.II, Paris, Hermann (1942).
29. S. Deser, *Ann Inst. Henri Poincare, Vol.XVI*, Paris, Gauthier-Villors, pp.79 (1972).
30. R. Lakes, *Phys.Rev.Lett.*, 80, (9), pp.1826ff (1998).
31. C. Itzykson, and J. Zuber, *Quantum-Field Theory*, Mc-Graw Hill, New York, pp.139 (1980).
32. T.W. Marshall, *New Developments on Fundamental Problems in Quantum Physics* edited by M. Ferrero, and A. Van der Merwe, A, Kluwer, Dordrecht, pp.231ff (1997).
33. T.W. Marshall, and E. Santos, in *The Present Status of the Quantum Theory of Light*, edited by S. Jeffers, et al., Kluwer Academic, Dordrecht, pp.67–77 (1997).
34. H.L. Armstrong, *Am.J.Phys.*, 51, (2), February 1983, pp.103 (1983).
35. M. Sachs, in *The Present Status of the Quantum Theory of Light*, edited by S. Jeffers, et al., Kluwer Academic, Dordrecht, pp.79–96 (1997).
36. R. Balescu, *Equilibrium and Non Equilibrium Statistical Mechanics*, John Wiley, New York, (1975).
37. B.G. Sidharth, *Chaotic Universe: From the Planck to the Hubble Scale*, Nova Science, New York, (2001).
38. arxiv.0708.2889

Average Exceptional Lie Group Hierarchy and High Energy Physics

M.S. El Naschie

King Abdullah Al Saud Institute of Nano & Advanced Technology, KSU, Riyadh, Saudi Arabia

Abstract. Starting from an invariant total dimension for an exceptional Lie symmetry groups hierarchy, we drive all the essential characteristic and coupling constants of the fundamental interactions of physics. It is shown in a most simplistic fashion that all physical fields are various transfinite scaling transformation and topological deformation of each other. An extended standard model on the other hand turned out to be a compact sub group H of a version of $E7$ exceptional Lie group $E7(-5)$ with $\dim H = 69$. Thus particle physics, electromagnetism as well as gravity and the bulk are all representable via modular spaces akin to the famous compactified version of F. Klein's modular curve.

Keywords: $E8$ exceptional Lie symmetry group-super strings-unification – E-infinity – standard model – high energy physics
PACS: 11.15.Tk, 10.10.Dm and 12.10.Kt

INTRODUCTION

In the last fifteen years or so a new spacetime model was developed almost from scratch using two of the simplest although most enigmatic mathematical objects, Cantor's triad set and the golden mean. In the case of a random Cantor set the two notions combine to produce an elementary Cantor set with a Hausdorff dimension equal to the golden mean [1,2]. From these modest initial ideas, E-infinity theory was developed and progressed steadily until clear contact was made with relatively more conventional subjects such as Kähler manifolds, Coxeter polytopes, strings and exceptional Lie symmetry groups [3]. In the last few weeks however unprecedented media attention was given to certain aspects of this theory although under different labels [4]. The present work is intended to clarify the basic idea of E-infinity theory which is that all fundamental fields of physics are derivable from each other by simple transfinite golden mean transformation, as indicated in Figs. 1 and 2.

THE BASIC THEORY

Super imposing all members of an extended exceptional Lie symmetry group family [2-4] an enormously complex structure is obtained. This structure is characterized by an invariant value measuring the complexity of this structure and given by 548.3281572 which turned out to be exactly equal to 4 multiplied with the inverse electromagnetic fine structure constant $\bar{\alpha}_0 = 137 + k_0$ where $k_0 = \phi^3(\phi^3 - 1) = 0.08203932$ and $\phi = (\sqrt{5} - 1)/2$ [2,3]. Realizing that $4\bar{\alpha}_0$ is a Hausdorff-like fractal dimension and may be used similarly to the Boltzmenian entropy as a measure of disorder, it is not very diffi-

CP1018, *Frontiers of Fundamental and Computational Physics*,
edited by B. G. Sidharth, F. Honsell, O. Mansutti, K. Sreenivasan, and A. De Angelis

cult to reason that topological transformation of the total structure may be economically expressed non-constructively via simple transfinite golden field transformation following the methodology of E-infinity spacetime theory [1]. Proceeding in this way and making use of the principle of maximum entropy, all fields of physics disclose themselves indirectly as interconnected transfinitely corrected modular spaces [1-3]. Thus similar to the well known $SL(2,7)$ holographic boundary, electromagnetism is essentially given by $SL(2,5)$ provided we take certain transfinite corrections into account. In the same vane the bulk is given by $PSL(2,9)$ and Einstein's gravity is subsequently fixed using a unified equation for all fundamental forces. Consequently starting from the dimension of the average symmetry of E-infinity $\sum_1^8 |E_i|$ one finds that [3]

$$|SL(2,7)|_c = (4\bar{\alpha})(\phi). \tag{1}$$

This fixes particle physics which means the weak and the strong interaction. Next we tackle electromagnetism by replacing the ϕ exponent by the inverse of the topological dimension, namely $\frac{1}{4}$ and finding that

$$\begin{aligned} |SL(2,5)|_c &= (4\bar{\alpha}_0)\left(\frac{1}{4}\right) \\ &= \bar{\alpha}_0 \\ &= 137.082039325 \end{aligned} \tag{2}$$

or

$$\begin{aligned} |SL(2,5)|_c &= |SL(2,7)|_c\left(\frac{1}{2}\right)\left(\frac{1}{1+\phi^3}\right) \\ &= \bar{\alpha}_0. \end{aligned} \tag{3}$$

Proceeding to the bulk one finds

$$\begin{aligned} |SL(2,10)|_c &= (\bar{\alpha}_0)(3+\phi) \\ &= |SL(2,5)|_c(3+\phi \\ &= |E_8E_8|_c \\ &= 496-k^2 \\ &= 495.9674775. \end{aligned} \tag{4}$$

Finally Einstein's gravity represented by $R^{(4)}$ of the Riemannian tensor is found from the transformation

$$\begin{aligned} R^{(4)} &= |SL(2,5)|_c\left(\frac{1}{2}\right)(\phi^4) \\ &= \frac{1}{2}(\bar{\alpha}_0)(\phi^4) \\ &= 20. \end{aligned} \tag{5}$$

Alternatively the unification equation [3]

$$B = H + G + E \tag{6}$$

gives

$$
\begin{aligned}
G &= B-[H+E] \\
&= |E_8E_8|_c - [|SL(2,7)|_c + |SL(2,5)|_c] \\
&= (496-k^2) - [(336+16k)+(137+k_0)]
\end{aligned} \tag{7}
$$

where $k=\phi^3(1-\phi^3), k_0=\phi^5(1-\phi^5)$ and $\phi=(\sqrt{5}-1)/2$. Inserting one finds $G=20$ as should be.

The preceding extremely simple transfinite scaling transformation shows clearly that all known physical fields are derivable from each other and that by starting from the topological invariant exceptional Lie symmetry group hierarchy we gave these transformations the status of a fundamental theory. In fact the well known dimensional Heterotic string hierarchy [1,2]

$$
\begin{aligned}
\frac{1}{2}(\bar{\alpha}_0)(\phi) &= \bar{\alpha}_g \cong 42 \\
\frac{1}{2}(\bar{\alpha}_0)(\phi)^2 &= \bar{\alpha}_{gs} \;;\; 26 \\
\frac{1}{2}(\bar{\alpha}_0)(\phi)^3 &\;;\; 16 \\
\frac{1}{2}(\bar{\alpha}_0)(\phi)^3 &\;;\; 10 \\
\frac{1}{2}(\bar{\alpha}_0)(\phi)^4 &\;;\; 6 \\
\frac{1}{2}(\bar{\alpha}_0)(\phi)^5 &\;;\; 4
\end{aligned} \tag{8}
$$

together with

$$
\begin{aligned}
\bar{\alpha}_0 &= (\bar{\alpha}_1)(\phi)+\bar{\alpha}_2+\bar{\alpha}_3+\bar{\alpha}_4 \\
&= (60)(\phi)+30+9+1 \\
&= 137+k_0
\end{aligned} \tag{9}
$$

rounds up the picture to a convincing theory. The root of this theory lies deep in history and was started by Coxeter's generalization of the golden mean proportioned Dodecahedra to a 120-cell polytopes and further still to the $E8$ Gosset and finally the exceptional Lie groups hierarchy and E-infinity theory [6].

In the next section we give various pictorial illustrations of the preceding analysis as well as the details of the transfinite extension of the modular Lie groups.

PICTORIAL ILLUCIDATION OF THE TRANSFINITE SCALING AND DETAILS OF THE TRANSFINITE MODULAR COMPUTATION

In Fig. 1 we see nonconstructively the initial and final state of the topological transformation from $E8$ Gosset to Klein's modular space $\Gamma(7)$. The dimension of the E_8E_8 bulk

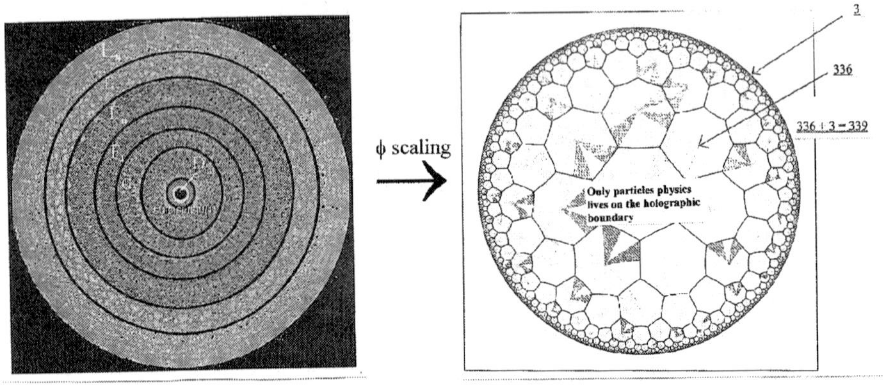

FIGURE 1. Transfinite scaling transformation from exceptional Lie group hierarchy to the holographic boundary represented by the compactified Klein modular curve $\Gamma(7)$.

is given in the crisp case by

$$\begin{aligned} |PSL(2,n=10)|+1 &= n(n^2-1)/2+1 \\ &= (10)(10^2-1)2+1 \\ &= 990/2+1 \\ &= 495+1 = 496. \end{aligned} \tag{10}$$

The corresponding exact transfinite (fuzzy) value is $496-k^2=495.96775;496$ where $k=0.18033989=\phi^3(1-\phi^3)$. The total invariant value of our exceptional Lie hierarchy representing the quasi average or fractal symmetry of our E-infinity spacetime may be found from either summing all the dimensions of the individual Lie groups of by noticing that $|E_8E_8|+|F_4|=496+52=548$. The transfinite value is obviously $(496-k^2)+(52+2k)=548+4k_0$. As for $\Gamma(7)$, we have in the crisp case

$$\begin{aligned} \dim\Gamma(7) &= |SL(2,7)| \\ &= (7)(7^2-1) \\ &= (7)(48) \\ &= 336. \end{aligned} \tag{11}$$

The exact fuzzy value is

$$\begin{aligned} |SL(2,7)|_c &= 336+16k \\ &= 338.88554382 \\ &; 339. \end{aligned} \tag{12}$$

Similarly $\bar{\alpha}_0$ may be found as a fuzzy modular $\Gamma(7)$. Thus starting from

$$\begin{aligned} |SL(2,5)| &= (5)(5^2-1) \\ &= 120. \end{aligned} \tag{13}$$

We make the following transfinite continuation

$$
\begin{aligned}
|SL(2,5+\phi^3)|_c &= (5)[(5+\phi^3)^2-(1+\phi^3)] \\
&= (5)[27.41640786-(1+\phi^3)] \\
&= (5)(26+k) \\
&= 137+k_0 \\
&= \bar{\alpha}_0. \qquad (14)
\end{aligned}
$$

The situation is shown in Fig. 2.

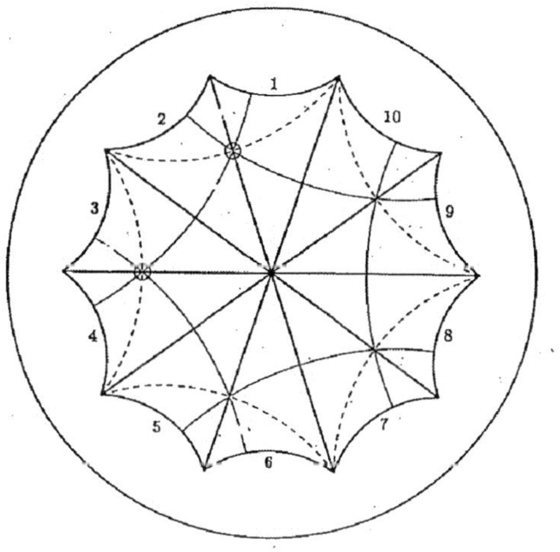

FIGURE 2. Modular curve $\Gamma(5)$ given by $(5)(5^2 \cdot 1) = 120$. On compactification we obtain $N = 120+17+k_0 = \bar{\alpha}_0$. It is a scaling of Figure 1a using $\lambda = \frac{1}{4}$ or Figure 1b using $\lambda = \frac{1+\phi}{4}$.

We may note that while 336 of the crisp $\Gamma(7)$ may be found from $(24)(7) = 168$ by doubling the automorphism $(2)(168) = 366$, we have similarly for $\Gamma(5)$ of electromagnetism $(24)(5) = 120$ and thus 240. For $\Gamma(10)$ one finds $(24)(10) = 240$ and subsequently $(2)(240) = 480$. It is vital at this point to remember that $120, 240$ and 480 are the roots numbers of $E7, E8$ and E_8E_8 exceptional Lie groups and at the same time, equal to the Kissing number in $D = 7$ and $D = 8$. By contrast there is no classical form for $E10$ and the Kissing number in $D = 10$ is 336. Equally noteworthy is that Ji-Huan He's polytopes has 496 elements in $D = 6$ and 17664 elements in $D = 10$ which reduces to the 69 elementary particles of the standard model when divided by 256 spin components [5,6].

Finally we give one of our most important and elegant results regarding the reduction of $E8$ to the standard model by noting that one of the real forms of $E7$, namely $E7(-5)$ has a maximally compact sub group H with dimension $|H| = 69$ equal to that predicted by E-infinity and Ji-Huan He's polytopes, namely 69elementary particles in the standard model. By contrast the 63 of Heterotic strings is found from $E7(+7)$subgroup.

CONCLUSION

All the fields of fundamental interaction are derivable from each other by transfinite scaling corresponding to topological transformation. Electromagnetism as well as the strong and weak fields may be given geometrical interpretation. This is in principle exactly as in Einstein's gravity but using hyperbolic modular spaces.

REFERENCES

1. M.S. El Naschie, *Elementary prerequisites for E-infinity: Recommended background reading in nonlinear dynamics, geometry and topology*. Chaos, Solitons & Fractals, 2006, **30**(3), pp. 579–605.
2. M.S. El Naschie, *High energy physics and the standard model from the exceptional Lie groups*. Chaos, Solitons & Fractals, 2008, **36**, pp. 1–17.
3. M.S. El Naschie, *The exceptional Lie symmetry groups hierarchy and the expected number of Higgs bosons*. Chaos, Solitons & Fractals, 2008, **35**, pp. 268–273.
4. R. Highfield, "Comment on Theory of Everything," Daily Telegraph http://www.Telegraph.co.uk/earth/main.jhtml?xml=/earth/ 2007/11/14/scisurfl114xml. (See also http://www.dailygrail.com/node/5770.)
5. M.S. El Naschie, Ji-Huan He's ten dimensional polytopes and high energy particle physics. *Int. J. Nonlinear Sci. & Numerical Simulation*, 2007, **8**(4), pp. 475–476.
6. M.S. El Naschie, Towards a quantum golden field theory. *Int. J. of Nonlinear Sci. & Numerical Simulation*, 2007, **8**(4), pp. 477–482.

Non-locality on the Time Axis

Toru Ohira

Sony Computer Science Laboratories, Inc., Tokyo, Japan 141-0022

Abstract. We consider the question of introducing the concept of "non-locality" in a time variable, which is opposite to the normal description of the concept in space. Simple delayed and anticipative dynamical models are proposed as representatives of non-locality in time. These models show rather complex behaviors as we increase the level of temporal non-locality. We further demonstrate that these models can induce a "resonance" when combined with spatial or temporal noise. Applications to physical and bio-physiological systems are also discussed.

Keywords: Non-locality, Delay, Advance, Dynamics
PACS: 05.40.-a,02.50.-r,01.55.+b

INTRODUCTION

"Non-locality" is a concept that has appeared in various phases and aspects of scientific contemplations. Earlier questions in physics were how two bodies separated in space, such as stars or electric charges, could influence each other. As we know, this question has led to the development of the concepts of forces, fields, mesons, and so forth. Also, in quantum mechanics, non-locality plays a key role in producing non-classical phenomena. The double slit experiment, quantum tunneling resonance, and EPR paradoxes are some examples. Naturally, non-locality has been considered as a concept in space.

In this paper, we consider the following question: what happens if we consider non-locality on the time axis rather than on the space axis? This is an analogous problem to the one that we discussed recently under the theme of "stochastic time"[1]. Namely, we have tried to map another "spatial" concept of noise or fluctuations onto the time axis. Together, they provide studies of non-locality and stochasticity in time, and might lead to a new point of view on differences and similarities between space and time[2].

As simple examples of non-locality on the time axis, we consider here a dynamical equation that depends on two separated time points. This approach leads to delayed and anticipative dynamical equations. We will show that as we change the parameters to control the level of temporal non-locality, we observe rather complex oscillatory behaviors.

These oscillations arising from non-locality on the time axis can be combined with "noise" to produce a "resonant" behavior similar to those investigated in the topic of stochastic resonance. We will see that both spatial and temporal fluctuations can lead to these stochastic resonances. We will close our paper by mentioning possible applications to physics and bio-physiological controls.

CP1018, *Frontiers of Fundamental and Computational Physics*,
edited by B. G. Sidharth, F. Honsell, O. Mansutti, K. Sreenivasan, and A. De Angelis

DELAYED AND ANTICIPATIVE DYNAMICS

We now discuss delayed and anticipative dynamics more specifically as simple examples of non-locality on the time axis. The general differential equation of the class of delayed dynamics is

$$\frac{dx(t)}{dt} = F(x(t), \bar{x}(\bar{t})). \tag{1}$$

Here, x is the dynamical variable, F is the "dynamical function" governing the dynamics, and $\bar{x}(\bar{t})$ is the state of x at some time point $t \neq \bar{t}$. Thus, the dynamics of x depends not only on its state at the current time t, but also on that at $\bar{t}$, which is separated from t the time axis. The level of temporal non-locality is the time interval between these two points.

If we take

$$\bar{t} = t - \tau, \quad \bar{x}(\bar{t}) = x(t - \tau) \tag{2}$$

we have a delayed dynamical system with "delay" $\tau > 0$.

On the other hand, we can take

$$\bar{t} = t + \eta \tag{3}$$

with "advance" $\eta > 0$. Given a suitable definition of $\bar{x}(t+\eta)$, this leads to "anticipative" or "predictive" dynamics, in which the dynamics of x depends on the state of x in the future. Here, we consider the case in which the current rate of change of x continues for the duration of the advance. This case is termed the "fixed rate prediction" and is given by

$$\bar{x}(\bar{t} = t + \eta) = \eta \frac{dx(t)}{dt} + x(t). \tag{4}$$

We would like to show how the change in the level of non-locality as represented by delay τ and advance η affects the dynamics. The following case is considered.

$$\frac{dx(t)}{dt} = -\alpha x(t) + f(\bar{x}(\bar{t})) \tag{5}$$

As is known from the study of delay differential equations, a temporal non-locality introduces non-linearity in the dynamics. Hence, our studies of the above equation were done with computer simulations of the following corresponding time–discretized equation.

$$x(t+1) = (1-\mu)x(t) + f(\bar{x}(\bar{t})), \tag{6}$$

where μ is a rate constant and $\bar{x}(\bar{t})$ are defined respectively for delayed and anticipative dynamics as

$$\bar{x}(\bar{t}) = x(t-\tau), \quad \bar{x}(\bar{t}) = x(t) + \eta(x(t) - x(t-1)). \tag{7}$$

Also, as the dynamical function, we choose the Mackey-Glass functions, which were first proposed for modeling the cell reproduction process [3].

$$f(y) = \frac{\beta y}{1+y^s}, \tag{8}$$

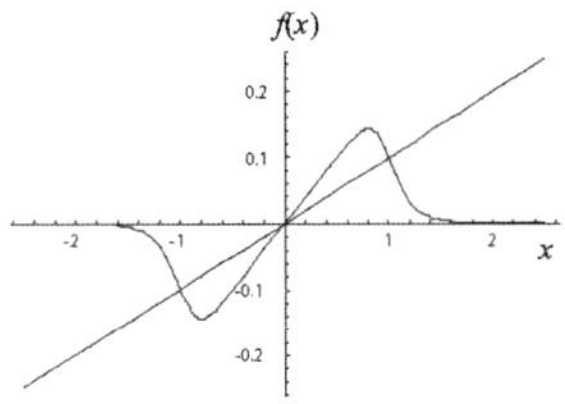

FIGURE 1. Mackey-Glass function $f(x)$ with $\beta = 0.8$ and $n = 10$. The straight line has a slope of $\alpha = 0.1$.

where β and s are parameters. The parameters of the equation are set so that without a delay or advance, $\tau = \eta = 0$, the dynamics of $x(t)$ monotonically approaches the stable fixed point (Figure 1).

Some examples of dynamics obtained by computer simulation are shown in Figure 2. The stability of the fixed point is lost as the level of temporal non-locality τ, or η,

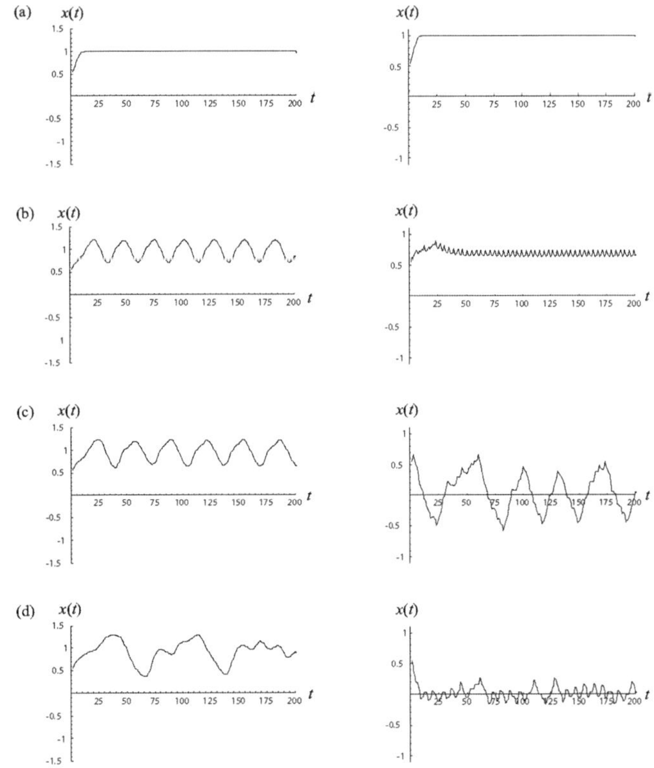

FIGURE 2. Examples of delayed (left column) and predictive (right column) dynamics for the Mackey-Glass map with $\mu = 0.5$, $\beta = 0.8$, and $n = 10$. For delayed dynamics, the initial condition is fixed at $x_0 = 0.5$ for the interval of $(-\tau, 0)$. For predictive dynamics, the initial condition is $x_0 = 0.5$ and $x_1 = (1-\mu)x_0 + f(x_0)$. The values of delay and advance $\tau = \eta$ are (a) 0, (b) 2, (c) 8, (d) 10, and (e) 20.

increases, leading to more complex dynamics. Unfortunately, analytical tools are not developed enough for us to fully understand these behaviors. It is interesting to note that the non-locality, which is in part responsible for the "strange" behaviors of quantum mechanics, can also induce a complex behavior in otherwise simple classical dynamical systems, when it is considered on the time axis.

DISCUSSION

We would like to discuss the non-locality on the time axis in this section.

Delayed and Anticipative Stochastic Resonance

A suitably adjusted combination of oscillatory dynamics and fluctuations or noise can lead to resonance-like behaviors, which have been investigated under the name of stochastic resonance[4, 5, 6]. Stochastic resonances have been studied in a variety of fields[7, 8, 9, 10, 11]. Normally, oscillations are brought in by external signals or forcing. Alternatively, we can take advantage of oscillatory behaviors associated with temporal non-localities. We have investigated possibilities in that directions and found such resonance phenomena theoretically[12, 13, 14]. They are termed as "delayed stochastic resonance" and "predictive stochastic resonance". There is also some experimental evidence of such phenomena [15]. More theoretical and experimental explorations are needed for further developments.

Application to Physiological Control: Stick Balancing

One of the fields of application of non-localities on the time axis is the area of biological and physiological controls, in which physiological delays and anticipative or predictive behaviors are ubiquitous. A recently investigated experimental paradigm is stick balancing control on the human fingertips[16, 17, 18]. Even in this simple task, control feedback delays and predictions of stick motions are present in an intricate mixture. For example, much of the time scale of stick fluctuations is smaller than the human physiological delay of a few hundred milliseconds, indicating that one cannot control the stick depending solely on feedback controls. Furthermore, more recent experiments suggest that moving an object with the other hand while balancing a stick could improve the performance[19, 20, 21]. This could be related to the stochastic resonance type effect we discussed above or to the importance of spatial non-locality in biological and physiological systems. Further experiments and theoretical research are currently underway.

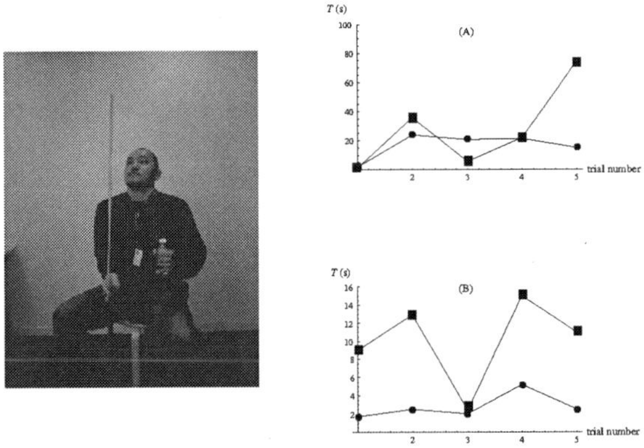

FIGURE 3. Left: Picture of subject balancing stick in one hand while moving object in other. Right: (A) Example of improvement on balancing tasks with (square) and without (dot) moving an object. The subject was given 5 trials without previous practice. By the 5th trial, the improvement was significant. (B) Another subject practiced for a few hours. Here, again improvement with moving the object was evident.

Other Applications

Delayed dynamical models, which are mathematically treated as delay differential equations and delayed random walks, have been studied in various contexts including physiological applications, as mentioned above[3, 22, 23, 24, 25]. Anticipative dynamics has been less investigated. However, mathematically similar equations are called "functional differential equations of an advanced type" or "advanced functional differential equations" [26, 27, 28].

In physics, anticipative dynamics and the concepts of "stochastic time", in which stochasticity is treated on the time axis, faces a problem related to the notion of "causality". It should be noted, however, that there are works such as the Dirac theory [29, 30, 31] and the Wheeler–Feynman theory [32] related to electrons, which include the effect of state in "future". It is left for further research to investigate the connection between the classical dynamical models with temporal non-locality presented here and those theories.

ACKNOWLEDGMENTS

I would like to thank Professor John G. Milton and Dr. J. L. Cabrera with NSF Grant 0617072 for their collaboration related to the delayed dynamical systems and the stick balancing. Thanks also go to Professor B. G. Shidarth for his suggestion about the Dirac theory of electrons.

REFERENCES

1. T. Ohira, "Stochastic Time," in *Frontiers of Fundamental Physics (FFP8), Eighth International Symposium*, edited by B. G. Sidharth, A. Alfonso-Faus, and M. J. Fullana, AIP Conference Proceedings 905, American Institute of Physics, New York, 2007, pp. 191–194.
2. T. Ohira, Physica A **379**, 483–490 (2007).
3. M. C. Mackey and L. Glass, Science **197**, 287–289 (1977).
4. K. Wiesenfeld and F. Moss, Nature **373**, 33–36 (1995).
5. A. R. Bulsara and L. Gammaitoni, Physics Today **49**, 39–45 (1996).
6. L. Gammaitoni, P. Hänggi, P. Jung, and F. Marchesoni, Rev. Mod. Phys. **70**, 223–287 (1998).
7. B. McNamara, K. Wiesenfeld, and R. Roy, Phys. Rev. Lett. **60**, 2626–2629 (1988).
8. A. Longtin, A. Bulsara, and F. Moss, Phys. Rev. Lett. **67**, 656–659 (1991).
9. J. J. Collins, C. C. Chow, and T. T. Imhoff, Nature **376**, 236–238 (1995).
10. F. Chapeau-Blondeau, Sign. Process **83**, 665–670 (2003).
11. I. Y. Lee, X. Liu, B. Kosko, and C. Zhou, Nano Letters **3**, 1683–1686.
12. T. Ohira and Y. Sato, Phys. Rev. Lett. **82**, 2811–2815 (1999).
13. T. Ohira, "Predictive Dynamical and Stochastic Systems," in *Cooperative Behavior in Neural Systems: Ninth Granada Lectures*, edited by J. Marro, P. L. Garrido, and J. J. Torres, AIP Conference Proceedings 887, American Institute of Physics, New York, 2007, pp. 182–189.
14. L. S. Tsimring and A. Pikovsky, Phys. Rev. Lett. **87** 250602 (2001).
15. C. Masoller, Phys. Rev. Lett. **88** 034102 (2002).
16. J. L. Cabrera and J. G. Milton, Phys. Rev. Lett. **89**, 158702 (2002).
17. J. L. Cabrera and J. G. Milton, Chaos **14**, 691-698 (2004).
18. J. L. Cabrera and J. G. Milton, Nonlinear Studies **11**, 305–318 (2004).
19. T. Ohira and T. Hosaka, "Control in Delayed Stochastic Systems," in *Proc. of DETC05, 2005 ASME Design Engineering Technical Conference*, (Long Beach, September 2005), DETC2005-84883.
20. T. Hosaka and T. Ohira, "Delayed Random Walks and Control," in *Flow Dynamics: The second international conference on flow dynamics*, edited by M. Tokuyama and S. Maruyama, AIP Conference Proceedings 832, American Institute of Physics, New York, 2006, pp. 487–491.
21. J. G. Milton, et al., Bull. of the APS (March Meeting 2006) **51**, 1529 (2006).
22. K. L. Cooke and Z. Grossman, J. Math. Anal. and Appl. **86**, 592–627 (1982).
23. J. G. Milton, et al., J. Theo. Biol. **138**, 129–147 (1989).
24. T. Ohira and T. Yamane, Phys. Rev. E **61**, 1247–1257 (2000).
25. T. D. Frank and P. J. Beek, Phys. Rev. E **64**, 021917 (2001).
26. T. Kusano, Hiroshima Math. J. **11**, 617–620 (1981).
27. V. B. Kolmanovskii, A. D. Myshkis, *Applied theory of functional differential equations*, (Kluwer Acad. Publ., Dordrecht, 1995).
28. R. P. Agarwal, M. Bohner and W. T. Li, *Nonoscillation and oscillation theory for functional differential equations*, (Marcel Dekker, New York, 2004).
29. P. A. M. Dirac, Proc. Roy. Soc. (London) **A 167**, p.148 (1938).
30. F. Rohrlich, *Classical Charged Particles*, (Addison-Wesley, Reading, Mass., 1965).
31. B. G. Sidarth, arXive:physics/0701237 (2007).
32. J. A. Wheeler and R. P. Feynman, Rev. Mod. Phys. **17**, p.157 (1945); **21**, p.425 (1949)

Theory of Time beyond the standard model

Eugene S. Poliakov

Institute of Physics of Time, Bolshaya Pionerskaya ul., 46-2, Moscow 115054, Russia

Abstract. A frame of non-uniform time is discussed. A concept of 'flow of time' is presented. The principle of time relativity in analogy with Galilean principle of relativity is set. Equivalence principle is set to state that the outcome of non-uniform time in an inertial frame of reference is equivalent to the outcome of a fictitious gravity force external to the frame of reference. Thus it is flow of time that causes gravity rather than mass. The latter is compared to experimental data achieving precision of up to 0.0003%. It is shown that the law of energy conservation is inapplicable to the frames of non-uniform time. A theoretical model of a physical entity (point mass, photon) travelling in the field of non- uniform time is considered. A generalized law that allows the flow of time to replace classical energy conservation is introduced on the basis of the experiment of Pound and Rebka. It is shown that linear dependence of flow of time on spatial coordinate conforms the inverse square law of universal gravitation and Keplerian mechanics. Momentum is shown to still be conserved.

Keywords: non-uniform time, generalized energy conservation
PACS: 04.50.Kd, 04.60.Kz, 06.30.Ft, 45.20.dh

INTRODUCTION

Let us assume that there is no universal time for all inertial frames to share. As such, we imply that time flows differently in different points of space. In other words, time dilation, which in General Relativity is an effect of gravity, is postulated in this work.

Axiom 1: The flow of time is not an invariant of space. In other words, time is non-uniform, i.e., every point of space (inertial frame) features its specific flow of time.

We borrow only one postulate from Einstein:

Axiom 2: The speed of light (c) is an absolute invariant, i.e. it does not depend on any variables and conditions. I.e., $c = \text{const}$.

The domain of velocity: $\forall u \in (0; c]$. That is, no physical entity can move faster than light, and no physical entity can be in the state of absolute rest.

Definition 1: By inertial frame, we mean a frame not affected by any external force.

Definition 2: The idea of physical entity encompasses an electromagnetic wave, a light quantum, a physical body, a mass point m, a neutrino-like elementary particle, i.e., anything featuring at least one of the following apparent physical characteristics: mass (m), energy (E), momentum (P).

Let us note that zero velocity $u = 0$ is excluded from the domain. This means that it is impermissible to use a notion of rest and its derivatives (e.g., *rest mass*).

CP1018, *Frontiers of Fundamental and Computational Physics*,
edited by B. G. Sidharth, F. Honsell, O. Mansutti, K. Sreenivasan, and A. De Angelis

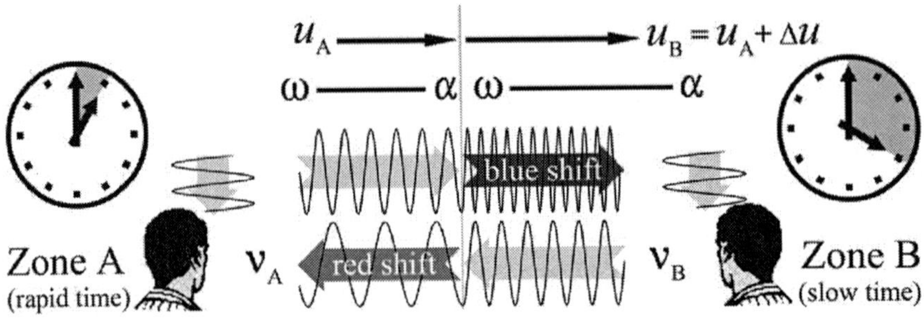

FIGURE 1. Time relativity and physical entities in the field of non-uniform time.

PRINCIPLE OF TIME RELATIVITY

Let us set the principle of time relativity in analogy to the Galilean principle of relativity. An observer attached to an inertial reference frame cannot detect any time flow variance within its own reference frame by the mean of any internal observation.

That is, an observer perceives the course of events in his own zone at a standard pace; all apparent physical processes occur with natural speed in his own zone. As a result, the observer is unable to find any anomalies of time flow within his own zone.

However, if the observer in zone B looks at the events occurring in zone A, he finds out that they go slower than he is used to seeing in his own zone (Fig. 1). E.g., the 'slow' observer B registers a longer half-life of a radioactive isotope in zone A.

On the contrary, the observer in zone A notices that all processes in zone B go faster than he expects (Fig. 1). Accordingly, for the 'rapid' observer A, it seems that the same radioactive isotope decays faster in zone B than in his own zone A.

If observer A sends his clock to zone B and after that it returns to him, then he notices that the clock has been fast as compared with the standard clock that has remained in zone A for the same period. Roughly speaking, the latter trial with clocks is reminiscent of the Hafele-Keating experiment [1].

A PHYSICAL ENTITY IN THE FIELD OF NON-UNIFORM TIME

Let us discuss a mass point (m) moving uniformly from zone A (rapid time) toward zone B (slow time) along a normal to the zone border. In the beginning, for the sake of simplicity, let us restrict our discussion to a one-dimensional model. Both zones are considered inertial, i.e., they are not affected by any external force.

It is easy to see that after the mass point crosses the border between zones A and B the observer in zone A will perceive the motion of the mass point in zone B in a rapid pace. Accordingly, he records a sudden increment of velocity, i.e., acceleration (Fig. 1). Similar observations will be registered by the 'slow' observer in zone B. Both observers suppose that the mass point was affected by some unknown force.

We draw attention to the fact that, although we have introduced mass m, accelerations detected by the observers in no way depend on mass point m, for they are completely conditioned by the rate of the flows of time in corresponding zones. Therefore, we have to refine the latter statement: the mass point is affected by some *fictitious* force.

It is appropriate to assume that time conditions vary continuously from point to point, rather than in discrete steps. So it is possible to calculate both acceleration (which will not be sudden), and time conditions to emulate a law of any fictitious force.

Now, let us consider a line segment $\alpha\omega$ travelling uniformly from zone A to Zone B along a normal to the border. It is easy to see that point α moves faster in zone B for both observers, while point ω moves slower as it remains in zone A. In the time between crossing the border by point α and by point ω the line segment $\alpha\omega$ becomes longer (Fig. 1). This makes the association with the tidal force pertaining to gravity almost obtrusive.

In other words, the fictitious force described is so close to resulting gravity that it is permissible to pose an equivalence principle.

The outcome of non-uniform time in an inertial frame of reference is equivalent to the outcome of a fictitious gravity force external to the frame. The zone with a slower time works as a 'centre of attraction' for the entities residing in zones of a faster flow of time.

From geometrical reasoning we can conclude that if time conditions vary continuously and monotonically, then the velocity of a mass point m will increase continuously and monotonically as well.

A mass point behaviour invariance under mass m also means that the point behaviour would change in no way, even if in the final analysis it turns out that mass itself depends on the flow of time (or on velocity, as it is in SR).

This permits the assumption that when accelerating continuously and monotonically, a mass point reaches the speed of light c. They are the axiom limitations, which do not allow a mass point to continue accelerating. As soon as the velocity of a mass point is defined only geometrically, there are no factors to prevent its velocity from reaching c.

Let us consider the behaviour of a light quantum (photon) as it travels from zone A (rapid time) to zone B (slow time). The velocity of a photon is c. According to axiom 2, any reasoning as to the kinematical acceleration of a photon is senseless. Let us consider the frequency of an electromagnetic wave (ν) instead.

Observer B registers light coming from zone A as 'faster,' having higher frequency ($\nu_B > \nu_A$). I.e., observer B notices a blue shift as compared to a standard source in his zone (Fig. 1). Observer A, on the contrary, registers light coming from zone B as 'slower,' with lower frequency ($\nu_A < \nu_B$). I.e., observer A notices a red shift compared to a standard source in his zone (Fig. 1). The blue shift is seen only by observer B, while the red shift is seen only by observer A. However, $\nu_A < \nu_B$ anyway.

The phenomenon described above is not solely our mental experiment, but it is a result of a number of researches, of which the foundation is Pound and Rebka's experiment [2]. The question is the interpretation of the outcome of the experiment.

STATEMENT OF PROBLEM

We will consider the behaviour of a physical entity in the field of non-uniform time. We assume that the flow of time varies continuously and monotonically. We consider the frame of reference to be empty and exclusive of any external force. We will consider a physical entity with initial conditions $E \to 0$, situated infinitely far and travelling in the direction of a decreasing flow of time according to the equivalence principle.

The initial condition $E \to 0$ (at $r \to \infty$) means an infinitely large red shift for a light quantum ($\nu \to 0$), and an infinitesimal velocity for a physical body ($u \to 0$). The coordinate r variation (travelling) for a physical entity in this frame of reference will, due to geometry, be unambiguously accompanied by continuous and monotonic increase of energy, i.e., frequency (ν) for a light quantum and velocity (u) for a physical body: $E = f(r)$.

As we introduce velocity u we imply a velocity, which the body acquires in point r under the action of non-uniform time occurring with the initial conditions set above. The physical sense of this value is the escape velocity from point r to infinity.

Let us introduce the dimensionless scalar variable χ so that

$$u_A \chi_A = u_A \chi_A \ \ or \ \ u\chi = const. \tag{1}$$

Special attention has to be drawn to the case of a mass point reaching c. According to Axiom 2, $c = \chi c$. In other words, the mass point reaches the speed of light at $\chi = 1$. That is,

$$u\chi = c, \tag{2}$$

where $\chi \geq 1$.

We point out that Equation (1) has no allusion to asymptotic behaviour near $\chi = 1$. The velocity of the mass point does not exceed c due only to the axiomatic principles.

In the quantum case, $\chi_A \Delta t_B = \chi_B \Delta t_B$. Therefore let us call variable χ flow of time.

Such a posing of the problem provides the possibility to establish unambiguous conformity between E (or u for a physical body) and χ; and, secondly, allows us to avoid the concept of potential energy, which essentially involves the idea of force that we escape by the equivalence principle. That is, in Fig. 1 the zone of 'rapid' time A features the higher flow of time χ_A, and the zone of 'slow' time B features the lower flow of time χ_B.

According to the equivalence principle, the further reasoning does not need an idea of 'gravitational' radiation, be it either in corpuscular or in wave form, nor does it need an idea of 'propagation' of gravity in space. Gravity has a nature pertaining to time.

Every time we talk about mass m, we imply the inertial mass, because the proposed principle does not require the idea of 'gravitational' mass, for according to the equivalence principle, the cause of gravity is not the mass of an attracting body, but a decrease of flow of time, invariant to both masses.

The latter proposition could be easily tested even in earthly conditions: according to the existent standpoint, acceleration due to gravity should (almost) linearly decrease from standard gravity g at the Earth's surface to zero in the centre of the Earth. The proposed theory, on the contrary, predicts monotonous increase of acceleration g along with the depth due to the decrease of the time flow toward the centre of the Earth that is what really causes gravity. Function $g(r)$ should obey the inverse square law both above the Earth and at its core. Such an increase of observable g along with depth is in strong agreement with experimental data (Table 1), although Stacey *et al.* interpret this outcome as 'the possible (but not probable) (*sic.*) systematic error arising from density inhomogeneity' [3] (and even 'possible violations of Newton's inverse-square law' [4]).

TABLE 1. Meter reading [3] comparisons with the increase of g under the inverse-square law.

Depth (m)	Δg theory $(0.001m/s^2)$	Δg test $(0.001m/s^2)$	Deviation (%)
267.23	0.824049	0.8252	0.14
487.11	1.502164	1.5038	0.109
599.94	1.850162	1.8522	0.11
720.13	2.22088	2.2233	0.109
834.14	2.572555	2.5753	0.107
935.29	2.884579	2.8877	0.108
948.16	2.924281	2.9274	0.107

It is worth noticing that the deviation is not only systematic, but also nearly fixed, which even further improves the inverse-square law compliance. An appropriate correction for the systematic bias (a decrement of $R_{\oplus}$ by only 1 m) results in a scandalous precision of 0.0003%. Similar results regarding the observable g at the Earth's core can be deduced from the work by Ander *et al* [4], although with bigger systematic error.

ENERGY AND MOMENTUM CONSERVATION

As we return to the case in Fig. 1, it is easy to see that according to Planck's quantum concept ($E = h\nu$), along with the increased frequency in zone B ($\nu_B > \nu_A$), energy increases ($E_B > E_A$) as well.

Thus the law of energy conservation is neither universal nor fundamental. Classical energy conservation is applicable only in the frames of reference featuring uniform time. This simply repeats a special case of Noether's theorem, according to which, energy conservation is effective only in the frames of uniform time. It is the generalized energy conservation law that should apply to the frames of non-uniform time:

$$E\chi = const. \tag{3}$$

Since it is absurd to suggest a specific energy conservation for material bodies, we extend it onto all physical entities.

If $\chi =$ const, then time is uniform and the classical energy conservation turns out to be correct. Since neither the red nor the blue shift have any physical limitations, so initially we can accept the domain of χ as $\forall\chi \in [0;\infty)$.

Accepting $E \sim mu^2$, it can be derived from Equations (2, 3) that

$$mu = const. \tag{4}$$

I.e., the conservation of momentum follows from the generalised energy conservation.

LAW OF UNIVERSAL GRAVITATION

To calculate the gravity force we differentiate Equation (3) with respect to spatial coordinate and get:

$$\chi \nabla E + E \nabla \chi = 0. \tag{5}$$

Since $\nabla E = F$ we derive,

$$F = \frac{C \nabla \chi}{\chi^2}. \tag{6}$$

It is easy to see that if χ depends on spatial coordinate linearly ($\chi = \mathfrak{P} r + X$), we get,

$$F = \frac{C \mathfrak{P}}{(\mathfrak{P} r + X)^2}, \tag{7}$$

where $\mathfrak{P} = \nabla \chi$ and X is the 'initial' flow of time at $r = 0$.

It is evident that at $\mathfrak{P} r \gg X$, Equation (7) becomes $F \sim r^{-2}$, i.e. Newton's inverse-square law of universal gravitation.

Since the law of universal gravitation is beyond doubt, our assumption that the flow of time depends linearly on coordinate, becomes a fact, which needs no experimental proof.

If so, Equation (1) turns into a more compact equivalent of the second Kepler's law:

$$v \chi = const, \tag{8}$$

where v is a velocity of any stable cyclic process.

The obtained result is another confirmation of the competence of the stated approach. However, within such a short paper, there is no sense to rewrite the entire mechanics.

REFERENCES

1. Hafele J.; Keating, R. (July 14, 1972). Around the world atomic clocks: predicted relativistic time gains. Science 177 (4044): 166-170.
2. Pound R. V.; Rebka Jr. G. A. (November 1, 1959). Gravitational Red-Shift in Nuclear Resonance. Physical Review Letters 3 (9): 439-441.
3. Stacey F. D. et al. (April 15, 1981). Constraint on planetary scale value of the Newtonian gravitational constant from a gravity profile within a mine. Physical Review D 23 (8): 1683-1692.
4. Ander M.E. et al. (February 27, 1989). Test of NewtonŠs inverse-square law in the Greenland ice cap. Physical Review Letters 62 (9): 985-988.

Quantum Information and Wave Function Collapse

S.N.Mayburov

Lebedev Inst. of Physics, Leninsky Prospect 53, Moscow, Russia, 117924

Abstract. Information-theoretical restrictions on the information transferred in quantum measurements, are regarded for the measurement of quantum object S performed via its interaction with information system O. This information restrictions, induced by Heisenberg commutation relations, are derived in the formalism of inference maps in Hilbert space. O restricted states ξ^O are calculated from Shroedinger S,O dynamics and the structure of O observables set (algebra); O decoherence by its environment is also accounted for some S, O systems. It's shown that this principal constraints on the information transfer result in the stochasticity of measurement outcomes. Consequently, ξ^O describes the random 'pointer' outcomes q_j, which correspond to the collapse of S pure state.

Keywords: Information Theory, Quantum Measurements
PACS: PACS: 03.65Ta, 03.67.-a

INTRODUCTION

Despite its significant achievements, Quantum mechanics (QM) still contains some open questions concerned with its internal consistency[1, 2, 3]. The most famous and oldest of them is the State Collapse or Quantum Measurement problem [4, 5]. In this paper this problem will be considered mainly within the framework of Information Theory [6, 7, 8]. Really, the measurement of some system S includes the transfer of information from S to the information system O (Observer) which processes and memorizes it. In Information Theory, any measuring system (MS) can be regarded as the information channel, which transfers the information about S state to O. Plainly, if some restrictions on the information transfer from S to O exist, they can influence the outcomes of measurements. It was shown that such constraints, induced mainly by Heisenberg commutation relations, result in the significant information losses for typical information channels [9]. Our calculations evidence that for MS such restrictions can induce the unavoidable stochasticity in the measurements S pure states, due to the loss of information about state purity [8].

In our approach the evolution of all physical objects can be described by the quantum state (density matrix) $\rho(t)$ which obeys Schroedinger-Lioville equation in arbitrary reference frame (RF). The information-theoretical formalism of systems' self-description or 'measurement from inside' is applied to the description of information acquisition by O in quantum measurements [7]. In Schroedinger QM framework, it can be realized via the formalism of inference maps in Hilbert space [5]. Overall, we shall demonstrate for simple MS model that such approach allows to construct the consistent measurement formalism, which describe the stochasticity of measurement outcomes without introduction of additional Reduction Postulate in Schroedinger (linear) QM scheme.

CP1018, *Frontiers of Fundamental and Computational Physics,*
edited by B. G. Sidharth, F. Honsell, O. Mansutti, K. Sreenivasan, and A. De Angelis

MODEL OF QUANTUM MEASUREMENTS

Here our measurement model will be described and some aspects of QM Measurements Theory, essential for our approach, will be discussed at semiqualitative level. In our model MS consists of the studied system S, detector D and the information system O, which acquires and processes the information about S state. The effects of D, O decoherence by their environment [10] will be considered in the final part of our paper. S is taken to be the particle with the spin $\frac{1}{2}$ and the measurement of its projection S_z will be studied. Its u,d eigenstates denoted $|s_{1,2}\rangle$, so that the measured S pure state is:

$$\psi_s = a_1|s_1\rangle + a_2|s_2\rangle$$

For the comparison, the incoming u, d 'test' mixture with the same $\bar{S}_z$ should be regarded also. This is S ensemble described by the gemenge[1] $W^s = \{|s_i\rangle, \mathrm{P_i}\}$, where $\mathrm{P_i} = |a_i|^2$ are the probabilities of $|s_i\rangle$ in this ensemble, its density matrix denoted ρ^s_m. Analogously to S, in our model D state in O RF is described by Dirac vector $|D\rangle$ in two-dimensional Hilbert space $\mathcal{H}_{\mathcal{D}}$. Its basis is constituted by $|D_{1,2}\rangle$ eigenstates of Q 'pointer' observable with eigenvalues $q_{1,2}$. The initial D state is: $|D_0\rangle = \frac{|D_1\rangle+|D_2\rangle}{\sqrt{2}}$. S, D interaction $\hat{H}_{S,D}$ starts at t_0 and finishes effectively at some t_1; for Zurek Hamiltonian $H_{S,D}$ with suitable parameters [10] it would result in S, D entangled final state:

$$\Psi_{S,D} = \sum a_i|s_i\rangle|D_i\rangle \tag{1}$$

in O RF, it follows that $\bar{Q} = |a_1|^2 - |a_2|^2$. The measurement of S eigenstate $|s_{1,2}\rangle$ results in factorized state $\Psi^{1,2} = |s_{1,2}\rangle|D_{1,2}\rangle$ in which Q has the eigenvalue $q_{1,2}$. If $a_{1,2} \neq 0$, then D is also described by the quantum state R_D, but due to S,D entanglement, it can't be completely factorized from S state, so it's instructive to use $\Psi_{S,D}$. In our model D, O interaction starts at $t > t_1$ and finishes at some t_2, during this time interval, the information about D state is transferred to O. Here we shall suppose that O can get all D information copiously, this assumption will be proved in sect. 3. It's supposedly true for human observer also, therefore some terms characteristic for human perception will be used in the description of our model.

Let's remind how the measurements are described in Information Theory [11]. The signal induced by the measured state and registrated by O in event n is characterized by the array of real parameters called information pattern (IP) $J(n) = \{e_1, ..., e_l\}$. The set of all possible J constitutes the independent 'information space', which define O recognition of measured states [7]. In quantum case, some parameters, in principle, can be uncertain, but it's unimportant for our problem. Consider, as the example, the measurement of S eigenstate $|s_{1,2}\rangle$, in such framework, O supposedly percepts $|D_{1,2}\rangle$ state in event n as IP: $J^D_{1,2} = q_{1,2}$. Q eigenvalues $q_{1,2}$ are D real properties [1], corresponding to the orthogonal projectors $P^D_{1,2}$. Hence for O the difference between this D states is the objective or Boolean Difference (BD) [3]. It is equivalent to the distinction between the logical operands Yes/No, or that's the same between the values $1/0$ of some parameter L_g. Yet in QM all measurable parameters are related strictly to the observables represented by Hermitian Operators on $\mathcal{H}$ (or POV in general formalism). For example, the parameter $L_g = 1/0$ for $|D_{1,2}\rangle$ corresponds to projector P^D_1.

Now let's regard the measurement in case when $a_{1,2} \neq 0$, i.e. ψ_s is $|s_i\rangle$ superposition. In this case, $\Psi_{S,D}$ state is objectively different from $\Psi^{1,2}$ but it doesn't mean automatically that O can discriminate them as the different signals. The standard or 'Pedestrian' interpretation [2, 3] (PI) of QM claims that without the inclusion of Reduction Postulate into QM formalism, O would percept S, D entangled state $\Psi_{S,D}$ as the superposition of $|D_{1,2}\rangle$ to which supposedly corresponds the simultaneous observation of $J^D_{1,2}$. More realistically, one can expect that if $\Psi_{S,D}$ is percepted by O as some IP J^s, it would differ from $J^D_{1,2}$. This assumption is the essence of famous 'Schroedinger Cat' Paradox [3]. No such exotic outcomes are registrated experimentally, in place of it $J^D_{1,2}$ are observed at random, so it is concluded that Reduction Postulate should be added to QM formalism. Yet the situation isn't so simple and doesn't favor such prompt jump to the conclusions. Really, given PI implications are correct, O should distinguish in a single event R_D or that the same $\Psi_{S,D}$ from both $|D_{1,2}\rangle$ states. Hence the relation of corresponding O IPs should be characterized by BD, i. e. $J^s \neq J^D_{1,2}$. So it should be at least one D parameter (observable) G^D which value g_0 for $\Psi_{S,D}$ is different from its values $g_{1,2}$ for $|D_{1,2}\rangle$. Roughly speaking, it should such D parameter G_D which is equal to $1/0$ depending on the presence/absence of D_i superpositions. To verify this hypothesis for $\Psi_{S,D}$ and $|D_{1,2}\rangle$, one should check the set (algebra) of D PV observables $\{G^D\}$ looking for the possible candidates. Let's suppose that such G^O - Hermitian operator exists, then it follows:

$$G^D\Psi_{S,D} = a_1|s_1\rangle G^D|D_1\rangle + a_2|s_2\rangle G^D|D_2\rangle = g_0\Psi_{S,D} \tag{2}$$

As easy to see, such equality fulfilled only for $G^D = I$. Any G^D, which is sensitive to the presence of superpositions, corresponds to the nonlinear operator on $\mathcal{H}_{\mathcal{D}}$, so the observation of such difference seems to be incompatible with standard QM formalism. Consequently, it seems impossible for O to distinguish $|D_i\rangle$ from R_D i.e. from $\Psi_{S,D}$ of (1) in a single event. From that it's logical to conclude that O would observe one of q_i outcomes in each event. For pure S ensemble it's reasonable to assume that QM expectation value $\bar{Q}$ will be obtained by O from the measurment of S ensemble with the number of events $N \to \infty$. To fulfill such relation, O should observe the stochastic $q_{1,2}$ outcomes with probabilities $P_{1,2} = |a_{1,2}|^2$. This considerations put doubts on the necessity of independent Reduction Postulate in QM; the similar hypothesis was proposed by Wigner [12]. Note that the obtained results don't mean that R_D is the probabilistic mixture of $|D_{1,2}\rangle$, rather R_D can be characterized as their 'weak' superposition, stipulated by the entanglement of S, D states.

MEASUREMENTS AND SYSTEMS' SELF-DESCRIPTION

Now the information system O will be considered as the quantum object, in this case, MS is supposedly described by quantum state ρ_{MS} relative to some external RF O'. In our model O pure state is a vector in two-dimensional Hilbert space $\mathcal{H}_{\mathcal{O}}$. Analogously to D, we settle O initial state $|O_0\rangle = \frac{|O_1\rangle+|O_2\rangle}{\sqrt{2}}$ where $|O_{1,2}\rangle$ are eigenstates of O 'internal pointer' observable Q_O with eigenvalues $q^O_{1,2}$. For suitable D, O Hamiltonian $H_{D,O}$ one can obtain at $t > t_2$:

$$\Psi_{S,D,O} = \sum a_i |s_i\rangle |D_i\rangle |O_i\rangle$$

As easy to see, D states only double S states for this set-up, so for the simplicity D can be dropped from further considerations. In such scheme S directly interacts with O by means of Hamiltonian $H_{S,O}$, which result in the final state:

$$\Psi_{MS} = \sum a_i |s_i\rangle |O_i\rangle \tag{3}$$

in some external RF O'. Our aim is to find the relation between this state and the information acquired by O. In Information Theory, the measurement of parameters of arbitrary system S' by an information system O^I is the mapping of S' states set N_S to the set N_O of O^I internal states [6]. In most general approach, which is generic for QM, the information acquisition by O^I can be described by the formalism of systems' self-description called also 'measurement from inside' [7]. In its framework, O^I considered as the subsystem of larger system $\Xi_T = S', O^I$ with the states set N_T. The information acquired by O^I about Ξ_T (including O^I itself) is described by O^I internal state R_O called also Ξ_T restricted state or restriction. For given Ξ_T system R_O is defined by the inference map M_O of Ξ_T state to N_O set.

The important property of inference map M_O is formulated by Breuer Theorem: if for two arbitrary Ξ_T states Γ, Γ' their restricted states R, R' coincide, then for O^I this Ξ_T states are indistinguishable, and for any nontrivial S', O^I at least one such pair of states exist [5]. In classical case, the origin of this effect is obvious: O^I has less degrees of freedom than Ξ_T and hence can't discriminate all possible Ξ_T states [7]. For quantum systems M_O ansatz should be derived from first QM principles, however, Schroedinger QM formalism only doesn't permit to derive $M_O(\Xi_T \rightarrow O^I)$ unambiguously and some additional inputs are needed for that. Breuer assumed phenomenologically that for arbitrary Ξ_T its restricted state is equal to the partial trace of Ξ_T individual state over S', i.e. is Ξ_T partial state on O^I. In our MS set-up for Ψ_{MS} of (3) it gives:

$$R_O^B = Tr_s \rho_{MS} = \sum |a_i|^2 |O_i\rangle \langle O_i| \tag{4}$$

Plainly, this ansatz excludes beforehand any kind of stochastic R_O behavior. For MS mixed ensemble, induced by incoming W^s gemenge, the individual MS states differ from event to event:

$$\varsigma^{MS}(n) = |O_l\rangle \langle O_l| |s_l\rangle \langle s_l| \tag{5}$$

where the frequencies of random $l(n)$ appearance in given event n are stipulated by the probabilistic distribution $P_l = |a_l|^2$. O restricted state for this mixed ensemble is also stochastic: in a given event

$$R_O^{mix}(n) = .\xi_1^O .or. \xi_2^O .$$

where $\xi_i^O = |O_i\rangle \langle O_i|$ appears with the corresponding probability P_i, so that the ensemble of O states described by the gemenge $W_{mix}^O = \{\xi_i^O, P_i\}$. $R_O^{mix}(n)$ differs from R_O^B in any event, hence for the restricted O individual states the main condition of cited theorem is violated. From that Breuer concluded that O can discriminate the individual pure/mixed MS states 'from inside', so the collapse of pure state can't be observed by O [5].

Meanwhile, we find that our IP formalism allows to calculate MS restriction to O unambiguosly, and the obtained results contradict to Breuer conclusion. Consider again the measurement of S eigenstate $|s_i\rangle$, it produces MS individual ς^{MS} states of (5), which restriction is $\xi^O_{1,2}$ with eigenvalues $q^O_{1,2}$. It's natural to expect that O identifies this states as IP $J^O_{1,2} = q^O_{1,2}$. Really, the difference between ξ^O_i states is boolean (classical), because in QM formalism ξ^O_i eigenvalues q^O_i are O real properties [1]. The formal difference of two O restricted states doesn't mean automatically that their difference will be detected by O. Such difference is the necessary but not sufficient condition, there should be also the particular O observation G^O, which indicate this difference between R_O and ξ^O_i. For example, for ξ^O_i states there are the eigenvalues q^O_i of observable Q_O, resulting in BD of $\xi^O_{1,2}$ for O. The calculations are analogous to the ansatz described by formulae (2), Really, suppose that such G^O exists, in QM framework, it should be Hermitian PV operator, from that G^O should obey:

$$G^O\Psi_{MS} = a_1|s_1\rangle G^O|O_1\rangle + a_2|s_2\rangle G^O|O_2\rangle = g_0\Psi_{MS} \tag{6}$$

Yet for O observables such equality fulfilled only for $G^O = I$. In the regarded case, only the parameters corresponding to nonlinear operators can establish BD between R_O and ξ^O_i states for O, but their observability contradicts to standard QM axiomatic. Consequently, O can't distinguish R_O and ξ^O_i states and resulting MS restriction is equal to:

$$R_O = .\xi^O_1 .or. \xi^O_2 . \tag{7}$$

i.e. it coincides with R^{mix}_O as the individual state. POV generalization of standard QM PV observables don't change this conclusions.

To reproduce the correct $\bar{Q}^l_O$ QM values for arbitrary l, O should observe the collapse of pure MS state to one of q^O_i at random with probability $P_i = |a_i|^2$, i.e. the ensemble of O states described by the gemenge $W^O = \{\xi^O_i, P_i\}$. It induces the corresponding O IP ensemble $Z^O = \{J^O_i, P_i\}$. In this approach we admitted *ad hoc* that the probabilities of q^O_i outcomes follow QM ansatz [13], however, they should be reconsidered in any new theory of measurements [1]. Here we note only that for our MS, which consist of spin-$\frac{1}{2}$ objects only, such relations can be derived from QM invariance relative to the space reflections and rotations. For example, for $a_1 = a_2e^{i\alpha}$ the equality $P_1 = P_2$ follows directly from QM reflections invariance. Note that the obtained results confirm our assumptions about the information perception by O, which were used in sect. 2.

The difference between the pure and mixed S states with the same $\bar{S}_z$ is indicated by 'interference term' (IT) observables. For our S states they are $S_{x,y}$ linear forms. For example, if $\frac{a_1}{a_2}$ is real, the maximal distinction corresponds to S_x, which gives $\bar{S}_x = \frac{1}{2}|a_1||a_2|$ for pure states and $\bar{S}_x = 0$ for the mixture. For MS entangled states such difference can be revealed only by joint S,O observables. As the example, consider the symmetric IT for MS:

$$B = |O_1\rangle\langle O_2||s_1\rangle\langle s_2| + j.c. \tag{8}$$

Being measured by external RF O' via its interaction with S, O, it gives $\bar{B} = 0$ for any $|s_i\rangle$ incoming mixture, but $\bar{B} \neq 0$ for entangled MS states of (4). For example, for incoming symmetric S state ψ_s^s with $a_{1,2} = \frac{1}{\sqrt{2}}$, the resulting MS state Ψ_{MS}^s is B eigenstate with eigenvalue $b_1 = 1$. Hence, in that case, S_x is mapped to B. However, B, as well as other ITs, can't be directly measured by O 'from inside', at least simultaneously with S_z, because they don't commute [5]. Note that, in general, the pure/mixed MS states with the same $\bar{Q}_O$ can be discriminated even by external O' only statistically, since their distributions of B values (or other ITs) overlap. Namely, for Ψ_{MS}^s the probability $P_B(b_{1,2}) = .5$ for such mixture, so its b distribution intersects largely with δ_{b1} distribution for B eigenstate Ψ_{MS}^s. Consequently, even O' can discriminate the pure/mixed MS states only statistically for MS ensemble with $N \to \infty$.

It's well known that the decoherence of pure states by its environment E is the important effect in quantum measurements [1, 10], we find yet that O decoherence by E doesn't play the principal role in our theory. However, its account stabilizes the described collapse mechanism additionally and defines unambiguously the used preferred basis PB $\{\xi_i^O\}$ of O final states. Really, for the specially chosen Hamiltonian of O,E interaction [10], it follows that Ψ_{MS} of (3) decoheres into MS,E final state:

$$\Psi_{MS,E} = \sum a|s_i\rangle|O_i\rangle \prod_j^{N_E} |E_i^j\rangle$$

where E^j are E elements (atoms), N_E is E^j total number. If an arbitrary O pure state Ψ_O is prepared, it will also decohere in a very short time into the analogous $|O_i\rangle$ combinations, entangled with E. Hence, of all pure O states, only $\xi_i^O = |O_i\rangle$ are stable relative to E decoherence. Consequently, it advocates the choice of such O states as O PB set, since in such environment O simply can't percept and memorize any other O pure state during any sizable time interval.

CONCLUSION

Any consistent physical theory should not contain the logical contradictions, it should be true also for the predictions of measurement outcomes. Yet as was noticed by Wigner: 'The simultaneous observation of two opposite outcomes of quantum experiments is nonsense' [12]. It seems from our analysis that the structure of QM Observable Algebra by itself excludes such observations without the exploit of independent Reduction Postulate.

We conclude that standard Schrödinger QM formalism together with the theory of systems' self-description permit to obtain the 'subjective' collapse of pure states without implementation of independent Reduction Postulate into QM axiomatic. In our approach the main source of stochasticity is the principal constraint on the transfer of specific information in S$\to$ O information channel. This information characterizes the purity of S state, because of its loss , O can't discriminate the pure/mixed S states. As the result of this information incompleteness, the stochasticity of measurement outcomes appear, which is the analog of fundamental 'white noise'. In addition, the formalism of systems' self-description permits to resolve the old problem of Heisenberg cut in quantum measurements, by the inclusion of the information system into quantum formalism properly and on equal terms with other MS elements.

The most exciting and controversial question is whether this theory is applicable to the observations made by human observer O, in particular, whether IP J^O describes the true O 'impressions' about their outcomes ? This is open problem, but at the microscopic level the human brain should obey QM laws, so we believe that the answer can be positive. Note that in our theory the brain or any other processor O plays only the passive role of signal receiver, the real effect of information loss, essential for collapse, occurs 'on the way', when the quantum signal passes through the information channel. The interesting feature of this theory is that the same MS state can be stochastic in O RF, but evolve linearly in O' RF. In particular, Ψ_{MS} restriction to O in O RF is stochastic state R_O of (7), yet in O' RF O partial state is R_O^B of (4), i.e. is the 'weak superposition'. The detailed explanation of this effect is given by the unitarily nonequivalent representations admitted in Algebraic QM [4]. Here we notice only that O and O' deal with different sets of MS observables, and so the transformation of MS state between them can be nonunitary. Obtained results agree well with our calculations in C^* Algebras formalism, in that approach the inference map M_O is the operator restriction of MS observable algebra to O (sub)algebra.

REFERENCES

1. P. Busch, P. Lahti, P. Mittelstaedt, *Quantum Theory of Measurements*, Springer-Verlag, Berlin, 1996, pp. 8–26
2. W. D'Espagnat, *Found Phys.* **20**, 1157–1169 (1990)
3. J. M.Jauch *Foundations of Quantum Mechanics*, Addison-Wesly, Reading, 1968, pp. 85–116
4. G. Emch, *Algebraic Methods in Statistical Physics and Quantum Mechanics*, Wiley, N-Y, 1972, pp. 71–89
5. T. Breuer, *Synthese* **107**, 1–9 (1996)
6. P. Mittelstaedt, *Interpretation of Quantum Mechanics and Quantum Measurement Problem*, Oxford Press, Oxford, 1998, pp. 67–109
7. K.Svozil, *Randomness and undecidability in Physics*, World Scientific, Singapour, 1993, pp. 46–87
8. S. Mayburov, *Int. J. Quant. Inf.* **5**, 279–287 (2007)
9. A. S. Holevo, R. F. Werner, *Phys. Rev.* **A63**, 032312–032326 (2000)
10. W. Zurek, *Phys. Rev.* **D26**, 1862–1876 (1982)
11. U.Grenander, *Pattern Analysis*, Springer-Verlag, N-Y, 1978, pp 12–37
12. E. Wigner, *Scientist speculates*, Heinemann, London, 1961, pp 47–59
13. J. B.Hartle, *Amer. J. Phys.* **36**, 704 (1968)

Hidden-Variable Theory versus Copenhagen Quantum Mechanics

Miloš V. Lokajíček

Institute of Physics, v.v.i., Academy of Sciences of the Czech Republic, 18221 Prague, Czech Republic

Abstract. The Copenhagen quantum mechanics has been newly analyzed and critical comments introduced already in literature have been gathered. It has been concluded that the hidden-variable theory should be preferred to the Copenhagen alternative, as well as to the standard model of classical physics. On the basis of given arguments it has been also concluded that the Copenhagen quantum mechanics should be declared as falsified.

Keywords: assumption bases, experimental tests, logical contradictions
PACS: 03.65.-w, 03.65.Ta

INTRODUCTION

In the following I should like to summarize the results of our long-years work. The final results were obtained in the last year and may be found in Refs. [1-3]. They have brought a new look at the physical theories of the 20th century, and especially at the Copenhagen quantum mechanics.

PHYSICS OF THE 20TH CENTURY

The physics of the 20th century was represented mainly by two theories: Copenhagen quantum mechanics and theory of special relativity. Both these theories have been applied to the same reality at the same time, even if they have exhibited some contradicting characteristics: chance versus causality, and quite different interpretations of the time parameter. And one must ask whether both the theories may be true. It is evident that it is necessary to conclude that something must be wrong.

We have performed the analysis of the Copenhagen quantum mechanics from this point of view. And we have started with the assumption basis of this theory. It is possible to say that it is built up on the four following main assumptions:

- first of all it is the time-dependent Schroedinger equation

$$i\hbar\frac{\partial}{\partial t}\psi(x,t) = H\psi(x,t), \quad H = -\frac{\hbar^2}{2m}\triangle + V(x) \tag{1}$$

  where Hamiltonian H represents the total (kinetic and potential) energy of a given physical system and x represents coordinates of all matter objects;
- the evolution of a physical system is given by function $\psi(x,t)$ and all physical quantities may be determined as expected values of corresponding operators:

CP1018, *Frontiers of Fundamental and Computational Physics,*
edited by B. G. Sidharth, F. Honsell, O. Mansutti, K. Sreenivasan, and A. De Angelis

$$A(t) = \int \psi^*(x,t) A_{op} \psi(x,t) dx \tag{2}$$

where A_{op} and functions $\psi(x,t)$ may be represented by operators and vectors of a suitable Hilbert space;

- in the case of Copenhagen quantum mechanics it has been required for the corresponding Hilbert space to be spanned on one set of Hamiltonian eigenfunctions $\psi_E(x)$:
$$H\psi_E(x) = E\psi_E(x); \tag{3}$$
- and in addition to, the mathematical superposition principle valid in any Hilbert space has been interpreted in physical sense, i.e., any superposition of two physical states has represented again another basic (pure) physical state.

It has been spoken often about two different interpretation alternatives of the Copenhagen mathematical model: orthodox (or Copenhagen) and statistical (or ensemble). However, it has been never mentioned that both the alternatives have corresponded to the different sets of assumptions. While Copenhagen alternative has involved all four introduced assumptions the ensemble alternative (called also as the hidden-variable theory) has corresponded to the first two assumptions only. And it is necessary to speak about two different theories, not interpretations. As to the Hilbert space in the latter alternative it must be chosen according to corresponding physical system; however, it must be always extended in comparison to the Hilbert space required by the third assumption.

And we should ask, first, what is the actual difference between the classical physics and the Schroedinger equation (or the hidden-variable theory).

SCHROEDINGER EQ. AND CLASSICAL PHYSICS

E. Schroedinger [4] was successful with his equation when he showed that for particles exhibiting inertial motion the identical behavior with classical physics was obtained. Let us express now the complex function $\psi(x,t)$ in Eq. (1) as

$$\psi(x,t) \;=\; \lambda(x,t)\, e^{\frac{i}{\hbar}\Phi(x,t)} \tag{4}$$

where both the functions $\lambda(x,t)$ and $\Phi(x,t)$ are real. Let us limit to time-independent potential $V(x)$. Eq. (1) may be then substituted by two equations for two real functions (see D. Bohm [5])

$$\frac{(\nabla\Phi)^2}{2m} + V(x) + V_q(x,t) \;=\; -\partial_t\Phi\,, \quad V_q(x,t) = -\frac{\hbar^2}{2m}\frac{\triangle\lambda}{\lambda} \tag{5}$$

$$\triangle\Phi + 2(\nabla\Phi)(\nabla\, lg\,\lambda) \;=\; -2m\,\partial_t\, lg\,\lambda \tag{6}$$

where $V_q(x,t)$ has been denoted as quantum potential.

Eq. (5) resembles Hamilton-Jacobi equation

$$\frac{1}{2m}(\nabla S(x,t))^2 + V(x) = -\partial_t S(x,t)\,. \tag{7}$$

There are, however, two differences: Hamilton principal function $S(x,t)$ is substituted by phase $\Phi(x,t)$ and the quantum potential is present in Eq. (5).

For inertial motion it holds: $V_q(x,t) = V(x) = 0$ and $\Phi(x,t) = S(x,t)$.

Let us assume now $V(x) \neq 0$ and let us limit to the basic solutions fulfilling the conditions (being characterized by one eigenfunction of H only)

$$\psi_E(x,t) = \lambda_E(x)e^{-iEt}, \quad H\lambda_E(x) = E\lambda_E(x).$$

In such a case $V_q(x) \neq 0$ is independent of t and $\Phi(x,t)$ and $S(x,t)$ are mutually different; $V_q(x)$ representing the numerical measure of such difference. There is not, however, any difference between the given results of Schroedinger equation and classical physics; see [1]. All basic solutions of Schroedinger equation are equivalent to classical solutions under the same conditions. However, it does not hold in opposite direction. In the case of discrete spectrum some solutions of Hamilton equations have not the corresponding counterparts in the Schroedinger equation. As to the other solutions of Schroedinger equation they must be interpreted as statistical combinations of basic solutions or of corresponding classical states. Our result is in the full agreement with results of Ioanidou [6] and Hoyer [7] who have shown that Schroedinger equation may be derived if classical physics is combined with a kind of statistical distribution (see also, e.g., [2]).

The advantage of Schroedinger equation consists then in obtaining a complete statistical result in one solution if a statistical distribution of basic initial state is given. Consequently, the Schroedinger equation is very suitable if some initial parameters (e.g., impact parameter in collision processes and, consequently, also total energy value) are not exactly known and only their statistical distribution may be established, as it occurs in measurement processes.

ARGUMENTS CONTRA COPENHAGEN ALTERNATIVE

Some arguments were raised against the Copenhagen quantum mechanics soon after it was proposed. It is necessary to mention Pauli in 1933 and Einstein in 1935. However, their arguments were not sufficiently reflected and taken into account; see [8,2].

In the preceding we have shown that the hidden-variable theory based on the mere Schroedinger equation and the Copenhagen alternative have represented two different theories. Now we should like to summarize four different kinds of critical arguments that should be interpreted as the falsification of the Copenhagen alternative:

(i) The first kind has concerned the mathematical model based on the four mentioned assumptions. Pauli showed that in such a case the corresponding Hamiltonian should have continuous energy spectrum from $-\infty$ to $+\infty$, which was in disagreement with the requirement of the energy being positive, or limited from below. In 1964 Susskind and Glogover showed then also that exponential phase operator was not unitary, which indicated that the given Hilbert space was not complete to describe fully a corresponding physical system. Many unsuccessful attempts have been done to solve these deficiencies in the past century.

The successful solution has been formulated only recently when it has been shown that it has been necessary to remove these deficiencies one after the other. Pauli's problem may be removed by extending the Hilbert space according to the earlier proposal of

Lax and Phillips and the other problem by defining the Hilbert structure according to the proposal of Fain. The whole approach has been described (with corresponding quotations) and explained in Refs. [8,9].

(ii) The preceding problems have concerned the mathematical structure of the Copenhagen model. Strong criticism based on physical features of the model was started by Einstein and his collaborators in 1935 [10]. They tried to show on the basis of a Gedankenexperiment that some non-physical (or rather anti-ontological) features were involved in Copenhagen quantum mechanics. They recommended to describe a microscopic object with the help of the hidden-parameter theory, which was refused, however, decisively by Bohr [11]. And the hidden-variable theory was refused also by physical community that accepted fully phenomenological ideology.

The situation changed partially after the paper of Bohm [5] and mainly the paper of J. Bell [12] were published. Bell derived some inequalities, the validity (or not) of which should have been tested experimentally and should have decided the controversy between Einstein and Bohr. Einstein's Gedankenexperiment was modified and corresponding experiments were realized; the coincidence probabilities of two photons transmitted through two polarizers lying in opposite directions were measured. The measurements were finished in 1982 with that the mentioned inequalities were violated [13]. The given result was interpreted as the refusal of hidden-variable theory and the victory of Bohr, which has lasted practically till now.

However, we have succeeded in the last time to show that the given conclusion concerning the EPR experiments was done on the basis of two mistakes. First, it has not been taken into account that Bell's inequalities have been based on one important assumption that has limited strongly the range of their validity. It has not been possible to apply them to the case of EPR experiment with two polarizers. In fact, their violation in experiment has refused the classical physics only and not the hidden-variable theory. Second, it was also the mistake of Belinfante (see [14]) that influenced the given conclusion. Belinfante argued mistakenly that the predictions of Copenhagen quantum mechanics and hidden-variable theory in the given experiment had to be necessarily different, which has not been true, either; see, e.g., [15].

(iii) The other critical comment concerns the existence of the discrete part of Hamiltonian spectrum. On one side one argues that Schroedinger equation predicts the existence of states belonging to discrete energy values. And on the other side, the model requires the existence of pure states belonging to all intermediate energy values (i.e., to all superpositions).

(iv) And finally we should like to add another experimental argument refusing Copenhagen alternative. We shall start from the standard theory of polarized light, on which the EPR experiment has been based. And we shall measure the transmission of light through three polarizers, as indicated by the following scheme

$$o---|---|^{\alpha}---|^{\beta}--->$$

where α and β denote axis deviations of the second and third polarizers from the first one. According to standard quantum-mechanical model it should hold for corresponding light transmission probabilities

$$P(\alpha,\beta) = \cos^2\alpha\, cos^2(\alpha-\beta).$$

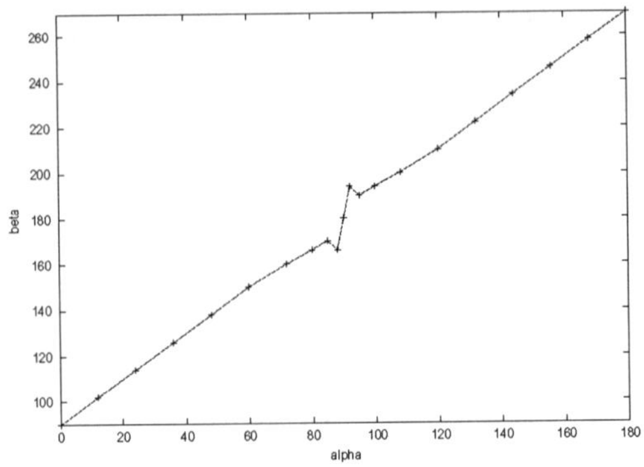

FIGURE 1. The pairs of angles α and β corresponding to minimal transmission values for chosen α; pair values used for the measurement shown in Fig. 2.

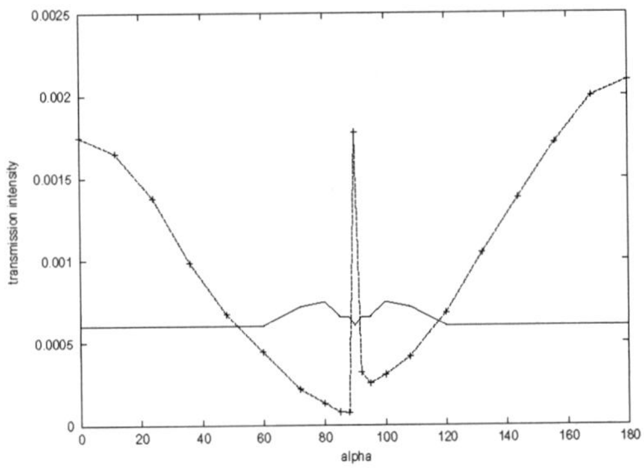

FIGURE 2. Light transmission through three polarizers (for $\alpha - \beta$ pairs shown in Fig. 1); experimental data - points on dashed line; quantum-mechanical (orthodox) prediction - full line.

The corresponding measurement was performed and the results were published in 1993-4; see [16]. Fundamental deviations from the given quantum-mechanical formula have been found, as it may be seen from Figs. 1 and 2.

The experiment was performed in two steps. First, for a chosen α the angle β was always established, so as the total light transmission to be minimal. The results are shown in Fig. 1. Then the values of total light transmission for individual angle pairs were measured; the results being represented by individual points on dashed line in Fig. 2. The quantum-mechanical predictions are represented by full line; the position of it being shifted in vertical direction somewhat arbitrarily as the values of the so called "imperfectness" of given polarizers were not available. In any case the standard quantum-mechanical prediction requires maxima at the positions where the experiment exhibits deep minima. The given discrepancy was not accented in [16] where the results were published for the first time as we were afraid reasonably that the paper would not have been accepted for publication in such a case.

CONCLUSION

And we are of the opinion that sufficient amount of arguments has been gathered that has enabled to denote Copenhagen quantum mechanics as falsified on both theoretical and experimental grounds, which should be regarded as certain knowledge obtained with the help of our reason (see [1]). Such a conclusion cannot be influenced by any other experimental data trying to bring a new support; see, e.g., results of delayed-choice experiment presented recently [17]. The same holds for the attempts of A. Leggett [18] trying to extend the Copenhagen model to macroscopic area. In all such cases the hidden-variable theory fulfills all necessary requirements.

All our analyzes performed in connection with Copenhagen quantum mechanics have led clearly to preferring causality and ontological approach to mere phenomenological description of measured values. One is forced to return to scientific methods ruling from Aristotle till the end of the middle age. It is in full agreement with the hidden-variable theory having decisive preference to other alternative physical theories at the present.

REFERENCES

1. M.V.Lokajíček: *New Advances in Physics* **1**, 69-77 (2007); also /arxiv/quant-ph/0611176.
2. M.V.Lokajíček: *Concepts of Physics* **4**, 317-39 (2007); also /arxiv/quant-ph/0611069.
3. M.V.Lokajíček: *Concepts of Physics* **4**, 657-72 (2007); also /arXiv:0710.3225v1 [phys.gen-ph].
4. E. Schrödinger: *Ann. Phys.* **79** (1926), 361-76; 489-527; 80 (1926), 437-90; 81 (109-39.
5. D. Bohm: *Phys. Rev.* **85**, 180-93 (1952).
6. H.Ioannidou: *Lett. al Nuovo Cim.* 34, 453-8 (1982).
7. U.Hoyer: "Synthetische Quantentheorie"; Georg Olms Verlag, Hildesheim, 2002.
8. P.Kundrát and M.Lokajíček: *Phys. Rev. A* **67**, art. 012104 (2003).
9. P.Kundrát and M.Lokajíček: "Irreversible time flow and Hilbert space structure" in *New Research in Quantum Physics*, edited by Vl.Krasnoholovets, F.Columbus, Nova Science Publishers, Inc., 2004, pp. 17-41.
10. A.Einstein, B.Podolsky and N.Rosen: *Phys. Rev.* **47**, 777-80 (1935).
11. N.Bohr: *Phys. Rev.* **48**, 696-702 (1935).
12. J.S.Bell: *Physics* **1**, 195-200 (1964).
13. A.Aspect, P.Grangier and G.Roger: *Phys. Rev. Lett.* **49**, 91-4 (1982).
14. F.J.Belinfante: *A survey of hidden-variable theories*, Pergamon, Oxford 1973, p. 283.
15. M.Lokajíček: *Controversy between Einstein and Bohr*, http:/arxiv/quant-ph/0601015.
16. J.Krása, M.Lokajíček and J.Jiřička: *Phys. Letters A* **186**, 279-82 (1994).
17. V. Jacques et al.: *Science* **315**, no. 5814, 966-8 (2007).
18. A. Leggett: "Does the everyday world really obey quantum mechanics?", http://online.itp.ucsb.edu/ online/colloq/leggett1/, accessed January, 2008.

Four qubit entangled pure States and K-way negativities

N. K. Sharma* and S. Shelly Sharma†

*Depto. de Matemática, Univ. Estadual de Londrina, Londrina, Parana, 86051-990, BRAZIL
†Depto. de Física, Univ. Estadual de Londrina, Londrina, Parana, 86051-990, BRAZIL

Abstract. It is known that by Stochastic local operations and classical communication (SLOCC) a pure state of four qubits can be transformed to a state belonging to one of a set of nine families of states. Using the eigen vectors corresponding to negative eigenvalues of global partial transpose and the K-way partial transposes, we construct measures of genuine four-partite, tri-partite, and bi-partite entanglement for all the nine families of states.

Keywords: Entanglement, four qubit canonical States, Global Negativity, K-way negativity
PACS: 03.65.Ud, 03.67.Mn

Negativity [1], based on Peres Horodecki PPT criterion, has been shown to be an entanglement monotone. Recently, we proposed a characterization of N-partite quantum states based on K-way negativities [2], where $2 \leq K \leq N$. The K-way partial transpose with respect to a subsystem of an N-partite composite system is constructed by partial transposition subject to specific constraints on transposed matrix elements for each value of K. A four qubit (ABCD) state may have genuine 4-partite, tri-partite, and bi-partite entanglement. It has been shown in ref. [3] that by Stochastic local operations and classical communication (SLOCC) a pure state of four qubits can be transformed to a state belonging to one of a set of nine families of states. We have shown in [2] that the partial transpose with respect to a subsystem can be written as a sum of matrices, where each matrix corresponds to a specific K-way partial transpose. In this contribution, we use the eigen vectors corresponding to negative eigenvalues of global partial transpose and the K-way partial transposes to construct measures of genuine four-partite, tri-partite, and bi-partite entanglement for all the nine families of states.

As an example, consider the set of states

$$\begin{aligned} G_{abcd} &= \frac{a+d}{2}\left(|0000\rangle + |1111\rangle\right) + \frac{a-d}{2}\left(|1100\rangle + |0011\rangle\right) \\ &\quad + \frac{b+c}{2}\left(|1010\rangle + |0101\rangle\right) + \frac{b-c}{2}\left(|1001\rangle + |0110\rangle\right), \end{aligned}$$

which represents one out of the nine families of four qubit states. These states are characterized by $E_3^A = 0$,

$$E_4^A = \frac{1}{2}\left(\left(a^2+d^2\right)^2 + \left(b^2+c^2\right)^2 + a^2d^2 + b^2c^2\right)$$

and $E_2^A = 1 - E_4^A$. The global Negativity N_G^A, and partial $K-way$ negativity E_K^A are defined in ref. 2]. Fig. 1 displays contour plots of N_G^A, E_4^A, and E_2^A for the normalized

CP1018, *Frontiers of Fundamental and Computational Physics*,
edited by B. G. Sidharth, F. Honsell, O. Mansutti, K. Sreenivasan, and A. De Angelis

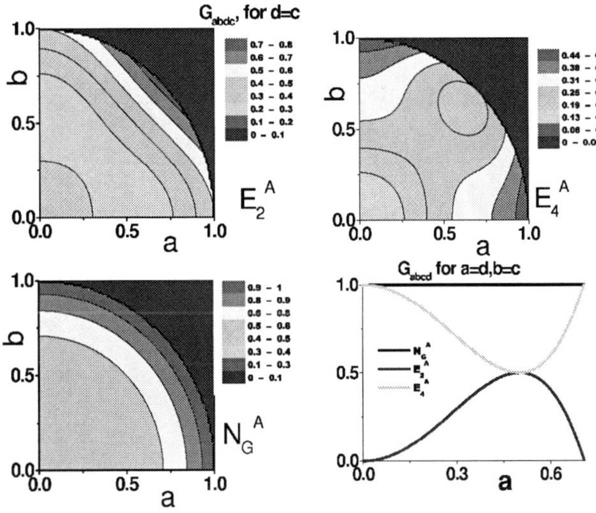

FIGURE 1. N_G^A, E_2^A, and E_4^A for the state G_{ABCD}.

states G_{abcd} for the case $c = d$ and for the special case $a = d$ and $b = c$.

Partial four way negativity E_4^A is a measure of genuine 4-partite entanglement of the state. Three-way partial negativity E_3^A determines the probabilistic entanglement that becomes available to the three parties after the fourth party measures the state of the qubit it holds. Similarly, two-way partial negativity E_2^A measures the bipartite entanglement between the entangled pairs of subsystems. As the Global negativity is a sum of partial K-way negativities plus a term coming from the overlap of eigenvectors of global partial transpose with the state operator, the local unitary rotations can be used to enhance a given set of partial K-way negativities at the cost of others. The coefficients in a canonical state being local invariants, the partial K-way negativities are proper entanglement measures satisfying the conditions of normalization, convexity and monotonicity. For a state that is not in canonical form, the negativities measure the coherences present in the composite system.

Financial support from Fundacao Araucaria, PR and FAEPE/UEL Brazil is acknowledged.

REFERENCES

1. K. Zyczkowski, P. Horodecki, A. Sanpera, and M. Lewen-stein, *Phys. Rev. A*, **58**, 883 (1998).
2. S. Shelly Sharma and N. K. Sharma, *Phys. Rev. A* **76** (2007); S. Shelly Sharma and N. K. Sharma, arXiv: quant-ph/0608062.
3. F. Verstraete, J. Dehaene, B. De Moor, and H. Verscheldel, *Phys. Rev. A* **65**, 052112(2002)

Controlled Manipulation of Three Cold Trapped ions in a Cavity

S. Shelly Sharma*, Eduardo de Almeida* and N. K. Sharma†

*Depto. de Física, Univ. Estadual de Londrina, Londrina, Parana, 86051-990, BRAZIL
†Depto. de Matemática, Univ. Estadual de Londrina, Londrina, Parana, 86051-990, BRAZIL

Abstract. In this work a scheme to generate maximally entangled three qubit W-states is proposed and entanglement dynamics of three two-level cold ions trapped in a high finesse cavity tuned to the red sideband of ionic vibrational motion, is investigated. Analytical expressions for the state of composite system for specific initial state preparations are obtained and used to calculate the probability of generating three ion W-state. We use the partial K-way negativities calculated from K-way partial transpose of the state operator, as a measure of K-way entanglement of the system. The composite system is found to have genuine four-partite, tripartite as well as bipartite entanglement, when number of center of mass vibrational quanta is greater or equal to three .

Keywords: trapped ions, W-state generation, entanglement, K-way negativities
PACS: 03.67.Bg,03.67.Mn

Interest in controlled manipulation of multipartite entangled states is motivated by the possible use in quantum communication, dense coding and quantum teleportation. We analyze the entanglement dynamics of spatially separated three two-level cold trapped ions in a high finesse cavity with the cavity tuned to the red sideband of ionic vibrational motion. Eigenstates of pauli operators model the internal states of ions. The composite system consists of ionic internal states, vibrational state in the trap, and the cavity field state.

Consider three cold trapped two level ions interacting with quantized field of a high finesse cavity. Each ion is placed at a node of the cavity field and ions are far enough to not have dipole interaction between them. With the cavity field tuned to red sideband of ionic vibrational motion, the relevant part of interaction Hamiltonian in the rotating-wave approximation and Lamb-Dicke limit ($\eta \ll 1$), in interaction picture is

$$\hat{H}_{II} = \hbar g\eta\,[\hat{\sigma}_{+}\hat{b}\hat{a} + \hat{\sigma}_{-}\hat{b}^{\dagger}\hat{a}^{\dagger}], \qquad (1)$$

where g is the ion-cavity coupling constant and η the Lamb-Dicke parameter. Analytical expressions for the state of composite system as a function of interaction time and interaction strength are obtained for specific initial state preparations and used to calculate the probability of generating three ion W-states. Figure (1) displays the probabilities $P_i(\tau)$ ($i = 0$ to 3) of finding the system in the states $|000,3,3\rangle$, $|W_1,2,2\rangle$, $|W_2,1,1\rangle$ and $|111,0,0\rangle$, respectively. Here $|W_1\rangle$ and $|W_2\rangle$ are W-like states with a single excited ion and two ions in excited state, respectively. The basis states are labelled as $|i_1,i_2,i_3,m,n\rangle$, where $i_{ion} = 0$ represent the ground and 1 the excited state of the ion. The number of vibrational quanta is m and the number of photons in the cavity is given by n. The system periodically returns to initial separable state with a period of $\approx \frac{2,3}{g\eta}$.

CP1018, *Frontiers of Fundamental and Computational Physics*,
edited by B. G. Sidharth, F. Honsell, O. Mansutti, K. Sreenivasan, and A. De Angelis

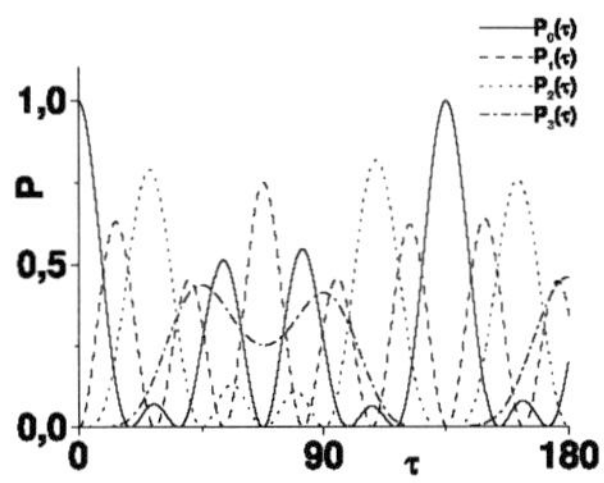

FIGURE 1. The probabilities P_i, $i = 0$ to 3 versus $\tau(= g\eta t)$ for the initial state $|000,3,3\rangle$.

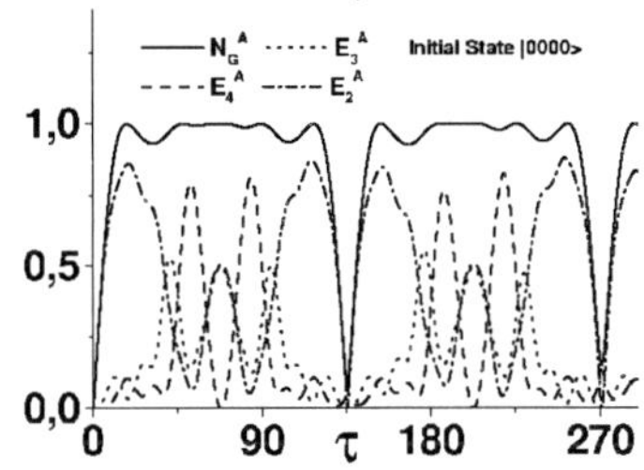

FIGURE 2. The global negativity N_G^A, and entanglement measures E_K^A, for $K = 2$ to 4 versus $\tau(= g\eta t)$ for the initial state $|0000\rangle$.

The number of initial state vibrational quanta (m) offers a control mechanism for manipulation of composite system states in the sense that whereas for m=1 only bipartite entanglement is possible, the system may have tripartite entanglement as well for m=2. For m greater or equal to three, the composite system may have genuine 4-partite, tripartite as well as bipartite entanglement. This is an interesting aspect unique to systems where vibrational motion of ions is coupled to cavity field in contrast to the ions coupled only to quantized cavity field or only to the vibrational modes. Entanglement dynamics is examined by plotting the Global [1], and partial 4-way, 3-way, and 2-way negativities [see ref. 2 for definitions], for the composite system state, as a function of interaction time. Figures (2) displays the global negativity and $K-$way entanglement measures for qubit A for initial state $|000,3,3\rangle$ of the system.

By using the ion-cavity coupling strengths achieved in experimental realizations, the interaction time needed for W-state generation is found to be of the order of 10μ sec.

Financial support from Fundacao Araucaria, PR, FAEPE/UEL and Capes Brazil is acknowledged.

REFERENCES

1. K. Zyczkowski, P. Horodecki, A. Sanpera, and M. Lewen-stein, *Phys. Rev. A*, **58**, 883 (1998).
2. S. Shelly Sharma and N. K. Sharma, *Phys. Rev. A* **76** (2007); S. Shelly Sharma and N. K. Sharma, arXiv: quant-ph/0608062.

Vortex-antivortex macroscopic quantum state

Alexey Okulov

A.M.Prokhorov General Physics Institute RAS, Vavilova str.38, 119991, Moscow, Russia

Abstract.
The macroscopic quantum state of ultracold atoms in laser optical trap is considered. A confining optical field is built as spatially periodic array of optical vortices. The optical angular momentum transfer induces the rotational motion of ultracold ensemble. A quantum motion in such a *labyrinth* trap is analyzed in the framework of the Gross-Pitaevsky equation.

Keywords: BEC, vortices, lattices, wavefunction, optical trapping
PACS: 42.50.-p, 42.50.Dv,42.50.Ex,03.67.Bg, 03.75.Gg,03.75.Lm

We analyze the pinning of quantized matter wave vortices by optical vortices in specially arranged optical dipole labyrinth traps [1]. Using previously obtained results for asymmetric elongated traps [2] and longitudinally modulated traps [3] we use the fact that $3D$ Gross-Pitaevsky equation[4]:

$$i\hbar\frac{\partial\Psi(\vec{r},t)}{\partial t} = -\frac{\hbar^2}{2m}\Delta\Psi(\vec{r},t) + V_{ext}(\vec{r})\Psi(\vec{r},t) + \frac{4\pi\hbar^2 a_s(\vec{B})}{m}\Psi(\vec{r},t)|\Psi(\vec{r},t)|^2, \qquad (1)$$

where

$$V_{ext}(\vec{r_\perp},z) = V_z + V_\perp = \frac{m\,\omega_z{}^2 z^2}{2} - Re[\alpha(\omega)]\,|E(\vec{r}_\perp)|^2 + \frac{m\,\omega_\perp{}^2|(\vec{r}_\perp)|^2}{2} \qquad (2)$$

and $\alpha(\omega)$ is atomic polarizability[5], admits the application of the standard method of separation of variables in pancake geometry[6].The equations for longitudinal and transversal parts of macroscopic wavefunction:

$$\Psi(\vec{r},t) = \Psi_{||}(z,t)\Psi_\perp(\vec{r}_\perp,t) \qquad (3)$$

are solved separately[1]. The first equation for 1D harmonic oscillator is solved exactly. The transversal two dimensional GPE is solved numerically by means of split-step FFT transform method on 512x512 computational mesh[7]. The vortex-antivortex "antiferromagnet"-like optical lattice of rectangular symmetry is found to support vortex-antivortex "antiferromagnet-like" matter waves.

In contrast to rotating "bucket" trap [8] and rotating "busket" trap [9], where superfluid vortex lattices rotate as a rigid body, our solution is static. The superfluid vortices in our "basket" trap are pinned at the nodes (i.e. zeros of amplitude or phase singularities) of optical interference pattern. The argument of $arg(\Psi_\perp)$versus transverse coordinates $\vec{r}_\perp$ shows clearly the loci of rectangularily spaced vortices with alternating circulations. The

CP1018, *Frontiers of Fundamental and Computational Physics*,
edited by B. G. Sidharth, F. Honsell, O. Mansutti, K. Sreenivasan, and A. De Angelis

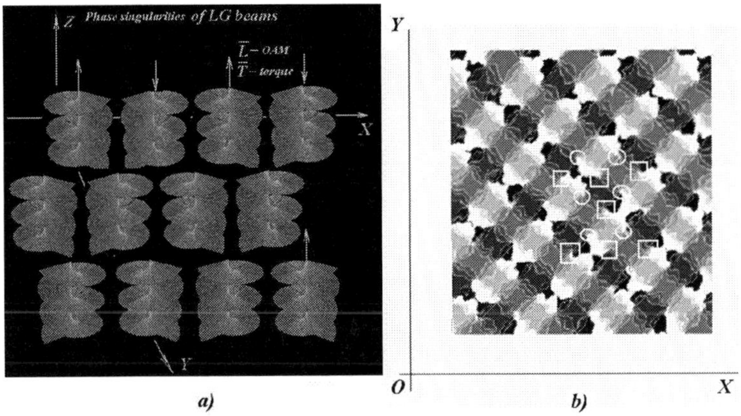

FIGURE 1. a) Multiply connected BEC optical trap composed of rectangular array of optical vortex beams. The optical phase dislocations are shown to match their phase fronts. Flip-flop orientations from site to site of local orbital angular momenta $\vec{\mathbf{L}}$ and corresponding torques $\vec{\mathbf{T}}$ are shown. **Z** - is beam propagation axis. b): The **2D** plot of the phase of macroscopic wavefunction **arg■(x, y)** in the **X, Y** - plane. **Z** - axis is normal directed to reader.

field of classical velocities obtained via Madelung transform is shown to form labyrinth pattern.

REFERENCES

1. A.Yu.Okulov, J.Mod.Opt.,**55**,N.2,p.241-257(2008).
2. V.M.Perez-Garcia, H.Michinel and H.Herrero, Phys.Rev. A **57**, 3837 (1998).
3. R.Fidele,P.K.Shukla,S.De.Nicola,M.A.Manko,V.I.Manko, F.S.Cataliotti, JETP Lett, **80**, 609 (2004).
4. F. Dalfovo, S.Giorgini, S.Stringari, L.P.Pitaevskii, Rev.Mod.Phys. **71**, 463 (1999).
5. R.Grimm, M.Weidemuller and Yu.B.Ovchinnikov, Adv.At.Mol.Opt.Phys. **42**, 95 (2000).
6. L.C. Crasovan,V.Vekslerchik,V.M.Perez-Garcia, J.P. Torres,D.Mihalache,L.Torner., Phys.Rev. A **68**, 063609 (2003).
7. A.Yu.Okulov, Opt.Comm., **99**, p.350-354 (1993).
8. J.R.Abo-Shaerr, C.Raman, J.M.Vogels and W.Ketterle, Phys.Rev.Lett.,**88**, 070409 (2002).
9. S.Tung, V.Schweikhard, and E.A.Cornell, Phys.Rev.Lett., **97**, 240402 (2006).

PARAQUANTUM YANG-MILLS THEORY AND PBRST TRANSFORMATIONS[1]

L. Maghlaoui *,† and N. Belaloui†

* *Université M'hamed Bouguerra, Faculté des Sciences, Département de Physique, Boumerdès, Algeria*
† *LPMPS, Faculté des Sciences Exactes, Département de Physique, Université Mentouri Constantine, Algeria*

Abstract. We investigate an extension of the non-abelian gauge theories in the paraquantum formalism of order 2 for both gauge fields and matter fields. The model,which consists in applying the Faddeev-Popov formalism on the Klein ordinary fields, gives a set of ParaBRST (PBRST) transformations satisfying the nilpotence property. This investigation conduct us to a discussion about the (im)possibility to reformulate an ordinary field theory solely in terms of parafields of order two. A modified Klein transformations which could give some progress in this direction is proposed.

Keywords: Yang-Mills Theory, Parastatistics, BRST Transformations
PACS: 11.10.Lm, 05.30.Pr, 11.10.Ef§

INTRODUCTION

In this work, we are interested in some questions related to the extension of the non abelian gauge theories in the paraquantum case of order two for both gauge fields and matter fields. In view of the complexity of the trilinear commutation relations satisfied by the parafields, we are lead to use the development in terms of Klein fields that verify normal bilinear relations and which permit the transition to the path integral formalism. We construct a set of transformations which leaves the theory invariant and which we will call PBRST transformations.

THE MODEL

It consists in a direct extention of the ordinary non abelian theory in terms of parafields. The model is described by a paraquantum Yang-Mills Lagrangian of order $r = 2$ satisfying a weak locality condition that permits to fix the structure of the observables and which describes a massive Dirac parafield $\Psi(x)$ and a massless parabosonic field A_μ.

$$L_{YM} = \frac{i}{2}\left[\overline{\Psi}(x), \not{D}\Psi(x)\right] + \frac{m}{2}\left[\overline{\Psi}(x), \Psi(x)\right] + \frac{\lambda}{2}(-1)^{N_\Psi} \vdots \left[\overline{\Psi}(x), \Psi(x)\right] \vdots - \frac{1}{4}F^a_{\mu\nu}F^{\mu\nu a} \tag{1}$$

---

[1] This work is supported by the Algerian Ministry of the Higher Education and Scientific Research under contract No.D2501/11/05.

CP1018, *Frontiers of Fundamental and Computational Physics*, edited by B. G. Sidharth, F. Honsell, O. Mansutti, K. Sreenivasan, and A. De Angelis

where, the second term of the mass term of $\Psi(x)$ is in accordance with the weak locality condition and imply that the masses of the ordinary Klein fields may be identified with $(m+\lambda)$ and $(m-\lambda)$. Here

$$F^a_{\mu\nu} = \partial_\mu A^a_\nu - \partial_\nu A^a_\mu + gC^a_{bc}A^b_\mu A^c_\nu,\ D_\mu\Psi_i = \partial_\mu\Psi_i + gT^k_{ij}A^k_\mu\Psi_j \tag{2}$$

$$N_\Psi = \int d^3x \left[\overline{\Psi}(x), \Psi(x)\right],\ \text{and}\ [T_k, T_m]_{ij} = iT^l_{ij}C^l_{km} \tag{3}$$

Different steps are followed, namely, the Green decomposition, the Klein transformations and the field redefinitions, then in the aim to have different couplings g_1 and g_2 respectively for the interaction fields Klein components $\mathbb{A}^{(1)}_\mu$ and $\mathbb{A}^{(2)}_\mu$ of the parafield A_μ, we postulale a new modified Klein transformation between the Green $\hat{\psi}^{(\alpha)}(x)$ and the Klein $\hat{\phi}^{(\alpha)}(x)$ components of any field $\Psi(x)$ as follows:

$$\hat{\psi}^{(\alpha)}(x) = (-i)^{1-\rho(\alpha)}\lambda_\alpha K_{\alpha-\rho(\alpha)}\hat{\phi}^{(\alpha)}(x) \tag{4}$$

then applying the Faddeev-Popov formulation by the introduction of the two terms $L^{(\alpha)}_{FP}$ and $L^{(\alpha)}_{GF}$, one obtain the final Lagrangian

$$L = \sum_{\alpha=1}^{2}\{i\overline{\varphi}^{(\alpha)}(x)\not{D}\varphi^{(\alpha)}(x) + (m+n_\alpha\lambda)\overline{\varphi}^{(\alpha)}(x)\varphi^{(\alpha)}(x) - \frac{\gamma\lambda_\alpha^2}{2}\left(\partial_\mu\mathbb{A}^{\mu(\alpha)}\right)^2$$
$$-\overline{\Omega}^{(\alpha)a}\partial^\mu\tilde{D}^{(\alpha)}_\mu\Omega^{(\alpha)a} - \frac{1}{4}\mathbb{F}^{\mu\nu(\alpha)}\mathbb{F}^{(\alpha)}_{\mu\nu}\} \tag{5}$$

where

$$D_\mu = \partial_\mu - ig_1T^k\mathbb{A}^{(1)k}_\mu - ig_2T^k\mathbb{A}^{(2)k}_\mu\ \text{and}\ \tilde{D}^{(\alpha)}_\mu\Omega^{(\alpha)a} = \partial_\mu\Omega^{(\alpha)a} + g_\alpha C^a_{bc}\mathbb{A}^{(\alpha)b}_\mu\Omega^{(\alpha)c} \tag{6}$$

$$g_1 = \frac{g}{\sqrt{2}}(\lambda_1 - K_2\lambda_2)\ \text{and}\ g_2 = \frac{g}{\sqrt{2}}(\lambda_1 + K_2\lambda_2) \tag{7}$$

One can easily verify that the following transformations are the nilpotent PBRST transformations of the theory

$$\delta\varphi^{(\alpha)}_i(x) = -iT^k_{ij}\left(a^1_{A_\mu}\Omega^{(1)k} + a^2_{A_\mu}\Omega^{(2)k}\right)\varphi^{(\alpha)}_j(x) \tag{8}$$

$$\delta\overline{\varphi}^{(\alpha)}_i(x) = iT^k_{ij}\left(a^1_{A_\mu}\Omega^{(1)k} + a^2_{A_\mu}\Omega^{(2)k}\right)\overline{\varphi}^{(\alpha)}_j(x), \delta\mathbb{A}^{(\alpha)a}_\mu = -\varepsilon\tilde{D}_\mu\Omega^{(\alpha)a} \tag{9}$$

$$\delta\Omega^{(\alpha)a} = -\frac{1}{2}\varepsilon g_\alpha C^a_{bc}\Omega^{(\alpha)b}\Omega^{(\alpha)c}, \delta\overline{\Omega}^{(\alpha)a} = 2\gamma_\alpha\varepsilon\partial^\mu(\partial_\nu\mathbb{A}^{\nu(\alpha)a}) \tag{10}$$

CONCLUSION

In this work, we investigate the (im)possibility to reformulate an ordinary field theory solely in terms of a parafields of order 2. Beyond the parafermion fields, we are particularily interested in the regrouping of the two gauge bosonic fields in term of a gauge parabosonic field of order 2. For lack of an answer, notice here that the modified Klein transformations proposed is not only intended to bring anomalous relations to normal relations (which is assured by the usual one), but this distinguish again the interaction fields by the introduction of a constant parameter λ_α

PART III

RELATIVITY AND COSMOLOGY

The redshift-distance relation extended to higher redshifts in Cosmological General Relativity

John G. Hartnett

School of Physics, the University of Western Australia,
35 Stirling Hwy, Crawley 6009 WA Australia

Abstract. The redshift-distance modulus relation, the Hubble Diagram, derived from Cosmological General Relativity has been extended to arbitrarily large redshifts. Numerical methods were employed and a density function was found that results in a valid solution of the field equations at all redshifts. The extension has been compared to 302 type Ia supernova data as well as to 69 Gamma-ray burst data. The latter however do not not truly represent a 'standard candle' as the derived distance modulii are not independent of the cosmology used. Nevertheless the analysis shows a good fit can be achieved without the need to assume the existence of dark matter.

The Carmelian theory is also shown to describe a universe that is always spatially flat. This results from the underlying assumption of the energy density of a cosmological constant $\Omega_\Lambda = 1$, the result of vacuum energy. The curvature of the universe is described by a *spacevelocity* metric where the energy content of the curvature at any epoch is $\Omega_K = \Omega_\Lambda - \Omega = 1 - \Omega$, where Ω is the matter density of the universe. Hence the total density is always $\Omega_K + \Omega = 1$.

Keywords: Cosmological General Relativity, high-z type Ia supernovae, gamma-ray bursts, cosmology
PACS: 04.20.-q, 04.50.-h

INTRODUCTION

Carmeli's Cosmological General Relativity (CGR) is a space-velocity theory of the expanding universe. It is a description of the universe at a particular fixed epoch of cosmic time t and involves a new dimension v, the velocity of the expansion of the fabric of space itself, in which the galaxies are sitting. For a full introduction to the theory, the reader should familiarize himself with either of the late Moshe Carmeli's books *Cosmological Special Relativity, 2nd Edition* [2] or *Cosmological Relativity* [4] .

Carmeli incorporated the Hubble law as a fundamental axiom, with $H_0^{-1} \approx \tau$, a universal time constant, into a general 4D Riemannian geometrical theory satisfying the Einstein field equations (Ref. [2], appendix A). The observables, redshift and distance (Hubble distance), are the coordinates of his theory. The expanding universe is then described by its *spacevelocity* "phase space" coordinates (v, r, θ, ϕ), which do not involve time t but this new dimension v. Cosmic time is measured from the present epoch, with $t_c = 0$, back toward the beginning with $t_c = \tau$, but determined from the expansion velocity of the fabric of space, i.e $t_c \to \tau$ as $v \to c$. In this case the galaxies act as tracers embedded in the expanding space. But in practice redshift, and not velocity, is what astronomers actually measure, as well as distance determined from the luminosity of the source.

CP1018, *Frontiers of Fundamental and Computational Physics*,
edited by B. G. Sidharth, F. Honsell, O. Mansutti, K. Sreenivasan, and A. De Angelis

PHASE SPACE EQUATION

Carmeli found a solution to his field equations, modified from Einstein's, (see [8] and [2, 5]) which led to the differential equation

$$\frac{dr}{dv} = \tau\sqrt{1+\left(\frac{1-\Omega}{c^2\tau^2}\right)r^2}, \tag{1}$$

where $\Omega = \rho/\rho_c$ the averaged matter density of the Universe expressed as a fraction of the critical density $\rho_c = 3/8\pi G\tau^2$. But the solution requires that

$$1+(1-\Omega)\frac{r^2}{c^2\tau^2} > 0. \tag{2}$$

In the above Carmelian theory it initially assumed that the Universe has expanded over time and at any given epoch it has an averaged density ρ. The solution of the field equations was sought on that basis. However because the Carmeli metric is solved in an instant of time (on a cosmological scale) any time dependence is neglected. But since we observe the expanding Universe with the coordinates of Hubble at each epoch (or redshift z) we see the Universe with a different density $\rho(z)$. In the following this fact is taken into account to extend the validity of the solution.

COMPARISON WITH OBSERVATION

In order to compare CGR theory with observations one requires a relationship that gives proper distance as a function of redshift. Equation (1) relates the velocity of the receding sources to their distance. Carmeli integrated (1) and found the redshift distance relationship

$$\frac{r}{c\tau} = \frac{\sinh\left(\beta\sqrt{1-\Omega}\right)}{\sqrt{1-\Omega}}, \tag{3}$$

where,

$$\beta = \frac{v}{c} = \frac{(1+z)^2-1}{(1+z)^2+1}, \tag{4}$$

and Ω the average matter density of the universe, a function of redshift z, must be found.

This equation was used in both Hartnett [8], which forms the basis of the work presented here, and Oliveira and Hartnett [13] where comparisons were made to the high redshift SNe Ia data from Riess *et al* [16], Astier *et al* [1] and Knop *et al* [12]. The proper distance is converted to magnitude as follows.

$$m(z) = \mathscr{M} + 5\,log\left[\mathscr{D}_L(z;\Omega)\right], \tag{5}$$

where $\mathscr{D}_L$ is the dimensionless "Hubble constant free" luminosity distance. Refer [14, 15]. Here

$$\mathscr{M} = 5\,log\left(\frac{c\tau}{Mpc}\right) + 25 + M_B + a. \tag{6}$$

The units of $c\tau$ are Mpc. The constant 25 results from the luminosity distance expressed in Mpc. However, $\mathscr{M}$ in (5) represents a scale offset for the distance modulus (m-M_B). It is sufficient to treat it as a single constant chosen from the fit. In practice we use a, a small free parameter, to optimize the fits.

In CGR the luminosity distance was found [10] to be slightly different to the expression used in the FRW theory. This fact, results from a relativistic effect on the emission times of the photons from the distant source. In CGR, times at cosmological distances add according to a relativistic addition law, [3] when referred to the observer at $t_c = 0$. When this is taken into account we get an additional factor $(1-\beta^2)^{-1/2}$, hence $\mathscr{D}_L$ is given by

$$\mathscr{D}_L(z;\Omega) = \frac{r}{c\tau}(1+z)\left(1-\beta^2\right)^{-1/2} \tag{7}$$

using (3), which is a function of Ω and z.

The parameter $\mathscr{M}$ incorporates the various parameters that are independent of the redshift, z. The parameter M_B is the absolute magnitude of the supernova at the peak of its light-curve and the parameter a allows for any uncompensated extinction or offset in the mean of absolute magnitudes or an arbitrary zero point. It also acts as a correction to the value of τ since $\mathscr{M}$ contains it. The absolute magnitude then acts as a "standard candle" from which the luminosity and hence distance can be estimated.

The value of M_B need not be known, neither any other component in $\mathscr{M}$, as $\mathscr{M}$ has the effect of merely shifting the fit curve (7) along the magnitude axis. However by choosing the value of the Hubble-Carmeli constant $\tau = 4.28 \times 10^{17}\,s = 13.58\,Gyr$, which is the reciprocal of the chosen value of the Hubble constant in the gravity free limit $h = 72.17 \pm 0.84$ (statistical) km s^{-1}Mpc^{-1}[13] $\mathscr{M} = 43.09 + M_B + a$.

EXTENDED REDSHIFT RANGE

The proper distance from (3) however has limited application because it has essentially been found by integrating (1) assuming constant density then using (4). The result being only valid over a limited redshift range. In order to improve on this, because the observables are distance (magnitude) and redshift not velocity, the differential equation (1) must first be converted to dr/dz using the chain rule as follows.

$$\frac{1}{\tau^2}\left(\frac{dr}{dz}\frac{dz}{dv}\right)^2 = 1 + \frac{1-\Omega(z)}{c^2\tau^2}r^2, \tag{8}$$

where it must be remembered that matter density Ω is also a function of redshift. The derivative dv/dz is then calculated from (4) and substituted into (8), which becomes

$$\frac{1}{c^2\tau^2}\left(\frac{dr}{dz}\right)^2 = \frac{16(1+z)^2}{(1+(1+z)^2)^4}\left(1+(1-\Omega(z))\frac{r^2}{c^2\tau^2}\right). \tag{9}$$

This differential equation is then solved for $r/c\tau$, but because it involves mixed terms on the rhs, it must be numerically solved. But we can employ a 'bootstrapping' technique.

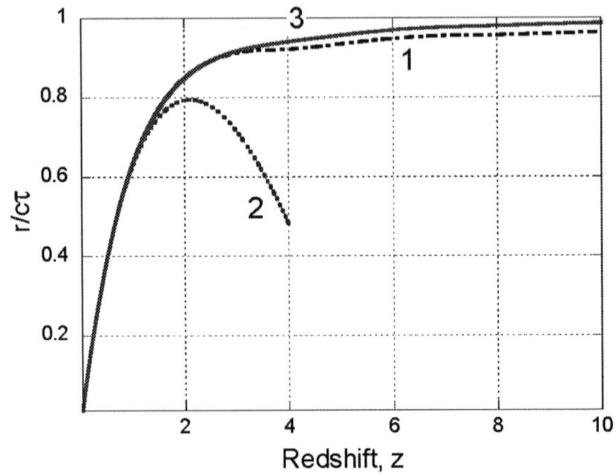

FIGURE 1. Redshift distance relation: Curve 1 represents $r/c\tau$ from the solution of (9) with $\Omega(z) = \Omega_m(1+z)^3$, curve 2 represents $r/c\tau$ from (3) with $\Omega(z) = \Omega_m(1+z)^3$ and curve 3 represents $r/c\tau$ from the solution of (9) but with the density taken from curve 1 of fig. 2.

By substituting the limited redshift solution from (3) into the rhs and assuming the averaged matter density in the universe

$$\Omega(z) = \Omega_m(1+z)^3, \tag{10}$$

(9) was numerically solved for $r/c\tau$. The result is shown in curve 1 of fig. 1. This is compared with $r/c\tau$ resulting from the initial equation (3) represented by curve 2.

Then $r/c\tau$ (of curve 1) was then taken as input on the rhs of (9) and iterated again to numerically solve for $r/c\tau$. However because of the requirement (2) on the solution from the field equations the result is only valid to about $z = 2.5$ with the assumed density dependence (10) on the present epoch density $\Omega_m = 0.04$. Dimensionally the matter density depends on the length cubed if we assume constant mass, but the precise density dependence is unknown at high redshifts. Therefore curve 1 of fig. 1 plotted beyond $z = 2.5$ can only be considered approximate.

Now in CGR there is the scale radius

$$R_0 = \frac{c\tau}{\sqrt{|1-\Omega|}}. \tag{11}$$

From (11) and the condition (2) the range of validity of the solution (1) is $0 \leq r/c\tau < 1 \leq R_0$ if $\Omega \leq 1$ and $0 \leq r/c\tau < R_0$ if $\Omega > 1$. This means the solution is valid for all values of the matter density $\Omega(z) < 2$.

Now if we again use the solution (curve 1 of fig. 1) and solve (9) for matter density Ω, it results in a density function such that (2) is always true. The density function is shown as curve 1 in fig. 2 and is compared with the initial density function (10) shown as curve 2. Finally using this as the density function in (9) and numerically solving for $r/c\tau$ results in curve 3 in fig. 1.

The new density function has some unexpected features, but it produces a smooth monotonically increasing function of distance on redshift as expected. The fluctuating

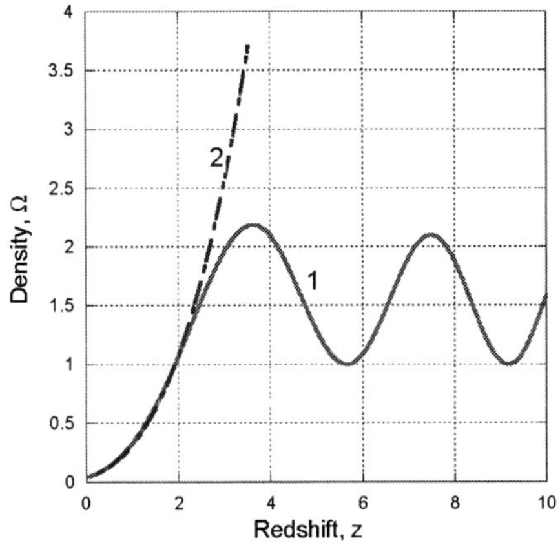

FIGURE 2. Density as a function of redshift: Curve 1 represents density derived from (9) and curve 2 represents the density function $\Omega(z) = \Omega_m(1+z)^3$ with $\Omega_m = 0.04$.

part of $\Omega(z)$, seen in curve 1 of fig. 2 has a decreasing period as a function of z, a minimum at $\Omega = 1$ and a maximum slightly greater than $\Omega = 2$. Any density function that remains within the range of this function for $z > 4$, even a constant density, will yield a smooth monotonic redshift-distance relation. However oscillating density as a function of redshift could represent episodic creation from the vacuum as the universe expands. Particle production at those epochs is a byproduct of the expansion process. Gemelli [11] found from a hydrodynamic solution of the 5D problem that particle production must occur from the expansion provided the Universe is not isentropic.

QUALITY OF CURVE FITS

The initial equation (3) has been curve fitted to SNe Ia data [13] and excellent fits resulted without any dark matter component. The best statistical fit resulted in a value of $\Omega_m = 0.0401 \pm 0.0199$, which is consistent with the observed baryonic matter density [7]. The fit was accomplished with the combined Gold and Silver SNe Ia data sets of Riess *et al* [16] with that of Astier *et al* [1] that extended to $z = 1.75$, the current limit of observational data. Inspection of the $r/c\tau$ curve (curves 1 or 3 shown in fig. 1) and the initial curve (curve 2) for the region $1.5 < z < 1.75$ indicates that only the last two SNe Ia data would be sensitive to the numerically determined extension here. See crosses in fig. 3.

Using $r/c\tau$ from the extended equation (9) I have curve fitted to the data set of Riess *et al* and also to combined data set of Riess *et al* and Astier *et al*. However before using Astier *et al* data I added 0.14 magnitudes to their quoted errors to align their errors with those of Riess *et al*. Otherwise Astier *et al* seem to have underestimated their magnitude errors, which unfairly bias the χ^2 calculations. The resulting χ^2s for the model here are shown in Table 1.

TABLE 1. Curve fit χ^2 parameter

Data set(s)	Ω_m	χ^2/N
SNe Ia	0.00	1.4120
Riess et al Gold & Silver	0.02	1.4047
$N = 185$	0.04	1.3987
$a = 0.2400$	0.06	1.3941
(a from $z < 0.23$)	0.10	1.3888
	0.27	1.4258
SNe Ia	0.00	1.2366
Riess et al + Astier et al SNLS	0.02	1.2341
$N = 302$	0.04	1.2326
$a = 0.1894$	0.06	1.2321
(a from $z < 0.24$)	0.10	1.2340
	0.27	1.2871
SNe Ia + GRBs	0.00	1.2191
Riess + Astier + Schaefer	0.02	1.2005
$N = 371$	0.04	1.1902
$a = 0.2113$	0.06	1.1860
(a from $z < 2$)	0.10	1.1870
	0.27	1.2538

Clearly more SN Ia data, at higher redshifts, are needed to test this prediction. Consequently GRBs have been used to get distance modulus and redshift as independent parameters (from Schaefer [17]) and are shown as dots in fig. 3. However the magnitude calculation is not really independent of the chosen cosmology as explained in the latter reference. Nevertheless using the data from columns 2 and 8 of Table 6 [17], and making the necessary changes to convert the magnitudes from the FRW Concordance Model to that of the Carmeli model, new distance modulii are obtained.

In order to use the data from Schaefer [17] first the specific details particular to a cosmological model he used are subtracted then the specific details from the CGR theory are applied. This means small corrections are made. This also means the magnitudes calculated this way are not really independent of our chosen CGR cosmology either. In fact, problems related to the cosmological dependence of the GRBs calibration, in principle, may invalidate such a procedure. Nevertheless we may get some benefit from this exercise. In fig. 3 both SN Ia and GRB data are represented on the same plot. However the SN Ia magnitudes are not dependent on any particular cosmology and present a much stronger constraint on the applied theory. Because it is shape of the curve fitting to the data that carries the most information if the theory fits the SN Ia data it should also fit the GRB data.

In order quantify the goodness of the least squares fitting I have used the χ^2 parameter which measures the goodness of the fit between the data and the theoretical curve assuming the two fit parameters a and Ω_m. The values of χ^2/N ($\approx \chi^2_{d.o.f}$) are shown in Table 1, calculated using published errors on the distance modulus data. The best fit value of a is found at low redshift and then applied to all fits with various values of Ω_m.

From Table 1 the smallest χ^2/N value are generally obtained where $0.04 < \Omega_m < 0.06$, especially when the GRBs are included. But the quality of the fits are not strongly depen-

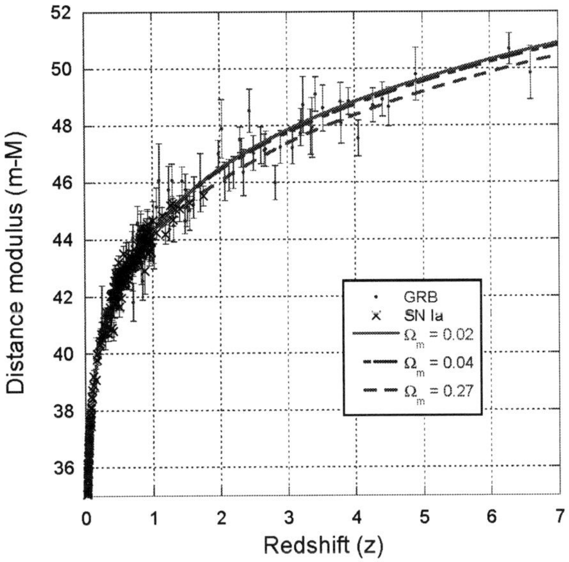

FIGURE 3. Distance modulus data from the combined SN Ia data set of Riess *et al* and Astier *et al* (crosses) and the GRB data (dots) derived from Schaefer [17] with curve fits using the extended equation (9). The top solid curve represents the distance modulus where $\Omega_m = 0.02$, the broken curve below it is where $\Omega_m = 0.04$ and the dashed curve below the latter is where $\Omega_m = 0.27$.

dent on the matter density. The locally measured baryonic matter budget has determined that $0.007 < \Omega_b < 0.041$ [7] where a Hubble constant of 70 km s^{-1}Mpc^{-1}was assumed. This region is consistent with the most probable values of Ω_m from this analysis. Definitely a matter density of $\Omega_m = 0.27$ is not necessary for the best fit. Therefore no exotic dark matter need be assumed. However to properly test this model much more SNe Ia data are needed where $z > 2$ and much better constrained magnitude errors. The GRB data are not determined independently of the cosmology tested and therefore by themselves don't provide a 'standard candle' test of the theory.

SPATIALLY FLAT UNIVERSE

In the earlier sections the time part of the metric $g_{\mu\nu}$ was ignored. Here we use coordinates $x^\mu = (ct, r, \theta, \phi, \tau v)$ where $\mu = 0-4$, but the full 5D description of cosmos has never been determined in the Carmelian theory. So the question must be answered, "What is the g_{00} metric component for the large scale structure of the Universe in CGR?"

First note the g_{11} metric component

$$g_{11} = -\left(1 + \frac{1-\Omega}{c^2\tau^2} r^2\right)^{-1}. \tag{12}$$

Using the scale radius we can define an energy density from the curvature

$$\Omega_K = \frac{c^2}{h^2 R_0^2} = \frac{c^2\tau^2}{R_0^2}, \tag{13}$$

which, when we use (11), becomes

$$\Omega_K = 1 - \Omega. \tag{14}$$

This quantifies the energy in the curved *spacevelocity*.

In the FRW theory the energy density of the cosmological constant is defined $\rho_\Lambda = \Lambda/8\pi G$ hence

$$\Omega_\Lambda = \frac{\Lambda}{3H_0^2}. \tag{15}$$

Even though the cosmological constant is not explicitly used in CGR, it follows from the definition of the critical density that

$$\rho_c = \frac{3}{8\pi G\tau^2} = \frac{\Lambda}{8\pi G}, \tag{16}$$

when the cosmological constant Λ is identified with $3/\tau^2$. Therefore in CGR it follows that

$$\Omega_\Lambda = \frac{\Lambda}{3h^2} = \Lambda\left(\frac{\tau^2}{3}\right) = 1. \tag{17}$$

In CGR $h = \tau^{-1}$. This means that in CGR the vacuum energy $\rho_{vac} = \Lambda/8\pi G$ is encoded in the metric via the critical density since $\rho_{eff} = \rho - \rho_c$ principally defines the physics. So $\Omega_\Lambda = 1$ identically and at all epochs of time. (The determination of Ω_Λ in [8] was flawed due to an incorrect definition.) Also we can relate Ω_Λ to the curvature density by

$$\Omega_K = \Omega_\Lambda - \Omega, \tag{18}$$

which becomes

$$\Omega_k = \Omega_\Lambda - \Omega_m, \tag{19}$$

at the present epoch ($z \approx 0$). Here $\Omega = \Omega_m(1+z)^3$ ($z \ll 1$) and hence $\Omega_K \to \Omega_k$ as $z \to 0$.

Finally we can write for the total energy density, the sum of the matter density and the curvature density,

$$\Omega_t = \Omega + \Omega_K = \Omega + 1 - \Omega = 1, \tag{20}$$

which means the present epoch value is trivially

$$\Omega_0 = \Omega_m + \Omega_k = \Omega_m + 1 - \Omega_m = 1. \tag{21}$$

This means that the 3D spatial part of the Universe is always flat as it expands. This explains why we live in a universe that we observe to be identically geometrically spatially flat. The curvature is due to the velocity dimension. Only at some past epoch, in a radiation dominated universe, would the total mass/energy density depart from unity.

I therefore write

$$g_{00}(r) = 1 + (1 - \Omega_t) r^2, \tag{22}$$

where r is expressed in units of $c\tau$. Equation (22) follows from $g_{00} = 1 - 4\Phi/c^2$ where Φ is taken from the gravitational potential but with effective density, which in turn involves the total energy density because we are now considering *spacetime*. The factor 4 result from a comparison of the CGR theory with Newtonian theory.

Clearly from (20) it follows that $g_{00}(r) = 1$ regardless of epoch. Thus from the usual relativistic expression

$$1 + z_{grav} = \sqrt{\frac{g_{00}(0)}{g_{00}(r)}} = 1, \tag{23}$$

and the gravitational redshift is zero regardless of epoch. As expected if the emission and reception of a photon both occur in flat space then we'd expect no gravitational effects.

Therefore we can write down the full 5D line element for CGR in any dynamic spherically symmetrical isotropic universe,

$$ds^2 = c^2 dt^2 - \left(1 + \frac{1 - \Omega}{c^2 \tau^2} r^2\right)^{-1} dr^2 + \tau^2 dv^2. \tag{24}$$

The θ and ϕ coordinates do not appear due to the isotropy condition $d\theta = d\phi = 0$. Due to the Hubble law the 2nd and 3rd terms sum to zero leaving $dt = ds/c$, the proper time. Clocks, co-moving with the galaxies in the Hubble expansion, would locally measure the same proper time.

However inside the Galaxy we expect the matter density to be much higher than critical, ie $\Omega_{galaxy} \gg 1$ and the total mass/energy density can be written

$$\Omega_0|_{galaxy} = \Omega_{galaxy} + \Omega_k \approx \Omega_{galaxy}, \tag{25}$$

because $\Omega_k \approx 1$, since it is cosmologically determined. Therefore this explains why the galaxy matter density only is appropriate when considering the Poisson equation for galaxies.[9]

As a result inside a galaxy we can write

$$g_{00}(r) = 1 + \Omega_K \frac{r^2}{c^2 \tau^2} + \Omega_{galaxy} \frac{r^2}{c^2 \tau^2}, \tag{26}$$

in terms of densities at some past epoch. Depending on the mass density of the galaxy, or cluster of galaxies, the value of g_{00} here changes. As we approach larger and larger structures it mass density approaches that of the Universe as a whole and $g_{00} \to 1$ as we approach the largest scales of the Universe. Galaxies in the cosmos then act only as local perturbations but have no effect on Ω_K. That depends only on the average mass density of the whole Universe, which depends on epoch (z).

Equation (26) is in essence the same expression used on page 173 of Carmeli [2] in his gravitational redshift formula rewritten here as

$$\frac{\lambda_2}{\lambda_1} = \sqrt{\frac{1 + \Omega_K r_2^2/c^2\tau^2 - R_S/r_2}{1 + \Omega_K r_1^2/c^2\tau^2 - R_S/r_1}}. \tag{27}$$

involving a cosmological contribution $(\Omega_K r^2/c^2\tau^2)$ and $R_S = 2GM/c^2$, a local contribution where the mass M is that of a compact object. The curvature (Ω_K) results from the averaged mass/energy density of the whole cosmos, which determines how the galaxies 'move' but motions of particles within galaxies is dominated by the mass of the galaxy and the masses of the compact objects within. Therefore when considering the gravitational redshifts due to compact objects we can neglect any cosmological effects, only the usual Schwarzschild radius of the object need be considered. The cosmological contributions in (27) are generally negligible. This then leads back to the realm of general relativity.

ACKNOWLEDGMENT

I would like to sincerely acknowledge the encouragement and support of the late Prof. Moshe Carmeli of Ben Gurion University, Beer Sheva, Israel.

REFERENCES

1. P. Astier, *et al* "The Supernova Legacy Survey: Measurement of Ω_M, Ω_Λ and w from the first year data set", *A & A* **447**(1):31Ű-48 (2006)
2. M. Carmeli, *Cosmological Special Relativity* (World Scientific, Singapore, 2002)
3. ibid. Sec. 2.15.4, p. 23.
4. M. Carmeli, *Cosmological Relativity* (World Scientific, Singapore, 2006)
5. M. Carmeli, "Accelerating Universe: Theory versus Experiment", [arXiv: astro-ph/0205396] (2002)
6. M. Carmeli, J.G. Hartnett, F.J. Oliveira, "The cosmic time in terms of the redshift," *Found. Phys. Lett.* **19**(3):277–283 (2006) arXiv:gr-qc/0506079
7. M. Fukugita, C.J. Hogan, and P.J.E. Peebles, "The cosmic baryon budget", *Ap. J.* **503**: 518-530 (1998)
8. J.G. Hartnett, "The distance modulus determined from Carmeli's cosmology fits the accelerating universe data of the high-redshift type Ia supernovae without dark matter," *Found. Phys.* 36(6): 839–861 (2006) arXiv:astro-ph/0501526
9. Hartnett, J.G. "Spiral galaxy rotation curves determined from Carmelian general relativity" *Int. J. Theor. Phys.* **45**(11):2147–2165 (2006) arXiv:astro-ph/0511756
10. J.G. Hartnett, F.J. Oliveira "Luminosity distance, angular size and surface brightness in Cosmological General Relativity," *Found. Phys.* 37(3): 446–454 (2007) arXiv: astro-ph/0603500
11. G. Gemelli, "Particle production in 5-dimensional Cosmological Relativity," *Int. J. Theor. Phys.* **45**(12): 2261–2269 (2006)
12. R.A. Knop *et al* "New constraints on Ω_M, Ω_Λ and w from an independent set of 11 high-redshift supernovae observed with the Hubble Space Telescope", *Ap. J.* **598**: 102-137 (2003)
13. F.J. Oliveira, J.G. Hartnett, "Carmeli's cosmology fits data for an accelerating and decelerating universe without dark matter or dark energy," *Found. Phys. Lett.* **19**(6):519-535 (2006) arXiv: astro-ph/0603500
14. S. Perlmutter *et al* "Measurements of the cosmological parameters Ω and Λ from the first seven supernovae at $z > 0.35$", *Ap. J.* **483**:565-581 (1997)
15. A.G. Riess, A. V. Filippenko, P. Challis, A. Clocchiatti, A. Diercks, "Observational evidence from supernovae for an accelerating universe and a cosmological constant", *Astron. J.* **116**: 1009-1038 (1998)
16. A.G. Riess, *et al* "Type Ia supernovae discoveries at $z > 1$ from the Hubble Space Telescope: Evidence for past deceleration and constraints on dark energy evolution" *Ap. J.* **607**: 665-687 (2004)
17. B.E. Schaefer "The Hubble Diagram to redshift > 6 from 69 Gamma-Ray Bursts" *Ap. J.* **660**: 16-46 (2007)

Topics in Cosmology

José De La Luz Montero García

Institute for Scientific and Technological Information (IDICT); National Capitol, Havana, Cuba.
E-mail: josedelaluzmg@gmail.com

Abstract. In a concise and simplified way, It was verified with the experimental dates of the Solar System and the 55 Cangri Extrasolar System, that the factor and the constant of planeticity, and the rotational geometric factor are nonlinear, volumetric a constant accelerations universal; it allow to enunciate the Kepler's Fourth Law and its relation with the Fourth Principle of Thermodynamics. An understanding of the Equivalence Principle is exposed and it proves that the Universal Gravitation is an inverse quadratic function of the energetic-longitudinal constant where all the fundamental interactions converge and universal rotation of the motion of matter.

Keywords: universal constant; equivalence principle; gravity; planetary system; space.
PACS: 06.20.Jr, 06.30.Bp, 11.10.Kk, 12.10.Dm, 98.80.Bp

THE SPATIAL CELLS BY ITSELF OF THE OWN REAL MATERIAL SPACE ARE THEIRS ATOMS

The process of essential convergence that has had and takes place in the sciences in the last centuries has allowed us to research the structure of the Real Material Space (RMS) and to prove qualitatively and quantitatively [1], that the World is one, is one; but it is one arithmetic-geometric-topologic-dimensional and dynamic-cellular-structural-quantum, simultaneously; with their corresponding intrinsic quantization of all material entity and the intrinsic discretion of all magnitude.

The results obtained and exposed here, more or less certain equations come neither from speculative interpretations nor of formulations, but from the universal constants which values and relations are unchangeable except to destroy the whole physical theory achieved by the science up to today, transcendental to the structure of a new scientific natural frame of the World.

The discovery [1] of that, the length, the physical time, the mass, the electrical charge, the thermodynamic temperature of energetic hierarchy, the natural grouping, the particle and the universal cycle, are connected and interchangeable between if, where only certain type of joining together of the eight will have, in rigor, independent existence as spatial properties of high meaning.

At the same time, the laws of equivalence and dimensional harmony, it constitute the inner structure of the Dimensional Primitive Complete System (DPCS) [1] of matter, and the Unified Theory of the Physical and Mathematical Universal Constants (UTPMUC) [1] supported by the Quantum Cellular Structural Geometry (QCSG) [1],

CP1018, *Frontiers of Fundamental and Computational Physics*,
edited by B. G. Sidharth, F. Honsell, O. Mansutti, K. Sreenivasan, and A. De Angelis

it fundamentality or generality or universality of a physical law or constant physics become necessary and sufficient criterion: Theorem of the Unicity of the World [1].

The unit of matter is, for essence, dimensional, being absolutely provable that, the RMS is the self-ensemble eight-dimensional and deployment ten-axial dimensional of the world [1], defines a own natural scale of matter and their level of discretion, allowing the calculation of any universal known constant or not.

The cells of the own RMS are their atoms of whose accretion whole, interactive and step-by-step everything existing arises to all the hierarchic levels, it do not have independent existence but always concatenated and connected, these cells below which it lacks sense to speak about structure and, therefore, about existence, as it proves the Fourth Principle of Thermodynamics [2]. This is characterized by a great dynamics. All the processes are by energy and order the cells of space are degenerated, the movement among cells does not cause any net change of energy and consequently, of entropy $\Delta S = 0$ [3]. In this sense, the movement turns out to be the translation with respect to a given system of reference of a perturbation of space itself. This is what we call object, system, entity, body, corpuscle, etc..

Such ultimate cells base the existence, uniqueness and harmony of matter, as well as its structural-formal and dynamic-functional diversity, do not depend on external conditions for their existence. Probably, the sense is this of absolute in its dimensional character, which will bequeath to us by I. Newton until today.

THE UNIVERSE HAS DIMENSIONAL BARRIERS THAT RECTIFY THE PLANCK´S UNITS

The matter has a measure or global connected scale or existential universal interval characterized by four types of dimensional barriers, which rectify Planck's Units [4], namely:

1. Micro unilateral barriers: longitude $\alpha = 3{,}3530 \cdot 10^{-34}$ m and time $\beta = 1{,}1184 \cdot 10^{-42}$ s;
2. Macro bilateral barriers: temperature $\gamma = 4{,}2911 \cdot 10^{31}$ K;
3. Barriers of step-like connection between the macro and the micro: mass $\theta = 4{,}5166 \cdot 10^{-7}$ kg, electric charges $\Delta = 1{,}0978 \cdot 10^{-17}$ C, grouping $\Omega = 1{,}1378 \cdot 10^{-25}$ kmol and particle $\phi = 6{,}8518 \cdot 10^{1}$ p;
4. Dynamic Systemic barrier: cycle $\varphi = 2\pi$ rad.

With this, the world of the universal constants:

- is infinite for their longitude and time: longitude of $(\alpha \rightarrow \infty)$ m and time of $(\beta \rightarrow \infty)$ s;
- is delimited for its temperature between: temperature $(0 \rightarrow \gamma)$ K;
- with a separation of two levels (2) whose border is characterized by: mass = θ kg and electric charges = Δ C;
- resulting in: grouping = Ω kmol;
- and the structures of particle = ϕ p;
- in which everything rotates global and locally $\varphi = 2\pi$ rad and everything is connected, from the cells of space up to groups of galaxies and beyond ones.

On scale of the ultimate cell of the space the gravitational radius ($R_G = \alpha$) the hexagono-icosahedral cell of the RMS, is connected and at short reach in whose successive linking it extends to all the universe, across the action at a distance, it connects it as a whole and it bases their unit and unicity. In the inner of R_G, they are theirs neighbors, and is a clear expression of the stability of the RMS, happens that:

$$R_G = \frac{2\cdot G}{c^2}\cdot\theta = 2\cdot\alpha \Rightarrow \frac{G}{c^2} = \frac{\alpha}{\theta} = 0{,}74236\cdot 10^{-27}\,\mathrm{m}\cdot\mathrm{kg}^{-1} \Rightarrow \frac{\theta}{\alpha} = 1{,}347\cdot 10^{27}\,\mathrm{kg}\cdot\mathrm{m}^{-1} \quad (1)$$

which is a general covariance principle dimensional. It is essential that the huge uniformity of the RMS, that all that exists is part of it, that it has no center, that cells cannot be distinguished one from the other, that the order is maximum and that cells are energetically degenerated disregard the RMS as a system of reference, which is simply not so. The system of absolute reference does not exit. Any point of it can be taken as reference and its kinetic relationship with the system is what decides it all.

This fact proves to us that physical relativity is universal and its manifestation, diverse. The physical bodies all and nobodies are these, are description of RMS, does not interest so much uniquely nor the state as the transition of these; which demands in addition, that the variation of a magnitude and all magnitude is not only employee of the given variables and the changes of these but, also, proportional to the magnitude itself, exception of its global existential invariants.

Having in consideration all previously expressed, applying to the relativist dependence of mass on velocity we will obtain an expression that is, from a state characterized by (m_1, β_1) towards another one state (m_2, β_2):

$$\Delta E = (E_2 - E_1) = (m_2 - m_1)\cdot c^2 = m_1\cdot\left(\exp\left\{\frac{f_0}{k}\left[e^{k\left(\beta_2^2-\beta_1^2\right)} - 1\right]\right\} - 1\right)\cdot c^2 = \theta\cdot c^2 \quad (2)$$

This one, unlike the invariant of Lorentz, allows to understand the Principle of Relativity as transition of states [5], where the relativist velocity of the physical body can be anyone, that is not prohibited that the system is accelerated which does not affect the invariant morphologically and; the mass will have, as inevitably it had to be, an increasing but finite and real value (θ).

HYPOTESIS NON FINGO!: THE UNIVERSAL ROTATION OF THE MOTION OF MATTER

Warned of everything before exposed, we, the authors of this work, realized that, for three decades, it would have been necessary to have begun "from before the beginning" and not "since the beginning of time" neither "on a finite universe with no beginning nor end."

We should make right to clear that the research object in cosmology is the RMS, on an unique consideration: the existential structural unit of the world, in correspondence with the Gödel´s incompleteness theorem, beginning us reason why it demanded to I. Newton to him to exclaim with great courage at his time and for all time: <<*Hypotesis non fingo*!>>. What is gravity?

Geometric Factor of the Planetary System or Planeticity

A simple analysis dimensional of the conical functions, similar to the made one by Newton allows us to obtain a dimensionless relation, between the means range ($\bar{d}$) of the foci of the central field of forces, the orbital period (T) and the average speed ($\bar{V}_0$), and that unlike the Kepler's Third Law does not depend on the mass of the star who generates the central field of forces [1].

Applying the Theorem of the Unicity of the World, we proved to the universality of the rotational motion $\varphi = 2\pi$, that it is formulated as follows:

$$\varphi \cdot (2\pi)^{-1} \dot{=} 2\pi \cdot (2\pi)^{-1} \dot{=} 2\pi \cdot \left[\bar{d}_i \cdot \left(T_i \cdot \bar{V}_{0_i} \right)^{-1} \right] \dot{=} 1 \qquad (3)$$

Kepler's Fourth Law or Fundamental Geometric Law of the Planetary Systems

"*The elliptical planetary motion takes place in such a way that their orbital average parameters behave uniformly like those of a body that move circular coplanar and in volumetric a central field of forces.*"

It outstanding that if the Planetary System, as the Galaxy System were not Euclidean geometry (flat space) such result it would be, theoretically impossible and very bad adjustment to the experimental dates; and between its characteristics it has the invariant plane of structural symmetry with orthogonal axis of rotation on their mass or inertial center; on the flat mentioned one the system is flatten.

The saying does not suppose that in the center of the galaxy a macro exists or mega stars but that the global galaxial motion and in its parts behaves as if such colossal star existed; sufficiently as considering it unique a body almost homogenous. Black Hole!

Longitudinal-Temporary Harmony of the Kepler's Third Law: Planeticity Constant

This, demands a structural geometric understanding to us of the longitudinal-temporary harmony of the Kepler's Third Law, it shows hexagonal structure deep of our Planetary System:

$$\left(\bar{V}_0^2\right) \cdot d = \left(\frac{2\pi \cdot d}{T}\right)^2 \cdot d = \left(\frac{\pi}{3}\right) \cdot \left(\frac{4}{3}\pi \cdot d^3\right) \cdot \frac{1}{T^2} = (\alpha_6) \cdot (E_0) \cdot \left(\frac{1}{T^2}\right) = (\alpha_6) \cdot \left(a_0^V\right) = J \quad (4)$$

if we defined sphere or orbital volume like E_0, volumetric acceleration of orbital sweeps out as a_0^V and central angle of the hexagonal space cell as α_6, we have the Constant of Planeticity J, an acceleration nonlinear, but volumetric one.

"*The constant of planeticity J is equal to the volumetric acceleration of the orbital sweeps out by the central angle of the hexagonal space cell inasmuch as the motion is compacted towards the symmetry plane.*" This is equivalent to say that:

"*All planetary motion in a central field of forces is so, that it happens with a unique volumetric acceleration of orbital sweeps out.*"

By simple algebraic transformations from the Kepler's Third Law (K), it is arrived:

$$\alpha_6 \cdot \left(d^3 \cdot T^{-2}\right) = (12\pi)^{-1} \cdot K = (3 \cdot \varphi)^{-1} \cdot K = R \qquad (5)$$

factor R, will be denominated: rotational geometric factor, the magnitude 12π makes see the hexagonal-icosahedrality of the volumetric space and its isotropy intrinsically quantized; an acceleration nonlinear, but volumetric one.

It was verified with the experimental dates for planets of the Solar System, for theirs asteroids and the 55 Cangri Extrasolar System with theirs five planets, obtaining itself a value average of $F = \left(2 \cdot 10^{-1}\right)^{2/3} = (2\pi)^{-1}$ and $R = 3{,}5082 \cdot 10^{-18}\,\mathrm{m}^3 \cdot \mathrm{seg}^{-2}$.

Perhaps, it is being the constant of acceleration that Einstein looked for and the reason for which today one considers that the universe is being accelerated. It is it?

In the conditions of free rotational motions as which it happen on scale of the Universe, the interaxial angle always is null inasmuch as the external gravity to the system is not significant; the rotation leads to the generation of the straight lines or material sides of the formation. In such material circumstances the formation that it happen is objectively volumetric, it their structural-formal and dynamic-functional diversity.

We have verified experimentally that if the angle of inclination of the axis of rotation with respect to the gravitational axis of terrestrial it is null, the material formation gravitationally flatten (icosahedrons) that is obtained has pentagonal form; if, on the contrary, the angle before is different from zero, then the form that is obtained is a perfectly delimited hexagon.

Understanding of the Equivalence Principle

The gravitation is, the fundamental spring into Universe, but this has to be understood beyond the conventional limit of Einstein's Inertial Principle and, extending it until the understanding of the phenomenon of anisotropic centripetal gravity (flat and orthogonal axis of symmetry) that proves, as non-homogenized aspect (structuring) of the material systems, where the rotation necessarily does not entail the existence of the centrifugal force but that it comes determined by the presence from inter-phases and the behavior of the accelerations.

This is the Global Structural Principle (GSP), it is a problem of the density of matter. It is not a mass problem.

The entities by this process it agglutinate losing their pseudo-homogeneity, and already to this manifestly discreet state and of inter-phases, the energetic concentration, the centrifuging and the catastrophes the non-agglutinate in a new cycle with permanent interchange of substances and radiation with the environs and other systems.

The Singularity is the Generality of the Real Material Space

The generalized cause of the gravitation is that in the world at local level always rotational processes are happening; to say that the gravity is universal is equivalent to express that the local rotation is generalized and characteristic of the global motion of matter $(\varphi = 2\pi)$.

The Universe is in constant generalized local rotation and, nevertheless it is structured; in effect, as the structuring and non-structuring are concatenated processes sequentially, that constantly is happening as original processes but in the sense of the fundamentality of the processes, it are the constant and eternal explosion of quantum cells of the RMS, are pronounced as fulfillment of the conservation laws, at the temperature of (2,735±0,06) K (according to COBE); the Universe pulsating, but locally of universal way, whose temperature is and has been always Natural Universal Radiations [3] and not and in a temporary sense historical as Big bang-Big crunch.

The Gravity is the Fundamental Force in Ultramicro and Megamacro

In ultramicro, the equivalent way can be obtained on the theoretical base of the laws of dimensional equivalence that are implicit in the fine structure constant α_F. Thus, the well-known equation:

$$\mu_0 \cdot \varepsilon_0 \cdot c^2 = 1 \Rightarrow G = \frac{\alpha}{\theta} \cdot c^2 = \frac{\alpha}{\theta} \cdot \frac{\alpha^2}{\beta^2} = \frac{\alpha^3}{\theta \cdot \beta^2} = 6{,}6721 \cdot 10^{-11} \mathrm{kg}^{-1} \cdot \mathrm{m}^3 \cdot \mathrm{s}^{-2} \qquad (6)$$

under the consideration as before we proved [1] of the connections gravitation-electromagnetism and equivalence with the expression (1):

$$\left(\frac{\mu_0}{G}\right)^{1/2} = 2 \cdot \phi \quad \Rightarrow \quad \mu_0 = (2 \cdot \phi)^2 \cdot G \Rightarrow \left(\frac{\mu_o}{G}\right) \cdot \alpha_F^2 = 1 \qquad (7)$$

If we calculated, in agreement with the Law of Universal Gravitation or the Newton's Second Law, the force between cells of the RMS, happens that:

$$F_G = G \cdot \frac{m \cdot m}{R^2} \Rightarrow F_G = G \cdot \frac{\theta^2}{\alpha^2} = F_N = m \cdot a = \theta \cdot \frac{\alpha}{\beta^2} \Rightarrow F_G = F_N = F_{Inertial} = \left[\mathrm{kg} \cdot \mathrm{m} \cdot \mathrm{s}^{-2}\right] \quad (8)$$

The Global Nexus Gravitatoric-Weak-Electrodynamics-Strong between the fundamental interactions convergent in the constant energetic-longitudinal L, a volumetric unique generalized force, as it were proved previously [2].

$$\vec{F}_{gener} = F_G = F_M = F_E = F_S = F_W = F_N = \frac{L}{d^2} = \frac{E \cdot \alpha}{\alpha^2} = \frac{G \cdot \theta^2}{\alpha^2} = \frac{\theta \cdot \alpha}{\beta^2} = \left[\mathrm{kg} \cdot \mathrm{m} \cdot \mathrm{s}^{-2}\right] \quad (9)$$

In megamacro, on Universal scale, usually it are expressed G in Astronomical Units (UA), where $M_S = M_\odot$, it is the Solar Mass and Y the Tropical Year [1], it allows:

$$\left(UA \cdot \alpha^{-1}\right)^3 \cdot \left(M_s \cdot \theta^{-1}\right)^{-1} \cdot \left(Y \cdot \beta^{-1}\right)^{-2} = (2 \cdot \pi)^{-2} \qquad (10)$$

This means that, the Nature changes of scale (hierarchic level of organization and measurement) when expanding the RMS of their basic cellular structure towards less dense disturbances of the same one like which we denominated Solar System, all the dimensions are modified, the diverse way conserving their main global invariants with their arithmetic-geometric-topologic-dimensional and dynamic-cellular-structural-quantum essence [1].

The Inertia is a Materialization of the Geometry of the Space Material Real

Previously expressed yet up to here, confirms the hypothesis on the nature of the noninertial forces enunciated by Ernst March of the following form:

"*The inertia of any system is the result of the interaction of that system and the rest of the universe. In other words, every particle in the universe ultimately has an effect on every other particle*" or "*the inertial mass is not an intrinsic characteristic of a motive, but a measure of its joining with the rest of the universe.*"

The gravitation is not another thing that the mechanical interaction between tie systems space always, inasmuch as it is not possible to conceive an object in rest, this does not exist simply. Equally, that from the triple equivalence of the mass (gravitational-inertial-energetic), it is not deduced that the gravitational mass is just as the inertial mass, like either, from the laws of dimensional equivalence and harmony of matter is not deduced that all their universal properties are equal.

REFERENCES

1. J. Novoa and J. Montero, *Are the Truly Constant Constants of Nature? How is the Real Material Space and its Structure?* Eighth International Symposium on Frontiers of Fundamental Physics (FFP8). AIP Conference Proceedings Volume 905 pp. 182-184. April 28, 2007. doi:10.1063/1.2737010; ISBN 978-0-7354-0412-0. Published June 2007.
2. J. Novoa and J. Montero, *The 4th Thermodynamic Principle?* Eighth International Symposium on Frontiers of Fundamental Physics (FFP8). AIP Conference Proceedings Volume 905 pp. 185-187. April 28, 2007. doi:10.1063/1.2737011. ISBN 978-0-7354-0412-0. Published June 2007.
3. J. Novoa and J. Montero, *Natural Universal Radiation?* Eighth International Symposium on Frontiers of Fundamental Physics (FFP8). AIP Conference Proceedings Volume 905 pp. 259-260. April 28, 2007. doi:10.1063/1.2737032. ISBN 978-0-7354-0412-0. Published June 2007.
4. J. Novoa and J. Montero, *Are Correct the Planck's Units? Is the Fine-Structure Constant Really Dimensionless?* Eighth International Symposium on Frontiers of Fundamental Physics (FFP8). AIP Conference Proceedings Volume 905 pp. 255-256. April 28, 2007. doi:10.1063/1.2737030. ISBN 978-0-7354-0412-0. Published June 2007.
5. J. Novoa and J. Montero, *Double Exponential Relativity Theory Coupled Theoretically with Quantum Theory?* Eighth International Symposium on Frontiers of Fundamental Physics (FFP8). AIP Conference Proceedings Volume 905 pp. 257-258. April 28, 2007. doi:10.1063/1.2737031. ISBN 978-0-7354-0412-0. Published June 2007.

Angular Power Spectrum in Modular Invariant Inflation Model

Mitsuo J. Hayashi*, S. Hirai, T. Takami,† and Y. Okame,
K. Takagi, T. Watanabe*

*Department of Physics, Tokai University, 1117 Kitakaname, Hiratsuka, 259-1292, Japan
†Department of Digital Games, Osaka Electro-Communication University, 1130-70 Kiyotaki, Shijonawate, Osaka, 575-0063, Japan

Abstract. A scalar potential of inflation is proposed and the angular power spectra of the adiabatic density perturbations are computed. The potential consists of three scalar fields, S, Y and T, together with two free parameters. By fitting the parameters to cosmological data at the fixed point $T = 1$, we find that the potential behaves like the single-field potential of S, which slowly rolls down. We further show that the inflation predictions corresponding to this potential provide a good fit to the recent three-year WMAP data, e.g. the spectral index $n_s = 0.951$.

The TT and TE angular power spectra obtained from our model almost completely coincide with the corresponding results obtained from the ΛCDM model. We conclude that our model is considered to be an adequate theory of inflation that explains the present data.

Keywords: Inflation; WMAP observations; Modular invariant potential.
PACS: 04.65.+e, 11.25.Mj, 11.30.Pb, 12.60.Jv, 98.80.Cq

1. INFLATION COSMOLOGY

Following three years of integration, the Wilkinson Microwave Anisotropy Probe (WMAP) data has significantly improved. There have also been significant improvements in other astronomical data sets[1]. The constraints on cosmological parameters, such as the spectral index and its running, as well as the ratio of the tensor to the scalar, have also been improved. The predictions of the theoretical inflation theories continue to find good agreement with these improved data sets. The ΛCDM model fits not only the three-year WMAP temperature and polarization data, but also small scale CMB data, light element abundances, large scale structure observations, and the supernova luminosity/distance relationship[1].

As a favored scenario to explain the observational data, it is customary to introduce a scalar field called inflaton into the theoretical models; there are, however, several problems in constructing successful theories: i) Precisely what is inflaton? ii) What kind of theoretical frameworks are most appropriate to describe the theory of particle physics, inflation and the recently observed accelerating universe? iii) How can the contents of the universe be explained?

These problems seem to require a far richer structure of contents than that of the standard theory of particles. Furthermore, phenomenologically, iv) Is the model consistent with the observed CMB angular power spectra? In particular, inflaton should satisfy the slow-roll condition in order that the model predicts the nearly scale-invariant spectral index as well predicting a sufficiently large number of e-folds and the angular power

CP1018, *Frontiers of Fundamental and Computational Physics*,
edited by B. G. Sidharth, F. Honsell, O. Mansutti, K. Sreenivasan, and A. De Angelis

spectrum.

In this talk, by assuming an inflationary potential[2], we investigate angular power spectra. This potential was originally derived by Ferrara *et al.*[3] in the context of T-duality and supersymmetry breaking in string theory via gaugino condensation. However, because we would like to explore the phenomenological implications of this model as an inflationary theory, we will suppose in the present study that the potential's parameters are not restricted by supergravitational backgrounds and gaugino condensation. We will show that our results agree with the three-year WMAP data, and are also consistent with the predictions of the ΛCDM model[4].

2. INFLATIONARY TRAJECTORY

We will compute the angular power spectra of the adiabatic density perturbations to compare with the results of 3 year WMAP. The inflaton field S and the gauge-singlet complex scalar fields Y and T are introduced and a scalar potential in the following form is considered[2, 4].

$$\begin{aligned} V(S,T,Y) &= \frac{3(S+S^*)|Y|^4}{(T+T^*-|Y|^2)^2}\Bigg(3b^2\left|1+3\ln\left[ce^{S/3b}Y\eta^2(T)\right]\right|^2 \\ &\quad +\frac{|Y|^2}{T+T^*-|Y|^2}\left|S+S^*-3b\ln\left[ce^{S/3b}Y\eta^2(T)\right]\right|^2 \\ &\quad +6b^2|Y|^2\left[2(T+T^*)\left|\frac{\eta'(T)}{\eta(T)}\right|^2+\frac{\eta'(T)}{\eta(T)}+\left(\frac{\eta'(T)}{\eta(T)}\right)^*\right]\Bigg), \end{aligned} \tag{1}$$

where η is Dedekind's η-function, η' is its derivative with respect to T, and c and b are the free parameters of this potential. For simplicity, S and Y are assumed to be real fields and the other matter fields are neglected.

First, we prove that the interrelation between S and Y gives rise to inflation. Upon minimizing V with respect to Y at the fixed point $T=1$, one eventually finds a single-field potential $V(S)$; the scalar field S plays the role of the inflaton field. Usually inflaton fields must satisfy the slow-roll condition in order for the corresponding inflation model to be successful. We have to choose the parameters c and b carefully.

We found that the potential $V(S,Y)$ at $T=1$ has a stable minimum at $(Y_{\rm min},S_{\rm min})\sim(0.00877,1.51)$ with $c=131$, $b=9.4$. These are the most suitable values for realizing the present experimental observations of three-year WMAP. (The value of $S_{\rm min}$ is consistent with $\langle S+S^*\rangle=\alpha' m_{\rm pl}^2$.) The inflationary trajectory can be well approximated by

$$Y_{\rm min}(S)\sim 0.00925e^{-0.0355S}, \tag{2}$$

which corresponds to the trajectory of the stable minimum for both S and Y. For large c, this potential is asymptotically approximated by

$$V(S)\sim\frac{3}{2}\left(\frac{e^{-S/3b}}{ac}\right)^6 S(2S^2+2bS-b^2), \tag{3}$$

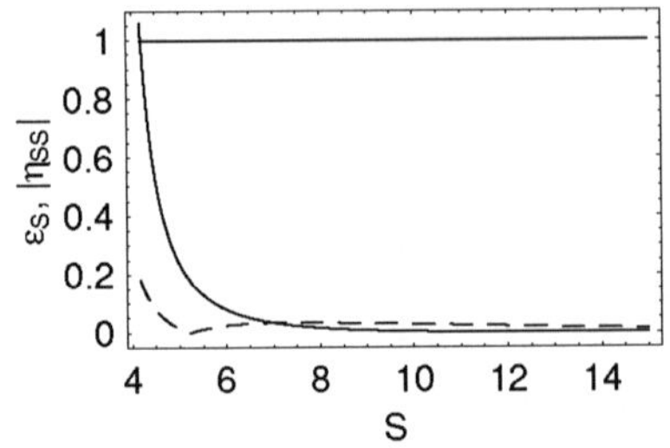

FIGURE 1. The evolution of the slow-roll parameters. The solid curve represents ε_S while the dashed curve denotes $|\eta_{SS}|$. The plots demonstrate that the potential $V(S)$ is sufficiently flat. Inflation ends at $S \sim 4.19$ in our model.

where $a \sim 0.824$ for $T = 1$. We can see inflation arises precisely due to the evolution of the scalar fields S and Y for the parameter values $c = 131$ and $b = 9.4$

Next, the slow-roll parameters are defined by

$$\varepsilon_\alpha = \frac{1}{2}\left(\frac{\partial_\alpha V}{V}\right)^2 \quad , \quad \eta_{\alpha\beta} = \frac{\partial_\alpha \partial_\beta V}{V}. \tag{4}$$

The slow-roll condition requires both these values to be less than 1. The end of the inflationary period is demarked by the slow-roll parameter ε_α approaching the value 1.

For the present potential the values of ε_S and η_{SS} are numerically obtained by fixing the parameters $c = 131$ and $b = 9.4$ as shown in Fig. 1; with these parameters the slow-roll condition is satisfied.

The potential V is stable at the self-dual point $T = 1$ in arbitrary points in the inflationary trajectory for our choice of the parameters $c = 131$ and $b = 9.4$.

3. INFLATIONARY PARAMETERS

Now we have to verify the amount of inflation. The number of e-folds which corresponds to our scale $k_* = 0.05\,\mathrm{Mpc}^{-1}$ must therefore be around 57. On the other hand, using the slow-roll approximation (SRA), N is also given by

$$N \sim -\int_{S_1}^{S_2} \frac{V}{\partial V} dS. \tag{5}$$

We could also have obtained the number of e-folds ~ 57, by fixing the parameters $c = 131$ and $b = 9.4$ and integrating from $S_{\text{end}} \sim 4.19$ to $S_* \sim 11.6$, i.e. our potential can produce a cosmologically plausible number of e-folds. Here S_* is the value corresponding to k_*.

We can also compute the scalar spectral index and its running that describe the scale dependence of the spectrum of the primordial density perturbation $\mathscr{P}_{\mathscr{R}} = (H/\dot{S})^2 (H/2\pi)^2$; these indices are defined by

$$n_s - 1 = \frac{d \ln \mathscr{P}_{\mathscr{R}}}{d \ln k}, \tag{6}$$

$$\alpha_s = \frac{dn_s}{d\ln k}. \tag{7}$$

These are approximated in the slow-roll paradigm as

$$n_s(S) \sim 1 - 6\varepsilon_S + 2\eta_{SS}, \tag{8}$$

$$\alpha_s(S) \sim 16\varepsilon_S\eta_{SS} - 24\varepsilon_S^2 - 2\xi_{(3)}^2, \tag{9}$$

where $\xi_{(3)}$ is an extra slow-roll parameter that includes the trivial third derivative of the potential. Substituting S_* into these equations, we have $n_{s*} \sim 0.951$ and $\alpha_{s*} \sim -2.50 \times 10^{-4}$. Because n_s is not equal to 1 and α_s is almost negligible, our model suggests a tilted power law spectrum. The value of n_{s*} is consistent with the recent observations; the best fit of three-year WMAP data using the power law ΛCDM model is $n_s \sim 0.951$[1].

Finally, estimating the spectrum $\mathscr{P}_{\mathscr{R}}$ in SRA,

$$\mathscr{P}_{\mathscr{R}} \sim \frac{1}{12\pi^2}\frac{V^3}{\partial V^2}, \tag{10}$$

we find $\mathscr{P}_{\mathscr{R}*} \sim 2.36 \times 10^{-9}$. This result is also in excellent agreement with the measurements derived from observations. It may also be noted that the minimum energy scale $V \sim 10^{-10}$ is also within the constrained range.

Gravitational waves are an inevitable consequence of all inflation models. The tensor perturbation (the gravitational wave) spectrum is given by

$$\mathscr{P}_{\mathrm{grav}} = 8\left(\frac{H}{2\pi}\right)^2 = \frac{2}{3\pi^2}V. \tag{11}$$

In SRA, the spectral index of $\mathscr{P}_{\mathrm{grav}}$ is given by the slow-roll parameters ε and η as

$$n_T = -2\varepsilon. \tag{12}$$

The ratio r' of the tensor quadrupole C_2^{grav} to the scalar quadrupole $C_2^{\mathscr{R}}$ is defined

$$r' = \frac{C_2^{\mathrm{grav}}}{C_2^{\mathscr{R}}} = 0.8625r = 13.8\varepsilon, \tag{13}$$

for the standard cold dark matter model. The gravitational wave spectrum does not evolve and remains frozen-in as a massless field, even after the horizon-exit, independent of the scalar perturbations.

4. THE ANGULAR POWER SPECTRUM OF THE MODEL

The angular power spectrum C_l^{TT} and the temperature-polarization cross-power spectrum C_l^{TE} will be given by

$$C_l^{TT} = 4\pi\int_0^\infty T_\Theta^2(k,l)\mathscr{P}_{\mathscr{R}}(k)\frac{dk}{k}, \tag{14}$$

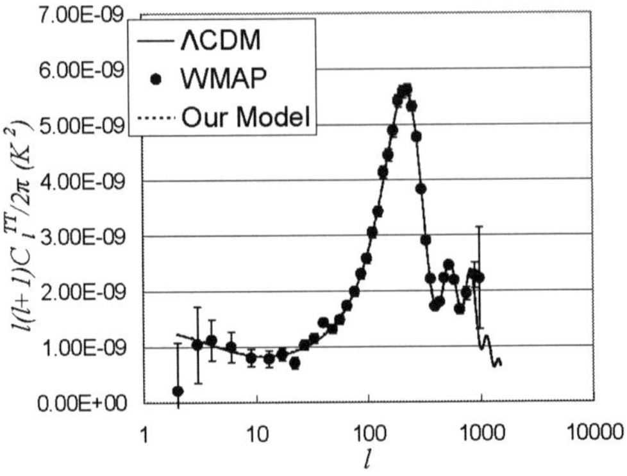

FIGURE 2. Temperature angular power spectrum (TT).

$$C_l^{TE} = 4\pi \int_0^\infty T_\Theta(k,l) T_E(k,l) \mathscr{P}_\mathscr{R}(k) \frac{dk}{k}, \tag{15}$$

where T_Θ and T_E are transfer functions and Θ is a brightness function.

Now we will describe the behavior of these power spectra according to our model. The scalar spectral index is $n_s(k_*) = 0.951$ and the running index is $\alpha_s(k_*) = -0.000250$ at $k_* = 0.05$ Mpc^{-1}, as has already been shown. We consider a tensor-to-scalar ratio $r' = 13.8\varepsilon = 0.00910$ ($\varepsilon = 0.000659$ at $k_* = 0.05$ Mpc^{-1}).

We will use the CMBFAST, where we have assumed the cosmological parameters to be: $\Omega_{\rm tot} = 1$ for the total energy density, $\omega = -1$ and $\Omega_\Lambda = 0.723$ for the dark energy, $\Omega_{\rm b} = 0.0426$ and $\Omega_{\rm cdm} = 0.2342$ for the baryonic and dark matter density, $h = 0.735$ for the Hubble constant[1]. The angular TT power spectra were normalized with respect to 11 data points in the WMAP data from $l = 65$ to $l = 210$, and the same values were used in the analysis of the angular TE spectrum.

The χ^2 values for the ΛCDM model with $\tau = 0.088$, which were given by Spergel *et al.*[1], were also calculated by the same method. (For the one-year WMAP data, $\tau = 0.17$ was favored.) On the other hand, the best fit of our model is realized at $\tau = 0.090$ for both TT and TE modes, which falls within experimental error, while the χ^2 values of our model seem to be better than those of the ΛCDM model.

The angular power spectrum of our model for the TT mode at $\tau = 0.090$ are presented in Fig. 2. Figs. 3 show the TE spectra for $l \leq 50$. Note that both the TT and TE spectra almost completely coincide with those of the ΛCDM model across the whole range of l. We would like to emphasize these two results as characteristic features of our model; within the area of scalar inflation models, our model can be regarded as an alternative to the ΛCDM model. Although both our model and the ΛCDM model perform on the whole satisfactorily in explaining the WMAP data, there remains one inconsistency in the suppression of the spectrum at large angular scales ($l = 2$). This problem is at present left to future investigations.

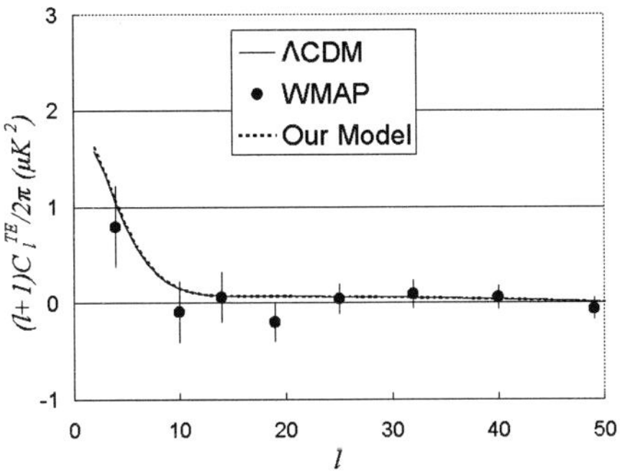

FIGURE 3. Temperature-polarization cross-power spectrum (TE) for $l < 50$.

5. SUMMARY AND DISCUSSIONS

In summary, the model we have here investigated is consistent with the present observational data and the ΛCDM model.

It appears that supergravity is one of the most plausible frameworks to explain new physics, including undetected objects, such as the inflaton, dark matter and dark energy. In particular, since the inflaton field is concerned with Planck scale physics, a dilaton field seems to be the most likely candidate for the inflaton. We have dealt with the parameters b and c as the completely free parameters. However, if we hope to consider our model as the one given by string-inspired supergravity with gauginocondensation of the hidden sector, the value of b = 9.4, which is defined from the coefficient of one loop beta-function of the renormalization group equation for E_8 or others, is too large to interpret it realistic. The construction of a realistic supergravity is another problem that must be tackled in the future, and would appear to be a fruitful approach. Because the agreement with the WMAP observations does not seem merely accidental, our next tasks include the investigation of an alternative derivation of an inflation potential of similar form to the present model using supergravitational theory.

Any successful theory of inflation should explain the mechanism of the preheating process. Reheating after inflation into radiation dominant big bang universe is under consideration.

REFERENCES

1. D.N. Spergel *et al.*, Wilkinson Microwave Anisotropy Probe (WMAP) Three Year Results: Implications for Cosmology, astro-ph/0603449, and reference therein.
2. M.J. Hayashi, T. Watanabe, I. Aizawa and K. Aketo, *Mod. Phys. Lett.* **A18**, 2785 (2003), hep-ph/0303029; M.J. Hayashi and T. Watanabe, Dilatonic inflation and SUSY breaking in modular-invariant supergravity, in *Proc. 32nd Int. Conf. High Energy Physics*, eds. H. Chen, D. Du, W. Li and C. Lu (World Scientific Publishing, Singapore, 2005), p. 423, hep-ph/0409084.
3. S. Ferrara, N. Magnoli, T.R. Taylor and G. Veneziano, *Phys. Lett.* **B245**, 409 (1990).
4. M.J. Hayashi, S. Hirai, T. Takami, Y. Okame, K. Takagi, T. Watanabe, *Int. J. Mod. Phys.* **A22**, 2223 (2007), astro-ph/0609657, and reference therein.

Weak Lensing on the CMB: Estimations Based on AP3M Simulations

Màrius Josep Fullana i Alfonso[1*], Josep Vicent Arnau i Córdoba⋆ and Diego Pascual Sáez Milán†

*Institut de Matemàtica Multidisciplinar, Universitat Politècnica de València
Camí de Vera s/n, 46022 València

⋆Departament de Matemàtica Aplicada and †Departament d'Astronomia i Astrofísica, Universitat de València, Dr. Moliner 50, 46100 Burjassot (València), Spain

Abstract. Lensed CMB maps are simulated with the Hydra4.2 code designed by the Hydra Consortium. This code has been modified to move CMB photons –through the simulation box– while the AP3M N-body is running. An appropriate ray-tracing procedure is used to do good weak lensing estimations with negligible periodicity effects (errors due to the periodic character of any universe covered by simulation boxes). The results of this paper improve on previous ones based on PM N-body simulations.

Keywords: Cosmic microwave background, Cosmology: theory, Large-scale structure of universe
PACS: 98.70.Vc, 98.65.-r, 98.65.Dx

INTRODUCTION

An appropriate ray-tracing procedure through N-body simulations was proposed in the basic papers [1] and [2]. Applications of this procedure, in the case of PM (Particle-Mesh) N-body simulations, can be found in [3] and [4]. Related talks presented by the authors in previous FFP Symposium editions were summarized in [5] and [6].

Recently, we have modified the sequential Hydra4.2 code, which was designed to perform hydrodynamical cosmological simulations. This code is freely available in the wave page of the Hydra consortium (see [7], [8], [9] and [10]). Our modifications allow us to move the Cosmic Microwave Background (CMB) photons through the simulation box. The N-body simulation performed by the code is an AP3M (Adaptative Particle-Particle Particle-Mesh) one, which can be used to reach resolutions better than those reachable with simple PM codes. The peculiar gravitational potential and its gradients – necessary to estimate the CMB weak lensing– are calculated and used at every time step of the N-body simulation. The photons are evolved under the action of this potential. Improvements based on parallel AP3M codes are in progress.

[1] E-mail: mfullana@mat.upv.es

CP1018, *Frontiers of Fundamental and Computational Physics*,
edited by B. G. Sidharth, F. Honsell, O. Mansutti, K. Sreenivasan, and A. De Angelis

RAY-TRACING PROCEDURE

N-body codes simulate the evolution of a distribution of dark matter halos in a box. Inside these halos, there are galaxy cluster with baryons. Since the baryonic component is subdominant, only the dark matter distribution is actually important to estimate gravitational effects as, e.g., weak lensing on the CMB. The simulation box can be considered as a piece of the universe, which is uniformly covered by identical pieces of this type (at any time). Such an universe is periodic by construction. The CMB photons are moved along null geodesics with appropriate directions, which have been chosen in such a way that all the regions of the simulation box crossed by these photons are statistically independent among them. Let us now give some details about this procedure. For a box size of size $L = 512\ Mpc$ and the preferred direction (PD) defined by the spherical coordinates $\theta = 73.34°$, $\phi = 16.96°$, any photon entering inside a certain box through the crossing point (i), passes to the next box at point (i+1) (see Fig. 1), and the distance, D, between points (i) and (i+1) is $\sim 104Mpc$; therefore, if the scales larger than $60Mpc$ are eliminated in the full potentials given by the AP3M code (only for lensing calculations, but not for evolving dark matter), photons cross uncorrelated regions in successive boxes, which makes negligible periodicity effects in our universe filled by boxes. For the PD shown in Fig. 1, CMB photons cross 16 boxes through independent regions from the initial redshift $z_{in} \sim 5(\sim 8000Mpc)$ until present time. It is assumed that lensing is produced at $z < z_{in}$, therefore the most extended lensed map we can built up is that corresponding to the image of a box face located at $z = z_{in}$. The angular size of this image is close to $\sim 4° \times 4°$. Scales greater than 60 Mpc are linear and the weak lensing they produce can be independently calculated (without N-body simulations); see, e.g., [3].

HYDRA SIMULATION

Our estimations are based on an unique simulation, which has been performed in the framework of the concordance model with the following parameters: $h = 0.71$, $\Omega_b = 0.04$, $\Omega_c = 0.23$, $\Omega_\Lambda = 0.73$, and $\sigma_8 = 0.9$ (see [11]). It is assumed that only adiabatic scalar energy density perturbations evolve in the universe. In other words, no tensor modes are considered at all. A reionization with optical depth $\tau = 0.1$ [11] has been assumed. The power spectrum of the scalar perturbations is calculated by using the code CMBFAST [12]. The number of dark matter particles is taken to be 256^3. This number is appropriate for the assumed box size ($L = 512Mpc$).

Figure 2 shows the correlation function $\xi(r)$ extracted from our unique simulation. Its form is that expected (see [13]) for the softening length (0.03 Mpc), the box size (512 Mpc), and the number of particles (256^3) we have chosen. The code modifications required by our CMB calculations have not altered the dark matter evolution. Our version of the AP3M N-body –including lensing estimations– works very well.

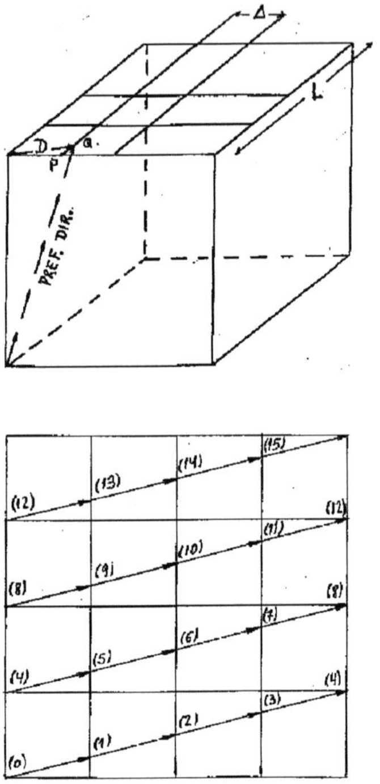

FIGURE 1. Top: Sketch of the photon motion along the PD (preferred direction) in the first box. Bottom: Representation of the point (i) where the PD crosses the upper face of the (i)-th box.

LENS DEFORMATIONS

Let us suppose a Gaussian unlensed map $T(\vec{n})$ of CMB temperatures. Hundreds of these maps have been simulated by using the CMB angular power spectrum of the so-called concordance model. In the absence of lensing, we would observe the unlensed temperature distribution; however, the CMB photons deviate and, consequently, the temperature observed in the direction $\vec{n}+\vec{\delta}$ is that corresponding to direction $\vec{n}$ in the unlensed map. The absolute temperature deformations due to lensing (photon deviations) are the quantities $\Delta T = T(\vec{n}+\vec{\delta}) - T(\vec{n})$. Ray-tracing through N-body simulations allow us to simulate the $\vec{\delta}(\vec{n})$ field of photon deviations. Then, the simulated unlensed maps may be deformed by using the resulting deviations and, finally, the maps of temperature deformations can be statistically analyzed. The solid line of Fig. 3 displays the power spectrum of the lens deformations obtained with the AP3M simulation described above.

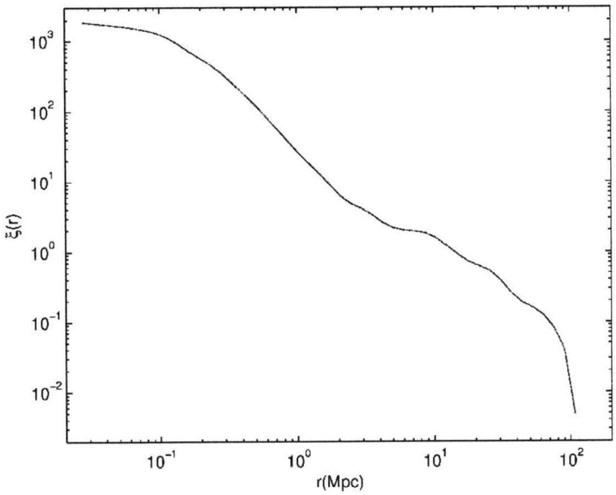

FIGURE 2. Two-point correlation function $\xi(r)$, at redshift $z = 0$, for the simulation described in the text.

In order to calculate the gradients of the peculiar gravitational potential (responsible for the lens effect), the PP corrections have been calculated at a set of points –on the photon null geodesics– separated by an uniform small distance of 0.1 Mpc. The dotted line of the same Figure shows the power spectrum of the lens deformations produced by the same AP3M simulation; nevertheless, in this case, the PP short range corrections have not been considered to calculate potential gradients on the null geodesics; that is to say, these corrections only have been used to evolve the dark matter, but not to calculate photon deviations.

DISCUSSION

The lens effect produced by scales greater than 60 Mpc can be easily studied by using linear approaches, e.g., see [3]. This effect is to be superimposed to that independently produced by scales smaller than 60 Mpc. This last effect requires N-body simulations to deal with small strongly nonlinear structures.

The use of N-body simulations constraints us to work in a periodic universe covered by simulation boxes. This periodicity affects lensing calculations (see [2], [3] and [4]).

If it is assumed that lensing is produced at $z < z_{in}$, the most extended lensed map we can built up is that corresponding to the image of a box face located at $z = z_{in}$. This fact motivates the use of large enough boxes leading to extended enough lensed maps.

The use of large boxes and high resolution (required by the strongly nonlinear scales) constraint us to run high resolution N-body codes as, e.g., Hydra and Gadget.

There are two alternatives: (1) we choose either pre-existent Hubble volume simulations or smaller simulations with higher resolutions as Millennium and, then, from the available time slices of the chosen simulation, we reconstruct the potential and its gra-

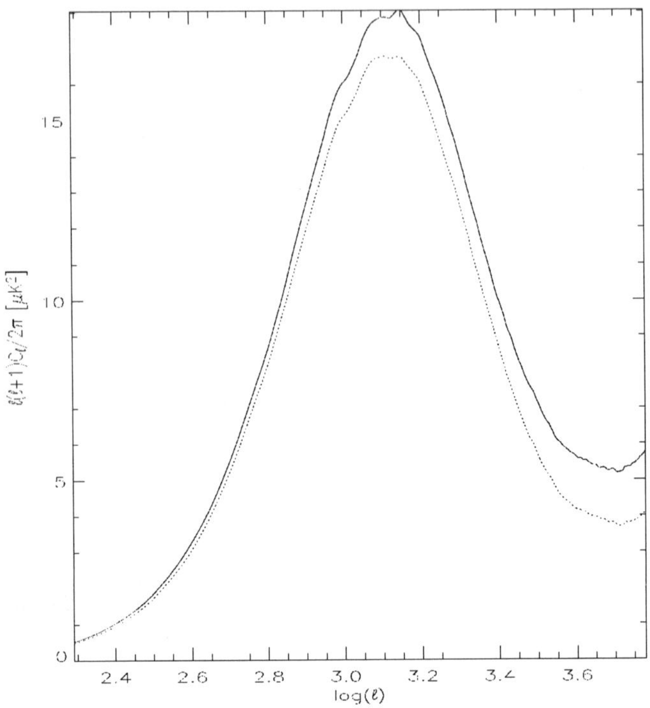

FIGURE 3. Solid (dotted) line shows the power spectrum of the temperature deformations –due to weak lensing in our AP3M simulation– for a estimation of the *photon deviations* which includes (does not include) local PP corrections. In both cases, these corrections are properly considered to evolve dark matter (AP3M code)

dients to estimate lensing effects on the CMB, or (2) we modify the chosen numerical code to move the CMB photons while the N-body simulation runs (all the time steps are then properly considered). We have verified that the reconstructed potential of method (1) has significant errors when the time slices are separated by $\sim$ 10 time steps of the N-body simulation.

We have successfully modified the AP3M Hydra4.2 code to move CMB photons according to the ray-tracing procedure displayed in [1] and [2].

Various results strongly suggest that our modified code is working well: (i) the good correlation function of Fig. 2, (ii) the power spectra of Fig. 3, which are similar, for $\log(\ell) < 3.4$, to those previously obtained with well tested PM codes (see [3]), (iii) the fact that the spectra obtained with simple PM codes (see [3]) continuously decrease for $\log(\ell) > 3.4$, whereas the spectra of Fig. 3 slightly grow for these large ℓ values, and (iv) the differences between the solid and dotted lines of Fig. 3, which depend on the angular scale as it is expectable. Points (ii) to (iv) are easily understood taking into account that local PP corrections (AP3M approach) power the growing of small scale structures and, consequently, these corrections contribute to the C_ℓ coefficients for large enough ℓ values ($\log(\ell) > 3.4$). The lack of resolution of the PM simulations implies an underestimation of these high multipoles. A good estimate requires a proper treatment of the PP corrections, which must be taken into account to deal with both dark matter evolution and lensing calculations.

PERSPECTIVES

Since maps of lens deformations are not Gaussian, they cannot be built up by using the angular power spectrum only. The bispectrum, trispectrum, and so on would be necessary. As a result of this fact, realistic deformations must be obtained by running N-body simulations (with high computational cost). Deviations from Gaussianity could be then estimated by analyzing the resulting deformation maps and the lensed maps (unlensed plus deformations).

Our modification of the Hydra4.2 code may be used to run simulations in big boxes with sizes between $\sim$ 1000 Mpc and $\sim$ 4000 Mpc (Hubble volumes). The lensed maps obtained from these simulations are extended enough (between $\sim$ 50 and $\sim$ 800 squared degrees) to allow the extraction of accurate angular power spectra for a wide range of ℓ values, for which, the ESA Planck Mission should lead to accurate C_ℓ estimates. The code is essentially ready and some improvements are being implemented. The computational cost is the main problem. An optimized parallel version of the AP3M code could help to deal with such a problem. One of these codes is being prepared in collaboration with some members of the Hydra consortium. Results will be presented elsewhere.

ACKNOWLEDGMENTS

This work has been supported by the Spanish Ministerio de Educación y Ciencia, MEC-FEDER project FIS2006-06062. Thanks to the Hydra Consortium for the free code Hydra4.2.

REFERENCES

1. P. Cerdá-Durán, V. Quilis, and D. Sáez, *Phys. Rev.* **69D**, 043002 (2004).
2. D. Sáez, N. Puchades, M.J. Fullana and J.V. Arnau. *Ray-Tracing through N-body simulations and CMB anisotropy estimations* in Proceedings of *CMB and Physics of the Early Universe*, Ischia, 2006.
3. L. Antón, P. Cerdá-Durán, V. Quilis and D. Sáez, *ApJ* **628**, 1 (2005).
4. N. Puchades, M.J. Fullana, J.V. Arnau and D. Sáez, *MNRAS* **370**, 1849-1858 (2006).
5. M.J. Fullana and D. Sáez. *Making Maps of the Rees-Sciama Effect* in Proceedings of *FFP6*, Udine, 2004.
6. M.J. Fullana and D. Sáez. *Status of CMB Radiation* in Proceedings of *FFP8*, Madrid, 2006.
7. P.A. Thomas and H.M.P. Couchman, *MNRAS* **257**, 11 (1992).
8. H.M.P. Couchman, *ApJ* **368**, L23 (1991).
9. H.M.P. Couchman, P.A. Thomas and F.R. Pearce, *ApJ* **452**, 797 (1995).
10. R.J. Thacker et al., astro-ph/9809221 (1998).
11. D.N. Spergel et al., *ApJS*, **170**, 377 (2007).
12. U. Seljak and M. Zaldarriaga, *ApJ*, **469**, 437 (1996).
13. K. Heitmann, P.M. Ricker, M.S. Warren and S. Habib, *ApJS*, **160**, 28 (2005).

The Speed of Light and the Hubble parameter: The Mass-Boom Effect

Antonio Alfonso-Faus[1]

E.U.I.T. Aeronáutica, Plaza Cardenal Cisneros s/n, 28040 Madrid, Spain

Abstract. We prove here that Newton's universal gravitation and momentum conservation laws together reproduce Weinberg's relation. It is shown that the Hubble parameter H must be built in this relation, or equivalently the age of the Universe t. Using a wave-to-particle interaction technique we then prove that the speed of light c decreases with cosmological time, and that c is proportional to the Hubble parameter H. We see the expansion of the Universe as a local effect due to the LAB value of the speed of light co taken as constant. We present a generalized red shift law and find a predicted acceleration for photons that agrees well with the result from Pioneer 10/11 anomalous acceleration. We finally present a cosmological model coherent with the above results that we call the Mass-Boom. It has a linear increase of mass m with time as a result of the speed of light c linear decrease with time, and the conservation of momentum mc. We obtain the baryonic mass parameter equal to the curvature parameter, $\Omega_m = \Omega_k$, so that the model is of the type of the Einstein static, closed, finite, spherical, unlimited, with zero cosmological constant. This model is the cosmological view as seen by photons, neutrinos, tachyons etc. in contrast with the local view, the LAB reference. Neither dark matter nor dark energy is required by this model. With an initial constant speed of light during a short time we get inflation (an exponential expansion). This converts, during the inflation time, the Planck's fluctuation length of 10^{-33} cm to the present size of the Universe (about 10^{28} cm, constant from then on). Thereafter the Mass-Boom takes care to bring the initial values of the Universe (about 10^{15} gr) to the value at the present time of about 10^{55} gr.

Keywords: gravitation, relativity, gravity quanta, Hubble, speed of light, anomalous acceleration, inflation, cosmology, Supernovae, big bang, Mach's Principle
PACS: 03.65. -W, 04.60. -M, 04.62. +V, 11.10-z, 98.80. -K

INTRODUCTION

In 1972 Weinberg [1] presented a relation that contained quantum parameters together with the Hubble cosmological parameter H. The order of magnitude of the mass of the fundamental particles was predicted and the relation is still catalogued as *unexplained*. It has also been classified as a coincidence of numerical numbers and some explanations [2], have been proposed, none of them gaining general acceptance. Here we prove it using Newton's universal law of gravitation, together with the momentum conservation law. We take a Machean approach considering the total momentum content of a mass M as Mc, a constant. The reality of the mass M, or the total momentum content Mc, is due to its interaction with the rest of the Universe. The total relativistic energy Mc^2 is also due to the interaction with the rest of the Universe. With these ideas in mind, together with a time varying speed of light as we will see, we derive the Weinberg relation as

---

[1] E-mail: aalfonsofaus@yahoo.es

CP1018, *Frontiers of Fundamental and Computational Physics*,
edited by B. G. Sidharth, F. Honsell, O. Mansutti, K. Sreenivasan, and A. De Angelis

a consequence of Newton's laws. By using a wave to particle interaction technique presented elsewhere [3] and [4], we derive a generalized red shift law that contemplates time variations in c and G (the gravitational *constant*). Then we prove that the speed of light c is proportional to the Hubble parameter H, and therefore time-varying. This time variation explains the anomalous acceleration found in space probes like the Pioneer 10/11 [5]. With the propagation equation of particles and photons we prove two very important constancies: all gravitational radii are constant. In particular the gravitational radius of the Universe is constant. Also the cosmological scale factor of the Universe is constant: neither expansion nor contraction is present in the Universe at a cosmological scale. The time variation of c implies a time variation of any mass. This is a result of the constancy of momentum: since the relation of any speed v to the speed of light c, v/c, has to be a constant to preserve special relativity, the constancy of momentum mv implies the constancy of mc. Then a decreasing speed of light implies an increasing mass, a Mass-Boom as we call it. Finally we make use of the Einstein cosmological equations to find the cosmological model that this theory predicts: a closed, finite, unlimited, and curved ($k = 1$) Universe with zero cosmological constant and a baryon matter content given by $\Omega_m = \Omega_k$. Neither dark matter nor dark energy is necessary in this model. However, an initial inflationary phase is also built in the model that brings the initial Planck fluctuation to the present size of the Universe.

WEINBERG'S RELATION

In 1972 Weinberg [1] discussed a relation obtained by using G (gravitation), c (relativity), $\hbar$ (quantum mechanics) and H (cosmology) to arrive at the mass m of a fundamental particle:

$$m^{\frac{1}{3}} \approx \frac{\hbar^2 H}{Gc} \tag{1}$$

We will follow a Machean approach by considering that the maximum momentum content of a mass M is Mc, a constant due to the rest of the Universe. It has acquired this momentum during a time t. Then the total force that can exert is Mc/t . The result of this heuristic argument can also be obtained by considering the rate of change with time of the total relativistic energy of the mass M, Mc^2, divided by c:

$$\frac{1}{c}\frac{dMc^2}{dt} = M\frac{dc}{dt} \tag{2}$$

We later will prove that the speed of light c is proportional to the Hubble parameter H, $c = HL$, where L is a constant. For a linearly expanding Universe the time derivative of H is $-H^2$ so that in absolute value we have for (2)

$$M\frac{dc}{dt} = MH^2L = MHc \tag{3}$$

Here we note that the Hubble parameter can be taken, in certain cases, as an operator that obtains the derivative, with respect to the cosmological time, of the function after it. To obtain the cosmological change with time we just put H in front of the function. This total force (3) can be thought of as an action fluxing over a spherical surface of radius r centred on the centre of gravity of M. It will act upon the small area represented by m, $m << M$, that has a size of the order of its Compton wavelength. Then m will feel a force exerted by the presence of M and given by

$$\frac{GMm}{r^2} = MHc\frac{(\hbar/mc)^2}{4\pi r^2} \tag{4}$$

and rearranging we arrive at

$$4\pi m^3 = \frac{\hbar^2 H}{Gc} \tag{5}$$

which is the Weinberg's relation (1) within a factor of less than 3.

GENERALIZED RED SHIFT LAW

In 1983 Adams [3] presented a wave to particle technique to derive the matter and photon propagation. Later, Alfonso-Faus [4], applied this technique to the case of a time-varying gravitational constant G. Now we do the same but allowing the speed of light c to vary with cosmological time t too. Since the factor in the Einstein's field equations in front of the stress-energy tensor is G/c^4, we have now the generalized red shift law equation as

$$G/c^4 N_\gamma \hbar \nu R = \text{constant} \tag{6}$$

Taking into account that the number of photons in a proper volume N_γ is constant (no particle creation), and using $n = c/\lambda$, we get eliminating G from (5) and (6)

$$HN_\gamma R(\hbar/mc)^3 = c\lambda \tag{7}$$

The wavelength λ is proportional to the cosmological scale factor R, and the Compton wavelength $\hbar/mc$ is constant. Then we have

$$c = H \times \text{constant} = HL \tag{8}$$

where $L \approx 10^{28} cm$, is the present size of the visible Universe ct. Since the Hubble parameter H is inversely proportional to the cosmological time t so is the speed of light c. Substituting (8) into (1) we get the new Weinberg's relation

$$m^3 \approx \frac{\hbar^2}{GL} \tag{9}$$

Here we see that we have obtained a relation that classically, with G, m, $\hbar$ and L all constants, has no time dependence in it. But allowing G and m to vary with time we have for $\hbar$ and L constant

$$Gm^3 = \text{constant} = \hbar^2/L \tag{10}$$

This is what is observed in the lunar range experiments with laser. Now, if we demand that the field equations of Einstein's general relativity continue to be derived from the action principle, we need to maintain constant the factors in front of the action integrals. This means

$$G/c^3 = \text{constant} \quad \text{and} \quad mc = \text{constant} \tag{11}$$

And these relations satisfy Weinberg's relation in the form given in (9).

EXPLANATION OF THE PIONEER 10/11 ANOMALOUS ACCELERATION

The fact that the speed of light varies with time as in (8) has enormous cosmological implications. We can get the photon acceleration as

$$\frac{dc}{dt} = L\frac{dH}{dt} \tag{12}$$

In the laboratory the speed of light is taken as constant c_0. Then we see a linear expansion of the Universe as $R = c_0 t$. Since H is R'/R we get for the photon acceleration a

$$a = \frac{dc}{dt} = -LH^2 = -Hc \tag{13}$$

Taking a range of values for H from 60 to 75 Km/sec/Mpc, the theoretical acceleration (13) for the photons is

$$a = -(7.13 \pm 0.3)\ cm/seg^2 \tag{14}$$

The observed anomalous acceleration in the Pioneer 10/11 case [5] is

$$a_p = -(8.74 \pm 1.33)\ cm/seg^2 \tag{15}$$

which is within the margin given in (14). From this point of view the observed acceleration is not satellite acceleration (as due to gravitational attraction). It is in the photons.

THE CONSTANCY OF GRAVITATIONAL RADII

With the wave to particle technique, [4], we get now for the matter propagation equation

$$\left(\left(\frac{G}{c^2}NP^\mu\right)_{;\alpha}\right)P^\alpha = 0 \tag{16}$$

Using $P^\mu = mU^\mu$ we get

$$U^\mu_{;\alpha}U_\alpha + \frac{d\ln\left(\frac{G}{c^2}mN\right)}{dt}U^\mu U^0 = 0 \tag{17}$$

and contracting with U_μ we finally get

$$\frac{d\ln\left(\frac{G}{c^2}mN\right)}{dt} = 0 \quad \text{i. e.} \quad \frac{GmN}{c^2} = \text{constant} \tag{18}$$

This means that all the gravitational radii of masses mN inside its proper volume are constant. The possible expansion or contraction of the Universe does not affect them.

THE CONSTANCY OF THE COSMOLOGICAL SCALE FACTOR

We have derived elsewhere, [6], the expression for the zero value of the right hand side of the Einstein's field equations

$$\nabla(\frac{G}{c^4}T^{\mu\nu}) = 0 \tag{19}$$

which is

$$\frac{\rho'}{\rho} + 3(\omega+1)H + \frac{\Lambda' c^4}{8\pi G\rho} + \frac{G'}{G} - 4\frac{c'}{c} = 0 \tag{20}$$

where we have allowed for the time dependence of G, c and Λ. The cosmological parameter Λ is a real constant of integration so that it disappears from (20). In fact we will later show that Λ is zero in the cosmological view. Here ρ is the energy density. Using the equation of state $p = \omega\rho$, integration of (20) gives

$$\frac{G\rho}{c^4}R^{3(\omega+1)} = \text{ constant} \tag{21}$$

or equivalently

$$\frac{GM}{c^2}R^{3\omega} = \text{ constant} \tag{22}$$

This result is very important. Since all the gravitational radii are constants and if w is not zero, we arrive at the conclusion that the Universe does not expand nor contracts, i.e.

$$R = \text{constant} \tag{23}$$

Then the gravitational radius of the Universe can be equated to its visible size with $R = ct = \text{constant} = L$. For M the mass of the Universe at present we have

$$\frac{GM}{c^2} = R = ct = L = \text{ constant } >> \ 10^{28}cm \tag{24}$$

No length in the Universe is expanding nor contracting. The dynamo paradox, that could ideally be constructed to obtain work from a rod attached to a galaxy that is going away from us, just goes away.

There is a way to interpret Mach's principle: the rest energy of any mass m is equal to its gravitational potential energy due to the rest of the mass of the Universe:

$$\frac{GMm}{R} = Rmc^2 \ \text{ or } \ \frac{GM}{c^2} = R \tag{25}$$

Comparing (24), (25) and (22) we get the value of ω as

$$\omega = -1/3 \ \text{ so that } \ p = -1/3\rho \tag{26}$$

We note that this is the same as the equation of state for photons, with a minus sign. We interpret it as the negative pressure of the gravity quanta.

MASS-BOOM AND THE COSMOLOGICAL EQUATIONS

From (11) and (24) we get again the Mass-Boom effect, [7], a linear increase of mass with cosmological time, $M \propto t$. We take the number of particles in the Universe N as constant so that in a certain system of units we have:

$$M = Nm = Nt \tag{27}$$

And from (24) we arrive at

$$G/c^3 = 1/N \quad \text{and} \quad m = t \tag{28}$$

The Einstein cosmological equations, with R = constant are then

$$\begin{aligned} 8\pi G\tfrac{p}{c^2} + \tfrac{Kc^2}{R^2} &= \Lambda c^2 \\ -\tfrac{8\pi}{3}G\rho_m + \tfrac{Kc^2}{R^2} &= \tfrac{\Lambda c^2}{3} \end{aligned} \tag{29}$$

Now with $\omega = -1/3\Lambda$ has to be identically zero. Then the two cosmological equations reduce to just one, i.e.

$$\frac{Kc^2}{R^2} = \frac{8\pi}{3}G\rho_m \tag{30}$$

We see that $K = 1$ is the solution to this equation. This cosmological model is spherical, closed, finite, unlimited, and static. Equation (30) can be interpreted as equilibrium between centrifugal forces, represented by the curvature term, and the negative pressure, the gravitational attraction. In the usual dimensionless parameters Ω we have the equivalent to (30)

$$\Omega_m = \Omega_k \text{ (undetermined)} \tag{31}$$

We can take the value 0.05 for both parameters. Then neither dark matter nor dark energy is necessary in this model.

THE INITIAL CONDITIONS OF THE UNIVERSE

During the first instants of time the speed of light must have been almost constant. This means that the Hubble parameter then was constant so that we have at that time:

$$H = R'/R = \text{ constant} \tag{32}$$

Integrating the above relation we get the solution:

$$R = R_0 e^{Ht} \tag{33}$$

This exponential expansion is equivalent to the very well known inflation phase at the initial stages of the Universe. During the inflation phase the first Planck's fluctuation of size 10^{-33} cm expanded very rapidly to the size 10^{28} cm, as of today, so that we have

$$e^{Ht} = 10^{61} \tag{34}$$

If t_1 is the first tic of the Universe, and t_i is the duration of the inflationary phase, then

$$t_i/t_1 = Ht = 61 \ \ln 10 >> 140 \tag{35}$$

Hence, only 140 tics were necessary to inflate the Planck's fluctuation to the present size of the Universe, and thereafter it remained at constant size. In our theory Planck's units have a time dependence as

$$\begin{aligned} \left(\frac{\hbar c}{G}\right)^{\frac{1}{2}} &= 10^{60}/c = 10^{20}t \approx 10^{-5}\ gr \\ \left(\frac{G\hbar}{c^3}\right)^{\frac{1}{2}} &= \text{ constant } = 1/10^{20} \approx 10^{-33}\ cm \\ \left(\frac{G\hbar}{c^5}\right)^{\frac{1}{2}} &= 1/(10^{20}c) = t/10^{60} \approx 10^{-44}\ s \end{aligned} \tag{36}$$

These units had an initial value of

$$\begin{aligned} t &= 10^{40} \\ t_1 &= 1 \\ \left(\frac{\hbar c}{G}\right)_1^{\frac{1}{2}} &= 10^{20}t_1 \approx 10^{-45}\ gr \\ \left(\frac{G\hbar}{c^3}\right)_1^{\frac{1}{2}} &\approx 10^{-33}\ cm = \text{ constant} \\ \left(\frac{G\hbar}{c^5}\right)_1^{\frac{1}{2}} &= t_1/10^{60} \approx 10^{-83}\ s \end{aligned} \tag{37}$$

Here we see that inflation by a factor of 10^{61} gives an initial stage of the Universe as

$$M_i = 10^{15} gr, \quad L = 10^{28} cm = \text{ constant}, \quad t_i = 10^{-23} s \tag{38}$$

From then on we have 5×10^{40} tics to arrive at

$$M_0 = 5 \times 10^{55} gr, \quad L = 10^{28} cm = \text{ constant}, \quad t_0 = 5 \times 10^{17} s \tag{39}$$

which is the present state of the Universe.

CONCLUSIONS

The wave-to-particle interaction technique is a powerful tool to determine very important relations in Nature. We have found the matter and photon propagation equations using the Einstein's general relativity field equations. We have found a generalized red shift law. We have allowed for time-varying G, m and c. We have found the new constants of Nature G/c^3, mc and c/H. We have defined a Mass-Boom cosmological model, a new frame of work for cosmology that has a decreasing speed of light and an increasing mass with cosmological time. The predicted photon acceleration agrees well with the anomalous acceleration of the Pioneer 10/11 satellites.

ACKNOWLEDGMENTS

I am grateful to Prof. Fullana for his help in preparing this paper for the FFP9 Proceedings.

REFERENCES

1. Weinberg, S., *Gravitation and Cosmology*, John Wiley and Sons, New York, (1972).
2. Sidharth, B.G., "The Universe of Chaos and Quanta", arXiv: qant-ph/9902028 8 Feb. (1999).
3. Adams, P. J. , Int. J. Theoret. Phys., "Large Numbers Hypothesis. I. Classical Formalism", Vol. 21, 603 (1982). and "Large Numbers Hypothesis. II. Electromagnetic Radiation" Vol. 22, 421 (1983).
4. Alfonso-Faus, A., Int. J. Theoret.Phys., "Cosmology with Time-Varying G", Vol. 25, No. 3, 293-316, (1986).
5. Anderson, J. D., et al., "Indication, from Pioneer 10/11, Galileo, and Ulysses data, of an apparent anomalous, weak, long-range acceleration". Phys. Rev. Lett., 81, 2258, (1998).
6. Belinchón, J.A. and Alfonso-Faus, A., Int. J. Mod. Phys., "A THEORY OF TIME-VARYING CONSTANTS", D10, 299-310 (2001) and arXiv: gr-qc/0404044 (2001).
7. A. Alfonso-Faus, V International Symposium "Frontiers of Fundamental Physics" Hyderabad, India 2003, and arXiv:physics/0302058 v1 - 18 February 2003.

Multiparticle Quantum Cosmology[1]

L. A. Glinka[2]

N.N. Bogoliubov Laboratory of Theoretical Physics, Joint Institute for Nuclear Research, Joliot–Curie 6, 141980 Dubna, Moscow Region, Russian Federation

Abstract. Fock space quantization of Hamiltonian constraints of General Relativity and thermodynamics of quantum states for flat Friedmann–Lemaître–Robertson–Walker metrics is presented.

Keywords: quantum cosmology, quantum gravity, tachyon, Bose condensates, 2nd quantization, thermodynamics of quantum states, canonical Dirac–ADM formalism of General Relativity, MEQS
PACS: 98.80.Qc, 11.25.-w, 04.60.-m, 14.80.-j, 05.30.Jp

INTRODUCTION. THERMODYNAMICAL EINSTEIN'S DREAM

General Relativity is a theory of some Riemannian manifold [1] with dynamics governs by the Einstein equations [2] thats can be deduced from the Hilbert dynamical action [3] by the Palatini variational principle [4]

$$S_{\rm EH} = \int d^4x\sqrt{-\det g_{\mu\nu}}\left(-\frac{1}{6}R+\mathscr{L}\right), \tag{1}$$

$$\frac{\delta S_{\rm EH}}{\delta g^{\mu\nu}} = 0 \Longrightarrow R_{\mu\nu}-\frac{1}{2}g_{\mu\nu}R=3T_{\mu\nu},\ \ T_{\mu\nu}=\frac{2}{\sqrt{-\det g_{\mu\nu}}}\frac{\delta\left(\sqrt{-\det g_{\mu\nu}}\mathscr{L}\right)}{\delta g^{\mu\nu}}, \tag{2}$$

where $R_{\mu\nu}$ is the Ricci tensor, $R=g^{\alpha\beta}R_{\alpha\beta}$ is the Ricci curvature scalar, $\mathscr{L}$ is a full Lagrangian of all physical fields, and $T_{\mu\nu}$ is the stress–energy tensor of these fields [3].

Let us to take up address an issue of a Riemannian manifold given by the flat Friedmann–Lemaître–Robertson–Walker metric [5] characterized by an interval[4]

$$ds^2 = g_{\mu\nu}dx^\mu dx^\nu = (dx^0)^2 - a^2(t)(dx^i)^2, \tag{3}$$

where a is the Friedmann conformal scale factor and x^μ, $\mu=0,1,2,3$ is a cartesian coordinate system. In the canonical Dirac–ADM approach to General Relativity [6] the metric (3) is characterized by vanishing shift function and lapse function N is given by

$$d\eta = N(x^0)dx^0 \equiv \frac{dt}{a(t)},\ \ \text{where } t=\tau+x^0,\ \ \tau=const, \tag{4}$$

---

[1] Given as a talk "Many Particle Quantum Cosmology" at the Ninth International Symposium *Frontiers of Fundamental and Computational Physics*, Udine University, Udine, Italy , and The Abdus Salam International Center for Theoretical Physics, Trieste, Italy, January 7-9, 2008

[2] E-mail to: laglinka@gmail.com

[3] In this paper the units $8\pi G/3 = c = \hbar = k_B = 1$ are used.

[4] For closed case $V_0 = \int d^3x < \infty$ the metrics (3) describe *cylindrical Einstein–Friedmann Universe.* [7]

CP1018, *Frontiers of Fundamental and Computational Physics*,
edited by B. G. Sidharth, F. Honsell, O. Mansutti, K. Sreenivasan, and A. De Angelis

where t is called cosmological time, and η is called conformal time. The metrics (3) in the Dirac-ADM approach are equivalent to the following Hamiltonian constraints

$$\mathrm{H} = p_a^2 - 4V_0^2 a^4 H^2(a) = 0, \tag{5}$$

where p_a is canonical momentum conjugated to a, and $H(a)$ is the Hubble parameter,

$$p_a = -2V_0 \frac{da}{d\eta}, \quad H^2(a) \equiv \left(\frac{1}{a}\frac{da}{dt}\right)^2 = \left(\frac{1}{a^2}\frac{da}{d\eta}\right)^2 = \frac{1}{V_0}\int d^3x\, \mathscr{H}(x) \tag{6}$$

where $\mathscr{H}(x)$ is summarized Hamiltonian of physical fields associated by the Legendre transformation with the Lagrangian $\mathscr{L}$ in the dynamical action (1). The Hubble law in the canonical approach is a result of integration of (5) and has a form

$$\int_{a_I}^{a} \frac{da'}{a' H(a')} = t_I - t, \quad \text{where } I \text{ means initial data.} \tag{7}$$

We propose call by the *Thermodynamical Einstein's Dream (TED)* a some formal road from General Relativity (1-2) to generalized thermodynamics of quantum states of a Riemannian manifold being a solution of this theory. The questions arise

- *Could TED really be realize?*
- *Can be exist a way that allows to get to know thermal properties of Spacetime?*
- *If these properties of Spacetime give new opportunities for Quantum Gravity?*

In this paper a proposal for realization of TED is demonstrated by using of second quantization in the Fock space and associated with this quantization formalism of the Bose condensation, where a *particle* is quantum state of a Spacetime, is discussed.

CONSTRAINTS, STRINGS, AND SECOND QUANTIZATION

Let us consider the Dirac–ADM primary constraints (5). Applying to this equation the formal identification

$$-4V_0^2 a^4 H^2(a) = m^2(a) \tag{8}$$

allows to equate the geometrical object (3) with free bosonic string with negative square of mass, called a *tachyon* [9], defined by the constraints

$$p_a^2 + m^2 = E^2, \quad E = 0, \quad m^2 = m^2(a) \le 0. \tag{9}$$

The Hubble law (7) can be treated for tachyon as definition of spatial volume V_0

$$\left(\int_{a_I^2}^{a^2} \frac{dy}{|m(y)|}\right)^{-1} \Delta t = V_0, \quad \Delta t = t - t_I. \tag{10}$$

First quantization of the constraints (9) gives the Wheeler–DeWitt equation [10] for (3)

$$i\,[\mathrm{p}_a, a] = 1 \implies \left[-\frac{\partial^2}{\partial a^2} + m^2(a)\right]\Psi = 0, \quad \Psi = \Psi^*, \tag{11}$$

that can be treated as the free-time Schrödinger equation [11]. However, this point of view has no physical sense. Equally well the Wheeler–DeWitt equation (11) can be understood as the $0+1$ Klein–Gordon–Fock equation, that treats a Riemannian manifold as a Bose field. Separation of variables method applied to the equation (11) gives

$$\left(i\Gamma^{\mu}\partial_{\nu} - \mathrm{M}^{\mu}_{\nu}\right)\Phi_{\mu} = 0, \ \ \mathrm{M}^{\mu}_{\nu} = \left[\begin{array}{cc} 0 & -1 \\ -m^2 & 0 \end{array}\right] \geq 0 \qquad (12)$$

where matrices $\Gamma^{\mu} = [\mathbf{0}_2, i\mathbf{I}_2]$ create Clifford algebra $\{\Gamma^{\mu}, \Gamma^{\nu}\} = 2\eta^{\mu\nu}\mathbf{I}_2$, and was introduced the state vector $\Phi_{\mu} = [\Psi, \ \Pi_{\Psi}]^T$ with Π_{Ψ} being classical canonical momentum field conjugated to wave function Ψ, here also $\partial_{\nu} = \left[0, -\frac{\partial}{\partial a}\right]^T$. Let us carry second quantization [8] of the equation (12) that is agreed with (11). This has a proposed form

$$\left[\begin{array}{c} \hat{\Psi}[a] \\ \hat{\Pi}_{\Psi}[a] \end{array}\right] = \left[\begin{array}{cc} 1/\sqrt{2|m(a)|} & 1/\sqrt{2|m(a)|} \\ -i\sqrt{|m(a)|/2} & i\sqrt{|m(a)|/2} \end{array}\right] \left[\begin{array}{c} G[a] \\ G^{\dagger}[a] \end{array}\right], \qquad (13)$$

and automatically fulfills principal rule of quantum field theory of the Bose systems

$$[\hat{\Pi}_{\Psi}[a], \hat{\Psi}[a']] = -i\delta\left(a - a'\right), \ \ a \equiv a(\eta), \ \ a' \equiv a(\eta') \qquad (14)$$

and lies in accordance with general algebraic approach investigated in papers of Von Neumann, Araki and Woods [12]. In result, considered Bose system is described by the dynamical basis in the Fock space

$$B_a = \left\{\left[\begin{array}{c} G[a] \\ G^{\dagger}[a] \end{array}\right] : \left[G[a], G^{\dagger}[a']\right] = \delta\left(a - a'\right), \left[G[a], G[a']\right] = 0\right\}, \qquad (15)$$

according to the evolution equation

$$\frac{\partial B_a}{\partial a} = \left[\begin{array}{cc} -m & \frac{1}{2m}\frac{\partial m}{\partial a} \\ \frac{1}{2m}\frac{\partial m}{\partial a} & m \end{array}\right] B_a, \qquad (16)$$

that can be diagonalized to the Heisenberg form by the Bogoliubov transformation

$$B'_a = \left[\begin{array}{cc} u & v \\ v^* & u^* \end{array}\right] B_a, \ \ \frac{\partial B'_a}{\partial a} = \left[\begin{array}{cc} -i\lambda & 0 \\ 0 & i\lambda \end{array}\right] B'_a, \qquad (17)$$

where

$$B'_a = \left\{\left[\begin{array}{c} G'[a] \\ G'^{\dagger}[a] \end{array}\right] : \left[G'[a], G'^{\dagger}[a']\right] = \delta\left(a - a'\right), \left[G'[a], G'[a']\right] = 0\right\}, \qquad (18)$$

is some new basis, and the Bogoliubov coefficients u and v are functions of a. This procedure gives simply $\lambda = 0$, and by this $G'[a] = \mathrm{w}_I = const$ defines stable vacuum $|0\rangle_I = \left\{|0\rangle_I : \mathrm{w}_I|0\rangle_I = 0, \ 0 = {}_I\langle 0|\mathrm{w}^{\dagger}_I\right\}$ in the static Fock basis $B'_a = B_I$,

$$B_I = \left\{\left[\begin{array}{c} \mathrm{w}_I \\ \mathrm{w}^{\dagger}_I \end{array}\right] : \left[\mathrm{w}_I, \mathrm{w}^{\dagger}_I\right] = 1, \left[\mathrm{w}_I, \mathrm{w}_I\right] = 0\right\}. \qquad (19)$$

In this manner, in proposed approach *quantum gravity is defined by monodromy between bases in the Fock space*

$$B_a = \begin{bmatrix} \left(\sqrt{\frac{|m(a)|}{|m_I|}}+\sqrt{\frac{|m_I|}{|m(a)|}}\right)\frac{e^{\lambda(a)}}{2} & \left(\sqrt{\frac{|m(a)|}{|m_I|}}-\sqrt{\frac{|m_I|}{|m(a)|}}\right)\frac{e^{-\lambda(a)}}{2} \\ \left(\sqrt{\frac{|m(a)|}{|m_I|}}-\sqrt{\frac{|m_I|}{|m(a)|}}\right)\frac{e^{\lambda(a)}}{2} & \left(\sqrt{\frac{|m(a)|}{|m_I|}}+\sqrt{\frac{|m_I|}{|m(a)|}}\right)\frac{e^{-\lambda(a)}}{2} \end{bmatrix} B_I, \tag{20}$$

where $\lambda(a) = \pm\int_{a_I}^{a} m(a')\,da'$. The field operator $\hat{\Psi}$ that represents the Spacetime (3) as boson can be concluded as

$$\hat{\Psi}[a] = \frac{1}{\sqrt{2m_I}}\left(e^{\lambda(a)}\mathrm{w}_I + e^{-\lambda(a)}\mathrm{w}_I^{\dagger}\right), \tag{21}$$

and has the properties

$$\left(\hat{\Psi}[a]\right)^n|0\rangle = \frac{e^{-n\lambda(a)}}{(2m_I)^{n/2}}\mathrm{w}_I^{\dagger n}|0\rangle \equiv |n\rangle, \tag{22}$$

$$\langle\, m|n\rangle = \frac{e^{(m-n)\lambda(a)}}{(2m_I)^{\frac{m+n}{2}}}\left\langle 0\left|\mathrm{w}_I^{m}\mathrm{w}_I^{\dagger n}\right|0\right\rangle, \tag{23}$$

$$\langle\, n|n\rangle = \sum_{p=0}^{n}\frac{e^{i\pi(p+1)}}{(2m_I)^n}C_{n-p}^{n}\left\langle 0\left|\left(\mathrm{w}_I^{\dagger}\mathrm{w}_I\right)^p\right|0\right\rangle = \frac{e^{i\pi}}{(2m_I)^n}\langle 0|0\rangle = \frac{e^{i(2n+1)\pi}}{(2m_I)^n}, \tag{24}$$

where for simplicity we omitted the index I near brackets of the expectation values, and $C_{n-p}^{n} = \frac{n!}{p!(n-p)!}$ is the Newton binomial coefficient, and was assumed that $\langle 0|0\rangle = 1$.

ONE-PARTICLE THERMODYNAMICS OF QUANTUM STATES

Initial data given by the basis (19) define thermal equilibrium state for ensemble of quantum states of the manifold (3). Let us consider thermal properties of this Spacetime - the final step of the Thermodynamical Einstein's Dream. Here we will study one-particle approximation [8] defined by density functional ρ_G in initial data basis

$$\rho_G = G^{\dagger}[a]G[a] = B_a^{\dagger}\begin{bmatrix} 1 & 0 \\ 0 & 0 \end{bmatrix}B_a = B_I^{\dagger}\begin{bmatrix} |u|^2 & -uv \\ -u^*v^* & |v|^2 \end{bmatrix}B_I \equiv B_I^{\dagger}\rho_{\mathrm{eq}}B_I. \tag{25}$$

Equilibrium entropy S and partition function Ω_{eq} of the system in initial data basis are

$$\mathrm{S} = -\frac{\mathrm{tr}\left(\rho_{\mathrm{eq}}\ln\rho_{\mathrm{eq}}\right)}{\mathrm{tr}\rho_{\mathrm{eq}}} \equiv \ln\Omega_{\mathrm{eq}},\ \ \Omega_{\mathrm{eq}} = \frac{1}{2|u|^2-1}. \tag{26}$$

Physical quantum states of considered manifold are assumed to be described in the Gibbs ensemble. One can compute directly the entropy of the system

$$\mathrm{S} = -\ln\langle \mathrm{n}\rangle,\ \ \langle \mathrm{n}\rangle = 2\mathrm{n}+1. \tag{27}$$

Here $\langle n\rangle$ is averaged occupation number n of quantum states determined by

$$n \equiv \langle 0|G^{\dagger}[a]G[a]|0\rangle = \frac{1}{4}\left(\left|\frac{m}{m_I}\right| + \left|\frac{m_I}{m}\right|\right) - \frac{1}{2}, \tag{28}$$

that together with natural conditions $n \geq 0$ and $|m| \geq |m_I|$ leads to the *energy spectrum*

$$\left|\frac{m}{m_I}\right| = \left|\frac{mc^2}{m_I c^2}\right| = \langle n\rangle + \sqrt{\langle n\rangle^2 - 1}, \tag{29}$$

and internal energy U and chemical potential μ computed by proper averaging are

$$\mathrm{U} = \left[1 + \left(2 + \frac{1}{\langle n\rangle}\right)(\langle n\rangle - 1)\right]\left(\langle n\rangle + \sqrt{\langle n\rangle^2 - 1}\right)\frac{|m_I|}{2}, \tag{30}$$

$$\mu = \left[1 + \frac{1}{\langle n\rangle^2} - \left(2 - \frac{1}{\langle n\rangle}\right)\sqrt{\frac{1}{\langle n\rangle^2 - 1}}\right]\left(\langle n\rangle + \sqrt{\langle n\rangle^2 - 1}\right)|m_I|. \tag{31}$$

From bosonic property one can assume in equilibrium the Bose–Einstein statistics

$$\Omega_{\mathrm{eq}} \equiv \left(\exp\left\{\frac{\mathrm{U} - \mu n}{\mathrm{T}}\right\} - 1\right)^{-1}, \tag{32}$$

and by this one can compute directly temperature of the system

$$\mathrm{T} = \frac{\langle n\rangle + \sqrt{\langle n\rangle^2 - 1}}{\ln(\langle n\rangle + 1)}\left[1 + \left(1 - \frac{1}{\langle n\rangle}\right)^2 + \left(\langle n\rangle - \frac{1}{2\langle n\rangle}\right)\sqrt{\frac{\langle n\rangle - 1}{\langle n\rangle + 1}}\right]\frac{|m_I|}{2}. \tag{33}$$

Maximal entropy quantum states (MEQS) of the Riemannian manifold are defined by

$$\mathrm{S}_{\max} = 0 \Longrightarrow n = 0,\ \langle n\rangle = 1,\ m = m_I,\ \mathrm{U}_{\max} = \frac{|m_I|}{2},\ \mu_{\max} = -\infty,\ \mathrm{T}_{\max} = \frac{\mathrm{U}_{\max}}{\ln 2}. \tag{34}$$

These allow conclude that *for the metrics MEQS are fully determinated by initial data.* The last equation in (34) is an equation of state for MEQS, that describes classical ideal Bose gas. Infinite value of chemical potential means that MEQS create closed system, and the point $n = 0$ is phase transition point for MEQS set. Initial data point can be interpreted as a birth-point of the Riemannian manifold. Measurement of MEQS formally defines initial data point, and it defines initial data reference frame. Furthermore, MEQS set have very interesting physical interpretation that is a conclusion of the reasoning

$$m_I = m = i2V_0 a^2 H(a) \Longrightarrow H(a) = \frac{Q}{a^2},\ Q = \frac{|m_I|}{2V_0} = \frac{\mathrm{U}_{\max}}{V_0} = a_I^2 H(a_I), \tag{35}$$

and for $\mathrm{U}_{\max} = \mathrm{p}_{\max} V_0$ it gives pressure $\mathrm{p}_{\max} = a_I^2 H(a_I)$. Form of the Hubble parameter $H(a)$ in (35) means that *set of MEQS for the manifold describes primordial radiation.* Initial data basis B_I is related with this light. In light of modern observational cosmology one can identify this primordial light with CMB radiation. One can see that λ in (20) is linear for MEQS and fulfills the d'Alembert equation

$$\lambda(a) = \int_{a_I}^{a} m(a')da'\Big|_{m=m_I} = m_I(a-a_I) \Rightarrow \Delta\lambda = 0, \ \ a(t) = a_I\sqrt{1+2Q(t-t_I)} \quad (36)$$

and by this can be treated as the Weyl scale [13] of the ensemble.

Quantum Gravity defined by the second quantization (13) used to the metrics (3) gives beautiful scenario for physics of the Universe. It allows to state that TED is realizable for considered manifold. From the String Theory point of view *tachyon is mass groundstate of bosonic string that describes some wider quantum theory of gravity.* Basing on this let one create the following conjecture: *for the Einstein–Hilbert (2) gravitational fields analogical deduction can be applied.* Moreover, multiparticle reasoning creates understanding of gravity by quantum field theory, and describes General Relativity by the Bose condensates. This approach can be called Multiparticle Quantum Gravity/Cosmology.

ACKNOWLEDGMENTS

The author benefitted from valuable discussions with B.G. Sidharth, A. De Angelis, J.G. Hartnett, and J. Lewandowski and is grateful to B.G. Sidharth for invitation to the 9th International Symposium *Frontiers of Fundamental and Computational Physics.*

REFERENCES

1. B. Riemann, *Nachr. Ges. Wiss. Göttingen* **13**, 133, (1920).
2. A. Einstein, *Sitzungsber. Preuss. Akad. Wiss. Berlin* **44**, N2, 778, (1915); *Sitzungsber. Preuss. Akad. Wiss. Berlin* **46**, N2, 799, (1915); *Sitzungsber. Preuss. Akad. Wiss. Berlin* **48**, N2, 844, (1915); *Ann. Phys.* **49**, 769, (1916), *Rev. Mod. Phys.* **20**, 35, (1948).
3. D. Hilbert, *Konigl. Gesell. d. Wiss. Göttinger, Nachr., Math.-Phys. Kl.* **27**, 395, (1915).
4. A. Palatini, *Rend. Pal.* **43**, 203, (1919).
5. A. Friedmann, *Z. Phys.* **10**, 377, (1922); *Z. Phys.* **21**, 326, (1924); G. Lemaître, *J. Math. Phys* **4**, 188, (1925); *Phys. Rev.*, **25**, 903 (1925); *Ann. Soc. Sci. Brux. A*, **47**, 49, (1931); *Ann. Soc. Sci. Brux. A* **53**, 51, (1933); H.P. Robertson, *Phil. Mag.*, **5**, 835, (1928); *Astrophys. Jour.* **82**, 248, (1935); *Astrophys. Jour.* **83**, 187, (1936); *Astrophys. Jour.* **83**, 257, (1936); A.G. Walker, *Proc. Lon. Math. Soc.* 2, **42**, 90, (1937).
6. P.A.M. Dirac, *Proc. Roy. Soc. Lond. A* **246**, 333, (1958); *Phys. Rev.* **114**, 924, (1959); *Proc. Roy. Soc. Lond. A* **246**, 326, (1958); *Can. J. Math.* **2**, 129, (1950); R. Arnowitt, S. Deser and C.W. Misner, in *Gravitation: an introduction to current research*, edited by L. Witten (John Wiley and Sons, New York, 1961), p. 227
7. A. Einstein, The meaning of relativity, Princeton University Press, Princeton (1922).
8. J.-P. Blaizot and G. Ripka, *Quantum theory of finite systems*, Massachusetts Institute of Technology Press, (1986).
9. D. Lüst and S. Theisen, *Lect. Notes Phys.* **346**, (1989).
10. J.A. Wheeler, in *Battelle Rencontres: 1967 Lectures in Mathematics and Physics*, Editors C.M. DeWitt and J.A. Wheeler, New York, 1968, p. 242; B.S. DeWitt, *Phys. Rev.* **160**, 1113, (1967).
11. A. Ashtekar, M. Bojowald and J. Lewandowski, *Adv. Theor. Math. Phys.* **7**, 233, (2003); J. Lewandowski *Private communication*; C. Rovelli *Private communication*
12. J. von Neumann, *Math. Ann.* **104**, 570, (1931); H. Araki and E.J. Woods, *J. Math. Phys.* **4**, 637, (1963); J. Dereziński, *Lect. Notes Phys.* **695**, 63-143, (2005).
13. Weyl H., *Raum-Zeit-Materie*, 6th ed., Springer-Verlag Berlin-Heidelberg, (1970).

5d relativistic hydrodynamics and cosmology

Gianluca Gemelli

L.S. B. Pascal, V. P. Nenni 48, Pomezia (Roma), Italy

Abstract. We consider the evolution of a perfect fluid in 5d relativity. This picture reveals a 4d particle production effect, the source of particle production originating from 5d geometry, and vanishing if the flow is isentropic. The result holds in 5d special relativity as well as in 5d general relativity, no matter if the cosmological fluid is a test-fluid or a self-gravitating fluid.

INTRODUCTION

As it is well known, 5d relativity is the framework of Kaluza-Klein theory of gravitation and electromagnetism (see e.g. [14]). Today, however, a 5d point of view is also perfectly legitimate in cosmology, since the introduction of Carmeli's theories [1, 2, 3], where the receding velocity of galaxies plays the role of fifth dimension. It is interesting to find out that this point of view has a link to open system cosmology, i.e. the cosmological scenario which includes particle production [12, 5].

Recently in fact some consequences of cosmological interest of the equations of hydrodynamics in 5d special relativity were considered in [6, 7]: it was found that such framework provides in a simple and natural way a mechanism for particle production.

5d equation of conservation in general corresponds to a 4-d balance equation (with source). Thus number density and stress-energy conservation in 5d imply particle creation and stress-energy balance in 4-d, respectively. This is interesting, since in cosmology particle production is usually introduced by hand (see e.g. [4]); the 5d-hydrodynamics approach instead has it as a natural consequence of 5d cosmology, and, rather unespectedly, even has it in the limit of negligible gravitation. One can find that the source of particle production vanishes if the fluid is isentropic [7].

Here we wish to show that such results directly extend to the case of 5d general relativity, no matter if the fluid is a test fluid or a self-gravitating fluid. Gravitation seems not to influence much such phenomenon, which instead seems to be due to presence of the fifth dimension, and again we have that particle production occurs only if the fluid is not isentropic. We mean to approach however the problem of constructing an explicit example of exact solution (even if of toy kind) of the 5d Einstein equations, interpreting a self-gravitating cosmological fluid with particle creation, in a forthcoming paper.

5D RELATIVITY

Let us consider a 5d manifold as the framework of 5d general relativity, with metric tensor g_{AB}. Capital Latin indices run from 0 to 4 while Greek ones from 0 to 3. We don't need to suppose a particular signature for the metric: we prefer to have the freedom to

CP1018, *Frontiers of Fundamental and Computational Physics*,
edited by B. G. Sidharth, F. Honsell, O. Mansutti, K. Sreenivasan, and A. De Angelis

consider the signature $- + + + -$ (as in Carmeli's cosmological relativity) as well as $- + + + +$. We thus introduce a parameter ε in the equations, which can assume the values $+1$ or -1. We introduce a special coordinate: ξ, for denoting the fifth dimension. In the particular case of Carmeli's cosmological general relativity ξ represents the receding velocity of the galaxies and $\varepsilon = -1$. Let us moreover denote $\xi_A = \delta_A^\xi$ (Ξ^A in [6, 7]). Let $\xi \cdot \xi = g^{\xi\xi} = \varepsilon$. We moreover denote derivative with respect to ξ by a prime.

Let $\mathscr{T}^{AB}$ be the 5d stress-energy tensor, involved in the 5d Einstein equations:

$$G^{AB} = \chi \mathscr{T}^{AB}, \tag{1}$$

where G is the 5d Einstein tensor and $\chi > 0$ the gravitational constant. Let us consider the decomposition of $\mathscr{T}$ with respect to ξ and its orthogonal complements:

$$\mathscr{T}^{AB} = T^{AB} + P^A \xi^B + P^B \xi^A + E \xi^A \xi^B \tag{2}$$

with $T^{\xi B} = T^{AB}\xi_A = 0$ and $P^\xi = P \cdot \xi = 0$ (i.e. such fields are in practice defined by their uppercase Greek indices components). From (1) it follows the 5d conservation of $\mathscr{T}$ (5d evolution equations for $\mathscr{T}$):

$$\nabla_A \mathscr{T}^{AB} = 0, \tag{3}$$

where obviously ∇ is the covariant derivative operator (with respect to the 5d metric g). Thanks to the constancy of the components ξ_A and of $\xi \cdot \xi = \varepsilon$, the splitting of (3) is rather simple:

$$\nabla_A P^A + E' = 0, \quad \nabla_A T^{AB} + \nabla_\xi P^B = 0. \tag{4}$$

5D HYDRODYNAMICS AND PARTICLE CREATION

The usual 4d set of equations of perfect hydrodynamics, for a fluid of proper density ρ, pressure p, number density r and co-moving frame U ($U \cdot U = -1$) is given by (see e.g. [9, 10]):

$$\nabla_\alpha (rU^\alpha) = 0, \quad \nabla_\alpha [(\rho + p)U^\alpha U^\beta] + \nabla^\beta p = 0. \tag{5}$$

Thus system (4) is a good candidate to represent a generalization of (5); the natural identifications to have a match then is:

$$P^\alpha = rU^\alpha, \quad T^{\alpha\beta} = (\rho + p)U^\alpha U^\beta + p g_4^{\alpha\beta} \tag{6}$$

where ${g_4}^{AB} = g^{AB} - \varepsilon \xi^A \xi^B$. If such identification is made, we have, in particular, from $(4)_1$:

$$\nabla_\alpha (rU^\alpha) + E' = 0 \tag{7}$$

which represents particle creation, with a source term $-E'$. And from the component of $(4)_2$ along U, we have:

$$(\rho+p)\nabla_\alpha U^\alpha + \dot{\rho} + r' = 0 \tag{8}$$

where we have denoted by a dot the time derivative $U^\alpha \partial_\alpha$. From (7) and (8) we can eliminate the divergence of U so to obtain:

$$-f(E' + \dot{r}) + \dot{\rho} + r' = 0, \tag{9}$$

with $f = (\rho+p)/r$. Now, from (9) and from the following thermodynamical relation (see e.g. [9, 10]):

$$d\rho = f dr + rT dS, \tag{10}$$

where S is entropy, we obtain the following relation for E':

$$E' = f^{-1}(rT\dot{S} + r') \tag{11}$$

It is exactly the same relation which was obtained in the flat case in [7]: contributions from covariant derivatives were eliminated by the splitting or were contained in the term $\nabla_\alpha U^\alpha$ and then canceled.

THE ISENTROPIC CASE

We have not assumed any prescribed form for the 5d stress-energy yet. If, by analogy with the 4d situation, we suppose:

$$\mathscr{T}^{AB} = (M+Q)V^A V^B + Q g^{AB} \tag{12}$$

for some M and Q, where the splitting of the 5d vector is $V^A = W^A + \mu \xi^A$, with $W \cdot \xi = 0$. It then follows from (6) that:

$$M + Q = r^2 \mu^{-2} (\rho+p)^{-1}, \quad E = r f^{-1} + \varepsilon p. \tag{13}$$

From comparison between $(13)_2$ and (11) we obtain:

$$r f^{-1} T(\dot{S} + f^{-1} S') = (\varepsilon - f^{-2}) p' \tag{14}$$

A first consequence of (14) is that if all fields are independent on ξ, like in ordinary 4-d hydrodynamics, we have $\dot{S} = 0$, which in fact is a well known consequence of system (5). Conversely, it is easy to check that if the flow is isentropic then particle production is absent. In fact if $dS = 0$ we have: $(\varepsilon - f^{-2})p' = 0$, and therefore there are two possible situations: $p' = 0$ or $\varepsilon = f^{-2}$. Now, if the fluid is isentropic the equation of state must be of the kind: $p = p(r)$ (see e.g. [9, 10]), thus from (10) if $p' = 0$ we have $r' = \rho' = 0$ and consequently $E' = 0$. On the other hand, if $\varepsilon = f^{-2}$, then we must have $\varepsilon = +1$ (in other words this cannot happen in Carmeli's cosmological relativity) and $r^2 = (\rho+p)^2$. We consequently have from (10): $r dr = (\rho+p)(d\rho + dp)$. Now if $dS = 0$ we have $d\rho = r^{-1}(\rho+p) dr$ so that in the end we have:

$$(\rho+p) dp = 0 \tag{15}$$

Thus, excluding the singular case $\rho + p = 0$ we conclude $dp = 0$ and consequently $dr = d\rho = 0$, which implies $dE = 0$ and thus again $E' = 0$.

Thus in any case, again just like in the flat situation (see [7]), the source of particle production vanishes if $dS = 0$.

REFERENCES

1. M. Carmeli (1995) *Found. Phys.* **25** 1029.
2. M. Carmeli (1996) *Found. Phys.* **26** 413.
3. M. Carmeli (2002) *Cosmological special relativity. The large scale structure of space, time and velocity* (second edition) World Scientific Singapore.
4. M. Cissoko (1998) *Gen. Rel. Grav.* **30** 521.
5. M. de Campos (2002) *Gen. Rel. Grav.* **34** 1393.
6. G. Gemelli (2006) *Int. J. Theor. Phys.* **45** 2226.
7. G. Gemelli (2007) "Hydrodynamics in 5-dimensional cosmological special relativity" *Int. J. Theor. Phys.* **46** 1431.
8. M. Greenhow, S. Moyo (1997) *Philos. Trans. Roy. Soc. London Ser. A* **355** 551.
9. A. Lichnerowicz (1967) *Relativistic hydrodynamics and magneto - hydrodynamics* Benjamin, New York.
10. A. Lichnerowicz (1994) *Magnetohydrodynamics: waves and shock waves in curved space - time*, Mathematical physics studies vol. 14, Kluwer academic publishers, Dordrecht - Boston - London.
11. S. Perlmutter et al. (1999) *Astrophys. J.* **517** 565.
12. I. Prigogine, J. Geheniau, E. Gunzig, P. Nardone (1989) *Gen. Rel. Grav.* **21** 767.
13. A. G. Riess et al. (1998) *Astron. J.* **116** 1009.
14. P.S. Wesson (1999) *Space, time, matter: modern Kaluza-Klein theory* World Scientific, Singapore.

Cosmology in the light of the Theory of Time

Eugene S. Poliakov

Institute of Physics of Time, Bolshaya Pionerskaya ul., 46-2, Moscow 115054, Russia

Abstract. Extreme cases of generalized energy conservation in the frame of non-uniform time (whose corollary reveals the nature of gravity) are discussed in detail. The inertial mass is shown to be substantially relative, infinitely growing as the flow of time tends to infinity. The nature of cold dark matter is explained as well as of the cosmic microwave background radiation, which turns out not to be a result of the big bang. Hubble's law is doubted. A new explanation for hydrogen and helium line broadening in the spectrum of the Sun is adduced as well as the way to solve the Solar neutrino problem. A thermonuclear-free star energy engine hypothesized by Kozyrev is proposed. The problem of the singularity is discarded. The theories of the big bang and black holes are disproved. A unified hypothesis of the nature of pulsars, quasars, and supernovae is proposed. A number of experiments are proposed to prove the theory of time under discussion.

Keywords: relativity of mass, CDM, CMBR, pulsars, quasars, supernovae
PACS: 26.65.+t, 95.35.+d, 97.60.Bw, 97.60.Gb, 98.54.Aj, 98.70.Vc, 98.80.-k

INTRODUCTION

In the previous work [1], we introduced a generalised energy conservation law to conceptualize non-uniform time:

$$E\chi = const, \tag{1}$$

from which the conservation of momentum ($mu = const$) follows automatically.

RELATIVITY OF MASS

Before we find a serious use for the generalised energy conservation law, let us note that keeping in mind definitions ($u\chi = const$) [1], it follows from the conservation of momentum that mass m is directly proportional to the flow of time χ:

$$m = \mathfrak{M}\chi. \tag{2}$$

(where $\mathfrak{M}$ is the reduced or 'basic' mass), while both energy and velocity are inversely proportional to χ. In fact, this means that mass m is inversely proportional to velocity u.

It is possible to express the generalised energy conservation law more certainly. In order to do so we take the gradient of the definition $u\chi = const$. Keeping in mind that Newton's second law ignores relativity of mass, we obtain:

$$F = mg = \mathfrak{P}\frac{\mathfrak{M}c^2}{\chi^2}. \tag{3}$$

CP1018, *Frontiers of Fundamental and Computational Physics*,
edited by B. G. Sidharth, F. Honsell, O. Mansutti, K. Sreenivasan, and A. De Angelis

Integration of this expression over variable χ yields:

$$E\chi = \mathfrak{M}c^2, \tag{4}$$

which becomes Einstein's well known formula when $\chi = 1$.

A NEUTRINO IN THE FIELD OF NON-UNIFORM TIME

First, we can estimate the flow of time for the Sun on the grounds that fictitious gravity is exhaustively determined in terms of the universal gravity law. The flow of time of the field of Sun in Earth's perihelion and aphelion gives us respectively:

$$\chi_{\odot}^{Peri} = 9895.35 \ and \ \chi_{\odot}^{Aph} = 10231.68. \tag{5}$$

Accordingly, it is opportune to predict that the flow of solar neutrinos at perihelion should be greater than at aphelion by 3.39887%.

The generalised energy conservation law forces another groundbreaking conclusion: the decreasing flow of time ought to generate neutrinos 'out of nothing.' This can be tested experimentally; due to the flow of time decreasing towards the centre of the Earth the stream of neutrinos should linearly increase in relation to depth.

EXTREME CASES

Now it is worth discussing the extreme cases of the generalised energy conservation: 1) $\chi \to \infty \Longleftrightarrow E \to 0$, 2) $\chi \to 0 \Longleftrightarrow E \to \infty$.

The 'infinite' flow of time: A corpuscular model. This case means that the flow of time tends to infinity ($\chi \to \infty$) while the energy of a body tends to zero ($E \to 0$). As it follows from Equation (2), the flow of time tending to infinity means an infinite increase of mass ($m \to \infty$) for a physical body. From the viewpoint of 'common sense' such a conclusion gives rise to grave doubts. But there is no place for doubt with respect to an experimental fact of the existence of invisible cold dark matter (CDM), discovered by Fritz Zwicky. According to the most modest estimations CDM makes up 75% of the mass of the whole Universe, while optimistic estimates reach 90%. CDM remains absolutely unobserved and is only inferred from circumstantial 'gravitational' indications.

The 'infinite' flow of time: A quantum model. In the case of a light quantum (a photon), the flow of time tending to infinity ($\chi \to \infty$), as the energy of a photon tends to zero ($E \to 0$), means an infinite red shift ($\nu \to 0$). However, a photon of infinitesimal energy $E \to 0$ at $\chi \to \infty$, having reached the zone of finite χ, acquires finite energy E due to the blue shift pertaining to the time flow. In other words, by means of the generalised energy conservation it would be possible to predict a certain degree of background radiation of the Universe, except that Penzias and Wilson already discovered this cosmic microwave background (CMB) radiation. This radiation is generally interpreted as a result of the 'big bang' radiation cooling due to the redshift, and therefore it is also called 'relic radiation.' Accordingly, it is considered to be the best evidence for the infamous 'big bang' theory. However, it is a question of interpretation, for it is permissible to interpret

CMB radiation as not a result of a 'bang,' however big it may have been, but as an effect of generalised energy conservation, because the sources of CMB radiation (which can be supposed to be CDM) are obviously situated in the zones of much faster time than the devices measuring CMB radiation that are situated on the nearby Earth ($\chi_{\oplus} \ll \chi_{\infty}$).

We can suppose that CDM radiates as a black body at $t^{\circ} \to 0$, and that due to the flow of time, CMB radiation is heated up to 2.725 K on average. Saying 'on average,' we should remember the difference between the flow of time in the Earth's aphelion and perihelion. Stefan-Boltzmann's law on the grounds of our estimations (5) permits the prediction that the temperature of CMB radiation measured at the Earth's perihelion should be 0.02 K higher than at the aphelion. This estimation ignores the flow of time 'caused' by the Earth. However, in any case, the CMBR temperature at the perihelion should be higher than aphelion temperatures. If this prediction is shown to be true, then case 'Expanding Universe' can be dismissed.

By the way, it requires little effort to show that the model of an expanding universe implies, in the terms discussed in our previous work [1], that the flow of time cannot remain constant at every point of space where $\chi \neq 0$; it should grow together with the Universe's expansion. Again, since the flow of time is not constant in any point of space, then energy conservation is altogether invalid. Thus, we have arrived at an absurdity.

Having stated this simple proposition, we cannot but cast a doubt on legitimacy of Hubble's law, which is considered the main basis for the expanding Universe hypothesis. According to Hubble's law, the red shift in light coming from distant objects is proportional to the object distance. However, the red shift can be explained by the difference in the flow of time, rather than by the Doppler effect alone. In other words, the red shift of a distant object is proportional to the time flow difference. From the above it does not at all follow that the more distant from the Earth an object is the faster it moves away. The red shift is observed even in the solar spectrum, yet, the Sun does not move anywhere.

'Zero' flow of time: A quantum model. The case of $\chi \to 0$ physically means a nearly complete stop of time. Energy of a photon tends, meanwhile, to infinity ($E \to \infty$). For a photon, it means an infinite blue shift ($\nu \to \infty$). In other words, however small a finite χ we take, there can always be found a smaller finite χ, which corresponds to a higher ν. This approach to the question does not impose any limits.

'Zero' flow of time: A corpuscular model. In the case of a physical body, the situation changes dramatically, because no physical body can travel faster than light. For matter, this imposes certain physical restrictions on the lower limit of the flow of time. For a material particle under the stated initial conditions, the range $0 \leq \chi < 1$ is a forbidden gap. In other words, Matter cannot exist at $\forall \chi \in [0;1)$.

KOZYREV ENERGY ENGINE

Let us return to the conclusion obtained from Equation (2):

$$m \to 0 \ at \ \chi \to 0. \tag{6}$$

Seeing the limitations of the domain of χ, it needs a refinement:

$$m \to \mathfrak{M} \ at \ \chi \to 1. \tag{7}$$

Nevertheless, quantum energy can exist no matter how low the χ value.

What does the range $0 \leq \chi < 1$ mean? What happens as a material particle of mass $\mathfrak{M}$ enters range $\chi < 1$? This mental experiment is worth setting, all the more so, as a fictitious force continues to act in the range $0 \leq \chi < 1$, and though trivial but necessary to repeat, is that a material particle travels absolutely regardless of its mass.

There is no other way than to assume that as a material particle enters the range $0 \leq \chi < 1$ its entire (kinetic) energy $E = \mathfrak{M}c^2$ transforms into a radiant energy according to the generalised energy conservation.

A reverse transition of radiant energy into matter is probably possible. But in the general terms, the reverse transition is non-symmetrical, for a reverse transition in which radiant energy is conserved entirely in a radiant form is quite possible as well. There is also the possibility for a reverse transition in which the radiant energy is partially conserved in a radiant form, while the rest of energy generates a material particle of mass m.

Rigourously, a reverse transition may generate several particles. It is quantum mechanics that entirely defines these processes. Yet, we can point out that if at least one of the particles carries an electric charge, then a reverse transition is necessarily accompanied by retaining a fraction of energy in a radiant form. In any case, transformation of a radiant energy into matter is probable. In other words, generation of matter is possible 'out of nothing,' literally 'out of the blue,' of energy.

The most remarkable thing is that even if first type transitions are in the state of energetic equilibrium with the reverse transitions and all the three types of reverse transitions are equally likely, then the neighbourhood of $\chi = 1$ works as an engine that transforms matter into radiant energy. This engine's efficiency compared to a thermonuclear reaction is like comparing the Sun to a candle.

Such an engine that produces radiant energy out of matter is a perfect match to describe a mechanism, which in hypothesis was expressed by N.A. Kozyrev [2], [3], [4], who considered thermonuclear fusion to be able to make up no more than 10% of the energy emitted by the Sun [2]. As for the Supergiants, according to him [2], their very existence cannot be explained by the hypothesis of thermonuclear reaction.

Kozyrev said, literally, the following: 'In ordinary conditions, space is passive and simply gives place for events. But time is the event *per se*, it can not only possess a passive characteristic of duration, but also represents a phenomenon of Nature. Then *time interacting with star matter, may turn out to be a source of its energy'* [2].

THE 'BLACK HOLE'

Let us discuss a model of a so-called 'black hole' from a standpoint of the generalised energy conservation. The 'black hole' is an object in space, where gravity is so powerful that even light cannot escape it. Any glow near a 'black hole' may be seen only from regions situated outside the border called the 'event horizon.' Beyond the 'event horizon,' 'singularity' occurs and any information on processes beyond the 'event horizon' is not accessible by any means.

At the 'event horizon,' the escape velocity is equal to the speed of light c. In terms of the flow of time, this means that at the 'event horizon,' $\chi = 1$. The 'black hole' itself is situated beyond the 'event horizon,' at $\chi < 1$.

However, as we have shown above, the matter of mass m cannot exist at $\chi < 1$. That is, the inertial mass of a 'black hole' matter situated beyond the 'event horizon' is equal to zero. In other words, 'black holes' exist only in fancy of their inventors or at best on paper as a result of some mathematical trickery outside the domain. Or, at the worst, 'black holes' become a devotional article. In reality, a 'black hole' cannot exist!

Nevertheless, objects, at whose border the escape velocity is equal to c and which are sources of huge fictitious forces of gravity, exist in nature. But they are not 'holes' and not 'black' anyway. Such an object is situated in the centre of every star.

TIME MODEL OF A STAR

Let us discuss stars in more detail. There are two reasons for doing so. Firstly, stars are the most evident sources of 'gravity' in the Universe. Secondly, stars are the biggest, if not the sole sources of energy in the Universe. In a time model of a star, its centre is of the greatest interest, as it is a centre of fictitious force, conditioned by the fact that the flow of time there is minimal ($X < 1$) (X is is the 'initial' flow of time at $r = 0$, which is rigorously not necessarily zero), and consequently, energy processes near the centre of star draw our specific attention.

As we have shown previously, matter cannot exist at $\chi < 1$. Inasmuch as the flow of time can be represented as $\chi = f(r)$, in terms of space we can introduce a sphere of radius R_1, within which matter does not exist. An estimation of R_1 yields 14865 km for the Sun. That is, the *inertial* mass of the R_1 neighbourhood of the centre of a star is zero (yet, the '*gravitational* mass' of this area is rather non-zero, lest we say immense).

This means that there is a perfect vacuum in the centre of a star. Maybe it is too bold a claim. Rigourously, neighbourhood R_1 is a zone of purely quantum effects. Hence, one cannot neglect the Heisenberg uncertainty principle ($\Delta P \Delta r \geq \hbar/2$). That is, the border of zone R_1 should be blurred. Nevertheless, it is simply inappropriate to talk about massive pressure in the centre of star (at $\chi < 1$, or $r < R_1$).

Partial vacuum of low pressure remains also nearby this area at $r > R_1$, as it is filled with elementary particles with maximal energy.

As distance from the centre grows, the energy of particles decreases until it becomes possible to combine hydrogen atoms and pressure appears.

At a certain R_T, pressure increases to a critical level. Then thermonuclear reaction begins producing helium. According to the redshift of helium lines it is possible to calculate the flow of time χ_T, where thermonuclear reaction takes place.

Because thermonuclear reactions take place in a range of χ, helium lines in the solar spectrum should be somewhat broadened and redshifted. For the same reason, hydrogen lines should be more broadened and more redshifted.

However, it requires significant time before a sufficient amount of hydrogen accumulates about point R_T for a thermonuclear reaction to begin. In the meantime, the energy engine of such a star along with particles emits huge radiant energy at R_1, where $\chi = 1$. This energy is not emitted continuously but discretely, as quanta. In the absence of hydrogen, this radiation spreads without hindrance and should be registered as regular scintillation in a narrow frequency range. This description fits pulsars well.

In time, this radiation starts to be absorbed and re-emitted by accumulated hydrogen so that regular scintillations give way to 'continious' radiation but now in a broad frequency range (characteristic of black bodies). At a distance (at high χ), this radiation will be registered as highly redshifted. The time conditioned red shift may move a part of the radiation to a radio frequency band. Again, this description fits quasars well.

As soon as the amount of hydrogen becomes sufficient for a thermonuclear reaction to begin, this object explodes as a supernova and quickly approaches the state of an ordinary thermonuclear star.

There is one final case to discuss: $X > 1$, although it does not refer to stars. It is evident that if $X > 1$, Kozyrev energy engine cannot work. Nevertheless, the energy nearby the centre of such an object is far higher than at its surface. Such a model is characteristic for planets. At best their energy engine is capable of volcanic activity.

The discussion of planets prompts another observation. Since volcanic activity was discovered on the Moon [5], any celestial bodies larger than the Moon ought to exhibit tectonism.

REFERENCES

1. E. S. Poliakov. Theory of time beyond the standard model. Ninth international symposium – Frontiers of fundamental and computational physics. Udine and Trieste, 2008.
2. Kozyrev N. A. Priroda zvezdnoy energii na osnove analiza nabludayemych dannych. Kozyrev N. A Selected works. Leningrad University, 1991. p.191 - 204.
3. Kozyrev N. A. On the Possibility of Experimantal Investigation of the properties of Time. Time in Science and Philosophy. Prague, 1971.
4. Kozyrev N. A. Astronomicheskiye nabludeniya posredstvom fizicheskich svoistv vremeny. Vspychivayuschie zvezdy. Trudy symposiuma priurochennogo k otkrytiyu 2,6-m teleskopa Burakanskoy astrofizicheskoy observatorii. Burakan, 5 - 8 Oct. 1976. Yerevan, 1977. p. 209 - 227. In Kozyrev N. A. Selected works. Leningrad University, 1991. p. 363 - 383.
5. Kozyrev N. A. Volcanism on the planets. Tectonophysics. 1964 - 65. Vol 1. N 5. P. 451 - 454.

New Concepts of Dimensions and Consequences

Philippe F. Journeau[a]

[a]*Aegeus, 5, rue Saint-Philippe du Roule, 75008 Paris*

Abstract. A Fourfold Co-necessity Principle (FCP) was introduced at FFP8, according to which our customary four dimensions, such as represented with (τ, Θ, φ, r), correspond to four covariant types of dimensions. The paper develops some of its estimated consequences, from the potential impact of minimal and maximal FCP sizes – beyond twofold space-time minimal length – to the derived existence of four types of singularities rather than only Black and White Hole, bringing geometric explanation and related values for ΛCDM density parameters: 0.04492 to 0.04516 for baryonic Ω_m, 0.21436 to 0.21727 for Ω_{CDM} and 0.73757 to 0.74048 for Ω_Λ.

Keywords: Principles, Cosmology, Causality, Dark energy, Dark matter, Dimension.
PACS: 01.55, 01.65, 01.70, 02.40, 03.65, 04.20, 04.40, 04.50, 04.60, 11.15, 14.80, 52.35, 80

INTRODUCTION

"The present knowledge of the elementary dynamical laws of physics is given by the application of QM to fields, namely Quantum Field Theory (QFT), by the particle-physics Standard Model (SM), and by GR (…) But the theories in this set are based on badly self-contradictory assumptions" summarizes Rovelli in [1].

Penrose scrutinizes this foundational issue in [2, p. 849] "Fundamental conflict with Einstein's principles": two Killing vectors at same location requires to "try to identify K_χ with K_φ, but that would certainly be committing a violation of a fundamental principle of Einstein's theory, since it would imply that we are thinking of these two Killing vectors as inhabiting the same space" and Penrose begins to "witness a clash between the fundamental principles of quantum mechanics and of general relativity."

The issue is solved if one considers the split between χ with φ states or forms as a part of the definition of this "same space" along with a length dimension co-necessary to it, and indeed covariant but 'change itself' will be seen as another co-necessary dimension. Reality – physicality – is space-time-field-particle rather than space-time among fields or than fields-particles 'in' space-time. GR for instance splits the fourfold reality, as required in order to be able to say something about it.

Therefore a Fourfold Co-necessity Principle (FCP), presented during FFP8, unites the two main sets of foundations – space and time dual dimensions versus field and particle dual phenomena – within a single framework of four co-necessary or interdependent types of dimensions that altogether "make the background and the reality itself [3]."

Indeed covariance is widely recognized as a foremost breakthrough of Einstein's: rather than being a dimension that can be grasped independently, wherever in space, time was then recognized as interdependent with it, and modeled by Minkowski. Furthermore, after GR, interactions between space-time and matter-energy were ever

CP1018, *Frontiers of Fundamental and Computational Physics*,
edited by B. G. Sidharth, F. Honsell, O. Mansutti, K. Sreenivasan, and A. De Angelis

better grasped as recalled by Cohen-Tannoudji [4]. Yet the idea prevails that each one, space and time or at least space-time versus matter, have independent ontology. This is a reason why the somehow stronger concept of co-necessity has been preferred, bearing sizeable consequences especially when widened to four types of dimensions.

As a consequence of FCP, where our own Kantian "power to know" meets with hypothesized external reality, time, space, change and form may not even be defined other than conventionally and relatively to the set of the others when any set of them is considered. Einstein refers to common experience of clocks and rods [5] even though, physically, clocks formally elapse also at length as in time and rods conversely.

Now the four-dimensionality considered in this paper is called time-form-change-length, knowing that the very meaning of each of these is itself subjected to the FCP. This doesn't imply that they have no underlying difference but that related severance is not possible in experience (even thought, as linguists such as Saussure and Peirce have devised it) even though ways to approach such goal through experiments addressing a quantum singularity of any of the four, are discussed below.

Space as length, as called from our customary related sense of locality, is therefore merely one of four dimensions. This will explain several small or cosmological scale paradoxes linked to causal connection usually seen to jump from one event to a spatial nearest. For a nearest according to time, or to form, or to change, may be not near in distance even though fourfold reality, basically limited with light ray, propagates and even bear this fourfold dimensionality and reality altogether within its own limits.

1. THREE DIMENSIONAL SUBSPACES

A consequence of FCP is that any split in it, although required for modeling purposes, decreases the level of external reality so represented. There are for instance four possibilities to sever three dimensions from the four co-necessity. Opposing a dimension called time to a subset of the three others called space was thought to depict the true nature of reality until many physicists, up to the Holography principle for instance, introduced other options. Smolin [6] mentions Kant about spatial three dimensions linked to several square distance laws but Bouton [7] also refers to Kant about a ternary dimensionality of "three modes of time": permanence, simultaneity and succession. The usual equivalence of Cartesian and Spherical polar coordinates shows that considering three dimensional length often works, if only because of the prevalence of underlying spherical form correlated to isotropy. A representation opposing reflexive simultaneity to non-commuting succession may also be developed as shown by Sidharth [8] through the split of the $g_{\mu\nu}$ in such components.

Similarly may be envisioned a three dimensional Change, continuous such as expansion, rotation and succession – when subjective imaginary 'i' – to discrete when punched in measured or actualized q-bits. And 3D form also, quite usual, is conformal easily confused, when configured, with 3D space, although formal logic and mathematics are length scale invariant: from length form S^0 to S^1, S^2 and S^3.

The (A, L, S, N) symbolism had been introduced in [3] for a FCP partaken by three worlds, such as Penrose's, where each type of dimensions corresponds with similar in the two others, yet with different characteristics. A set such as (τ, Θ, φ, r) brings a provisional application of it, a four dimensional measurement customary formalism.

A further consequence of the FCP is that there are four types of 3D combinations, which are also singularities since the fourth dimension reduces there to a quantum otherwise reality per se vanishes. Hawking and Penrose have agreed about Weyl curvature different from zero by a quantum at Big Bang, hence White Hole.

These subspaces may be symbolized as SH for (A, L, N) Spontaneous change singularity or Hole, (A, S, N) or LH (Law singularity or Hole), (A, L, S) or WH (White Hole) and (L, S, N) for BH Black Hole. (A, L, N) will encompass all logical, invariant order levels, as opposed to 'S' spontaneous form or order level breaking, which is not by itself entropic. SH is related to Penrose's "sea of entangled states" [2] or Hawking's universe wave function [9, 15] and LH to the sub-reality that Penrose calls "rock of un-entangled states" or rather 'disentangling' as subspace of transformations inasmuch as occurring, of which the dynamics however necessarily succeeds with at least one law – thence alone, singular – as required by FCP.

(A, L, N) is a 3D sub-reality like this BH (L, S, N) 3D space through Schwarzschild radius while (A, L, S) is conjectured to be everywhere – below Planck scale and up to light boundary – with several characteristics such as proposed by Penrose for White Holes. (A, L, N) subset of formal sub-realities is composed of forms of progressively encompassing order levels of which sphere stacking and enclosing gives an image. Logical or formal is often opposed to physical but FCP suggests that length is per se no more physical than form: it is rather the distance from FCP, or reality, that measures the decrease of physicality and therefore each of the four subsets discussed here is more physical, like a split opposing (A, L) to (S, N), than any sole dimension. Furthermore, Conformal theory is among the directions selected by physicists, together with several other major theoretical breakthroughs ranging from String theories minimal surface, DSR theories minimal length, Holography principle which already considers two spatial dimensions rather than three, Loop quantum gravity and spin networks from spin behavior and Non-commutativity, where change may occur apart from length distance, Twistor theory already mentioned and Causal set theory. FCP appears compatible, eve intersecting well with them (for example minimal area).

Opposed to length distance WH produces it reciprocally to BH, time swallowing (actually 'at length'). It is here conjectured that the Fourfold co-necessity combined four minima and maxima, which logically derive solely from universe being discrete – and, after a beginning, therefore finite – suffice to produce and inflate thereafter radiation, then mass and step by step the universe as we see, deduct and represent.

Such emergence seems ongoing a Big Bang, possibly such as the mass boom mechanism envisioned by Afonso-Faus [10], since it has to be fourfold a Big Bang rather than localized solely in (2D) space-time, which would be contradictory with FCP. Such so-called vacuum of modern physics is event boarded by rays born from it, contributing to space expansion like 4D Fourfold spheres added one after another through increasingly heightened order level in (A, L, N) whereby formal, logical world is relatively present at all form scale levels in the universe, where entropy growth leads from each outer edge 'A' to local 'N' where Killing vectors turn space-like, vanishing as says Penrose. To S^1 to S^2 rotation axe changing vs. disk at any scale.

(A, L, S) yields length scale invariance as defined. Such a rule is followed by each monomial law, such as Newtonian (hence GR), and some polynomial and by exponential series in circle or reciprocal logarithmic spiral through rotations such as

Wick (t to iτ) or similarly what may be called Weyl's (t to iΘ) with here t taken as angle or form, simultaneity or reflexivity of time shape, rather than its permanence. Or 'Wake' (t to iφ), active or successive or efficient causal, that several authors associate with non-commuting rotation or spinning. Require diverse signatures (Witten, cf. [3]).

Truth logical formal condition imply forms don't intersect otherwise deformation, hence symmetry, density or coherence breaking, has to occur, whether form or field sphere is atomic or galactic. All fourfold spheres in (A, L, N) subspace obey such rule, suggesting that SH fills so-called 'dark energy' share of the universe according to Hales theorem out of Weyl's rotation as evocated above. Likewise (A, S, N) occupies the 'dark matter' subset of acting upon in relation to above so-named 'Wake' rotation.

2. FORM OR ORDER LEVEL SCALING

Transition from some form or phase order level to the next is important matter in physics, which gives a scaling along the 'form' dimension as each dimension has its own range: between minimal and maximal universe radius for length scale but [0, 2π] for continuous Θ and [0, π] for continuous φ changes with again none of them met in phenomena independently from the three others. Difference therefore comes in experience as performed, from choices input, such as resolution, hence form level. Experience implies form level breaking to new entanglement and casting (cf. 3 below) to next, lower level – "the generic feature of all phase transitions" says Liu [11] – or containment base, such as Pauli exclusion law. Then length widening opens more distance opportunities for forms, compelling Second Law of thermodynamics (Balian).

Required universal neguentropy law (cf.3) thence adds ever bigger fourfold spheres while conversely entropy grows with space within a given form. This maximal scale consequence of FCP operates from WH edges opposite to BH to draw neguentropy versus entropy fourfold path, similarly to Feynman allowing paths from both ends at particle level. And suggesting entropy flow conserved and related to $\Omega=1$ preserved, possibly phase space volume also at universe or between two order levels, through form flow from WH neguentropy into BH entropy: approaching Schwarzschild radius, entropy locally grows as transformations break most complex and sizeable forms towards more simple symmetry, a decay down to radiation itself finally absorbed in a process indeed reciprocal to Big Bang or WH. Yet their shared neguentropy to entropy flow fits with Lachèze-Rey and Luminet [12] depicting BH "frontier of space-time."

Reciprocally to Hawking-Beckenstein formula, 'A' may be defined as neguentropy, proportional to temporal WH area, linked to both Weyl curvature quantum and yielding a rather filled vacuum in SH, of which a layer is Dirac's electrons "sea".

Neguentropy conjecture is that evolution of the universe, from initial point to final, yields order level increase with space differentiation conversely to Schwarzschild boundary approach spatially breaking order level as it decreases time evolution rate.

Hawking brings characteristics of a three dimensional Big Bang Euclidian half-sphere subspace geometry [9, p. 190]: Probability of change is e^{-I} with action I for instance $2\pi m^2/qE$, experimentally verified, for electron pair creation. His "e^S manners to form a Black Hole" yield e^{-S} probability of one change in the set, its ∃/∀ ratio.

This will support evaluations of Ω_Λ and Ω_{CDM} proposed below in relation to FCP.

3. NEGUENTROPY, DARK ENERGY AND DARK MATTER

Formal world logically allows no interpenetration. Hence (A, L, N) as SH prevents change, and therefore 'real', to occur within density parameter Ω_Λ maximal 74% (Hales) set of spheres of formal exclusion since SH is subspace of invariance like Black Hole (L, S, N) is singularity subspace where time, better defined from neguentropy, stops, hence entropy flow becomes maximal with Weyl curvature.

Likewise change, whatever the kind of rotation involved, encompasses all possible face centered - both spherical and cubic connection – spinning in universe fourfold geometry, hence amid $(2n+1)\pi/2$ imaginary i^{2n+1} to $(2n)\pi/2$ real i^{2n}. It includes any possible level of symmetry breaking, from neguentropic Higgs to entropic 'Weak' rotation in (γ, W^+, W^-, $Z°$) and the rotations mentioned above, measured as time 'ticks'. Any such rotation angle is a multiple of a minimum change. At order 2 it is bispinor 'change of change' (or rotation axis), which induces minimum mass or equivalently slight deformation from one face center to the con-next one, as compared to order 1 U(1) where spin axis doesn't change. Related Group levels representation have much progressed from Galois to most recent summaries [13, 14]. Kepler conjecture Face Cubic Centered fits phenomena both Cartesian and Spherical compliant. It unites 'Least action' with isotropic or 'Least form' for a fourfold sphere to get in touch with another. The difference between gravity and $\pi/2$ or π changes or events is therefore that the minimal rotations of each of the (1,2,0,1) vs. (1,0,2,1) level 2 order representations mentioned in [3] are not 'real', since not fourfold, by themselves, even though involving a correspondence between two of the four 3D subspaces, of which Hawking had guessed the WH to 4D glue (below) $S = A/4 = \pi/2$.

Separating the imaginary and real parts of these discrete multiples rotations within SO(3) will yield the two series of $e^{-(2n+1)\pi/2}$ and $e^{-2n\pi/2}$ as probabilities of corresponding change. Each n corresponds to a form, order level. There is no infinitesimal Lie rotation involved here since this paper considers discrete physical reality K_Σ, moreover at cosmological scale and therefore without much K_i to K_Σ [3] subjective relationships where, apart from other singularities, smoothness may be involved.

Obviously neither these spheres nor all these rotating face centered qubits are fully real since the first ones, within (A, L, N), like the seconds as (A, S, N), don't obey the FCP requirement. However their intersection does, which is the real $\Sigma e^{-2n\pi/2}$ series, roughly equal to 0.04516 for Ω_m, the contribution of real matter, while the total of both series from n=1, i.e. $\Sigma e^{-n\pi/2}$, is close to 0.26243 for Ω_M including $\Omega_{CDM} = 0.21727$.

The sum of maximal $\Omega_\Lambda = \pi$ sqrt(2)/6 from Hales theorem and $\Omega_M = 0.26243$ is roughly equal to 0.74048 + 0.26243 = 1.00291. None directly involves time and postulating Ω equal to 1 gives Ω_m possibly at $\pi/4 - \Omega_\Lambda = 0.04492$ from S^1 disk rather than S^2, since real belongs to both (A, L, N) formal scalar and (A, S N) oriented non-commutative sub-principles, taking into account thence such reduced tridimensionality, which scale invariant disk versus rotating axis exhibits at all scales. But if Ω is capped at 1, then Ω_{CDM} bottom value is 0.21436, or similar bracket for Ω_Λ.

Indeed the FCP picture of the universe derived from form or order scale part of its dimensions is that phase space, like other dimensions and spaces sometimes added to a classical space-time four dimensionality, rather has to be considered part of it. Form as a dimension is extensive, rather than intensive, as 'phase space' volume also suggests.

The proximity of Kepler-Hales' orange stack analogy to each nearest connection is as much formal as space or time ones, Conformal a z axis with nearest form level either similar, decreased or increased depending isentropy, entropy or neguentropy with change. Obviously the umbrella called 'isolated system', or z capped evolution, imply entropic decays to lesser order level. In this image it means a filled orange decaying in many sub-oranges, which will replicate Kepler filling rule as homogeneity is reached. It requires hot WH source continuously irradiating neguentropy from nowhere, possibly such as cosmic rays and high energy pairs of particles, which may be seen contributing to ever bigger oranges along z axis, being spatially quite tiny, actually Planck long. This (A, L, S) subspace seems inducing this "Hartle-Hawking-Israël wavefunction" that Maldacena [15] reminds "arising from gluing half the Euclidean geometry at t=0 to half the Lorentzian geometry" then as scalar belonging (A, L, N) which is cosmological constant at its widest level and as operator part of (A, S, N).

The balance within $\Omega=1$ may result from neguentropy-entropy flow consistency in a universe less spatially inflating a bubble than FCP reality in which space length is single and co-necessary complement to three other types of dimensions. It is clear that the conjectured mechanism exposed above neither refute nor support strongly FCP but it goes along with as spherical coordinates used form FLRW to particles comforts it even though φ should itself be referred to an axis of change that is change of axis (as computed above) rather than a fixed axis. Considering form and change as the other two amid the four types of dimensions may achieve the metric equivalently dynamic conception of outer reality with a neguentropy versus entropy flow globally invariant.

ACKNOWLEDGMENTS

The author is grateful to Pr. B Sidharth and to FFP9 Symposium Committee for their kind invitation and to Pr. J. Madore, colleagues and friends for encouragements.

REFERENCES

1. C. Rovelli, *Unfinished Revolution*, arXiv:gr-qc/0604045v2, 2006
2. R. Penrose, *The road to reality*, Alfred A. Knopf, New York, Jonathan Cape, London, 2004
3. P. Journeau, *Evolution of the concept of dimension*, AIP proceedings # 905 2007, FFP8 Madrid 2006
4. G. Cohen-Tanoudji, *Les constantes universelles*, Pluriel, Hachette, Paris, 1998
5. A. Einstein, *On the electrodynamics of moving bodies*, tr. From Annalen der Physik, 17:891, 1905
6. L. Smolin, *Rien ne va plus en physique*, trad. Alexeï Grinbaum, Dunod, Paris 2007
7. C. Bouton, *Au-delà de la représentation Kant et le problème de l'idéalisme*, Philosophie, n°81, Editions de Minuit, Paris, 2004
8. B. Sidharth, *The Common Origin of Gravitation, Relativity, Electromagnetism and Quantum Theory* in American Institute of Physics AIP conference proceedings # 905, 2007 for FFP8, Madrid 2006
9. S. Hawking & R. Penrose, *The Nature of Space and Time*, Princeton University Press, 1996
10. A. Alfonso-Faus, *Galaxies: kinematics as a proof of the existence of a universal field of minimum acceleration*, EUIT, Madrid, 2006
11. C. Liu, *Spontaneous Symmetry Breaking and change*, University of Florida, 2002
12. M. Lachèze-Rey & J. P. Luminet, *De l'Infini*, Dunod, Paris, 2004
13. M. El Naschie, *Deterministic quantum mechanics versus classical mechanical indeterminism and nonlinear dynamics*, in AIP proceedings # 905, 2007 for FFP8, Madrid 2006
14. G. Lisi, *An exceptionally simple Theory of Everything*, arXiv: 0711.0770v1 [hep-th], 2007
15. J. Maldacena, *Eternal black holes in Anti-de-Sitter*, Institute for Advanced Study, hep-th/0106112v6

Connecting Fundamental Constants

D. Di Mario

Viale Zara 114, 20125 Milan - Italy

Abstract. A model for a black hole electron is built from three basic constants only: h, c and G. The result is a description of the electron with its mass and charge. The nature of this black hole seems to fit the properties of the Planck particle and new relationships among basic constants are possible. The time dilation factor in a black hole associated with a variable gravitational field would appear to us as a charge; on the other hand the Planck time is acting as a time gap drastically limiting what we are able to measure and its dimension will appear in some quantities. This is why the Planck time is numerically very close to the gravitational/electric force ratio in an electron: its difference, disregarding a $\pi\sqrt{2}$ factor, is only 0.2%. This is not a coincidence, it is always the same particle and the small difference is between a rotating and a non-rotating particle. The determination of its rotational speed yields accurate numbers for many quantities, including the fine structure constant and the electron magnetic moment.

Keywords: Constant of gravitation, Planck black hole, black hole electron, fine structure constant.
PACS: 12.10.-g, 04.60.-m

Planck Black Hole and Planck Permittivity. If we devise a particle with a Planck time $t_p = \sqrt{\pi hG/c^5}$ and a Planck mass $M = h/t_pc^2$ we have created the basis for the Planck black hole. The gravitational force is GM^2 but the severe constrain of time t_p outside the black hole will allow us to measure only an energy which depends on this time. It behaves like a very fast shutter that allows us to see only an infinitesimal fraction of the energy. We experience only this minute energy and we would see it as generated by a mass $M_0 = M\sqrt{t_p}$. Now it should come with no surprise that if we use M_0 we will not get a black hole because this mass will not satisfy the Kerr-Newman condition. It is a black hole after all, but the measured mass is M_0 and not M as expected. We are still talking about the very same particle but its mass will undergo the limitation of time t_p. We would still feel the effect of mass M, but we do not see it as a gravitational force but as an electric force generated by what we would call the Planck charge $Q = M\sqrt{4\pi\varepsilon_p G}$ where ε_p is the Planck permittivity. We can calculate ε_p if we think that mass M_0 would not stand still but would move at a velocity v with the minimum quantum of action $\hbar$:

$$M_0v^2 = \hbar \tag{1}$$

The dimension for v will be clear later. As M and M_0 are the same particle, we could think that the kinetic energy applied to mass M is the same as the energy of mass M_0:

$$M_0c^2 = 2\pi Mv^2 \tag{2}$$

If we square both terms and multiply by the constant of gravitation we have:

$$GM_0^2/GM^2 = 4\pi^2(v/c)^4 = t_p \tag{3}$$

CP1018, *Frontiers of Fundamental and Computational Physics*,
edited by B. G. Sidharth, F. Honsell, O. Mansutti, K. Sreenivasan, and A. De Angelis

We define the Planck permittivity ε_p as the ratio v/c and if we substitute the gravitational force of mass M with the equivalent force given by the Planck charge Q we have:

$$GM_0^2/(Q^2/4\pi\varepsilon_p) = t_p \tag{4}$$

From Eq. 4 we see that the ratio of the measurable gravitational force and the electric force in a Planck particle is exactly t_p. Both charge Q and mass M_0 are an order of magnitude larger then the expected electron charge and mass while permittivity is only 0.3% off the known value. This is due to the fact that this is a non-rotating particle where the dynamic interaction of the electric and magnetic force is not yet taken in account. We have an additional time dimension in M_0 because time t_p is acting as a fast shutter. The same happens with velocity v whose actual dimension is $(m/sec)\sqrt[4]{sec}$ and numerically is identified with the vacuum conductance. The additional time dimension will have an effect on other quantities as well and must be always accounted for.

Rotating Particle and Magnetic Anomaly. A suitable physical model is the ring model with some sort of toroidal force field around the tiny black hole. A rotating charge will set up a magnetic field opposing its own rotation until a stability point is reached where the resulting electron force would be the difference between the electric and the induced magnetic force. The rotational speed u_0 is a fundamental parameter and with a slightly modified equation originally proposed by Sutton et al. in 1992, we define a relativistic connection with what we would call the initial fine structure constant α_0:

$$\alpha_0 = 2(1-u_0^2/c^2) \tag{5}$$

In order to calculate u_0 we try to relate α_0 to some properties of the particle and specifically how charge Q compares with a ring of unitary charge Q_u. We define α_0 as the ratio $\sqrt{W_u/W_p}$ where $W_u = 16\pi^4 Q_u^2/t_u$ would be the energy of the unitary charge in the unitary time applicable to a toroidal particle and $W_p = Q^2/t_p$ would be the energy of charge Q within time t_p. This initial fine structure constant gives the initial rotational speed, Eq. 5, and is an intrinsic property of the Planck black hole and can be calculated directly from basic constants: $\alpha_0 = (2\pi^2/c)\sqrt{\pi/c}\,\sqrt[4]{2G/h}\,\sqrt[16]{c/\pi hG}$. This particle is rotating at relativistic speed and will increase its energy to a level corresponding to a charge $Q_0 = Q(c/u_0)$. It is on charge Q_0 that we must calculate the effect of the electric force and the induced magnetic force in order to get the initial electron charge e_0:

$$e_0 = \sqrt{Q_0^2 - Q_0^2 u_0^2/c^2} = Q_0\sqrt{\alpha_0/2} = Q/\sqrt{2/\alpha_0 - 1} \tag{6}$$

We have the initial electron mass m_b from M_0 in the same way as we had e_0 from Q_0:

$$m_b = M_0\sqrt{1-u_0^2/c^2} = M_0\sqrt{\alpha_0/2} \tag{7}$$

If numerical calculations are carried out we find a difference ranging from 0.01% to 0.1% compared to known quantities. Total agreement was achieved by changing the rotational speed u_0. There is indeed a slightly lower speed u_e yielding all expected data although, for the mass, we have to define an additional term that takes care of the

radii variation. The hypothesis is that the interaction of this basic electron with virtual particles in the vacuum has the effect of lowering its rotational speed.
There is a direct relationship between permittivity and the charge originating it: for a static particle we have $\varepsilon_p = Q^2/4hc$ that will increase to $Q_0^2/4hc$ for a rotating particle. Our current value for permittivity is $\varepsilon_0 = 10^7/4\pi c^2$ and is an exact quantity so we could lower speed u_0 until we have ε_0. The result is a speed decrease of $111 m/sec$ and a slightly lower value for Q_0, Q and e_0 which now coincides with the electron charge e.
As no other constant was used but h, c and G, it is possible to elaborate all the available data and write two equations showing the relationship among all these quantities:

$$\alpha^2 - 2\alpha + (4\pi^2/e\,c)^2\sqrt{\pi hG/c} = 0 \tag{8}$$

$$\alpha^3 - 2\alpha^2 + 10^{-7}(2\pi)^5\sqrt{\pi G/c^3 h} = 0 \tag{9}$$

So far we have not considered the radius variation due to the change of the rotational speed. As $M_0 = 8h^3/\pi Q^4$ we have that the black hole radius variation of M_0 is related to the fourth power of the variation of Q hence to the fourth power of the fine structure constant variation. We would now consider the radius variation of the ring and the radii variation of the ring section, most likely an ellipse. These three radii variation would influence the electron mass m_e with the factor $C_e = (\alpha/\alpha_0)^{12}\sqrt[8]{(1-\alpha/2)^3}$. Eventually we would have:

$$m_e = M_0\sqrt{\alpha/2}(\alpha/\alpha_0)^{12}\sqrt[8]{(1-\alpha/2)^3} \tag{10}$$

Based on certain assumptions still under investigation, it is possible to describe the ring speed variation. This will originate a factor $C_a = \sqrt[4]{1-\alpha/2}(\alpha/\alpha_0)^4(2-\alpha_0)^8/(2-\alpha)^8$. The quantity $C_e/C_a - 1$ would be the electron magnetic anomaly a_e:

$$a_e = [(2-\alpha)\alpha/\alpha_0(2-\alpha_0)]^8\sqrt[8]{1-\alpha/2} - 1 = (e_0/e)^{16}\sqrt[8]{1-\alpha/2} - 1 \tag{11}$$

While the ratio C_e/C_a will give us the electron magnetic moment $\mu_e = (C_e/C_a)e\hbar/2m_e$.

Gravitational and Electric Force. From Eq. 3 and 4, for a static particle, we have:

$$GM_0^2 = \pi\varepsilon_p^3 Q^2 \tag{12}$$

We could write Eq. 12 in terms of known quantities providing we take in account the rotation and slowdown of the particle. The change in each quantity has been calculated and placed in a single term $C_r = (\alpha/\alpha_0)^{32}\sqrt[4]{(1-\alpha/2)^{19}}$ thus getting:

$$Gm_e^2 = \pi\varepsilon_0^3 e^2 C_r \approx \pi\varepsilon_0^3 e^2 \tag{13}$$

As C_r is close to unity we could ignore this term if we are satisfied with a difference of about 1% between the right and the left term. From Eq. 4 we see that the gravitational to the electric force ratio is exactly t_p. This ratio is slightly modified because of the rotation of the particle and if no corrective factor is introduced we will be left with a difference of about 0.2%. Now we can calculate this corrective factor and eventually we have:

$$F_g/F_e = t_p(\alpha/\alpha_0)^{24}\sqrt[4]{(1-\alpha/2)^3} \approx t_p \tag{14}$$

The electric force F_e can be calculated directly from the rotating Planck mass M:

$$F_e = e^2/4\pi\varepsilon_0 = GM^2(\alpha/2) \tag{15}$$

The electric force appears to be a gravitational force but with a fundamental difference: it has a polarity, positive or negative. We could possibly explain the existence of a polarity if we think that the Planck mass is a black hole trying to get larger and larger or smaller and smaller. This tendency will be experienced as a polarity: a variable gravitational field over an extremely long time. Time dilation in a black hole causes the variation of the gravitational field to take place over a long period of time, probably the age of the universe, and we would be left only with a polarity: negative or positive and in our everyday experience we would see the Planck mass as a steady and ageless particle, a charge in fact. A fitting application could be the creation of an electric field directly from the gravitational field variation taking place during the materialization of an electron-positron pair from a gamma ray. The materialization time $t_e = h/m_e c^2$ originates a variable gravitational field Gm_e/t_e and once we introduce the quotient $K_p = Q/M(\alpha/\alpha_0)\sqrt{1-\alpha/2}$ and consider the rotation of the particle, we have the electric field E_e generated by the electron without the prior knowledge of its charge:

$$E_e = (Gm_e/t_e)\sqrt{2/\alpha}\,/K_p(\alpha/\alpha_0)^{24}\sqrt[4]{(1-\alpha/2)^3} \tag{16}$$

In our world there would be a very small additional tug between two bodies whose distance changes with time. This could be important, among other things, in the measurement of the gravitational constant: any experiment entailing a moving mass could be flawed and give different results depending on the mass distance and relative motion.

Conclusion. Planck time t_p is the time window through which we experience the presence of the Planck mass and its energy. Time t_p is extremely short and we will detect only a minute fraction of the Planck mass. This small mass has an additional time dimension and is identified with the electron. The precise determination of its rotational speed yields all known electron parameters and a link with the gravitational constant.

Elaboration of existing equations allows a given quantity to be written in many interesting ways and the electron charge e below is just an example. $4\pi^2$ in the right term is part of W_u which includes the unitary dimension of charge and time:

$$e = Q/(\alpha/\alpha_0)\sqrt{2/\alpha-1} = e_0\sqrt{(2-\alpha_0)\alpha_0/\alpha(2-\alpha)} = 4\pi^2\sqrt{t_p/\alpha(2-\alpha)} \tag{17}$$

It should be mentioned that Eq. 9 gives also a negative fine structure constant, $-\alpha+\delta$, where $\delta = 1+\alpha/2-\sqrt{1+\alpha-(3/4)\alpha^2} = 2.653\times10^{-5}$. Such negative value would imply a speed faster than light by 0.18%. It would not involve any material particle but would substantially add to the complexity of vacuum. The other solutions are α, common to both Eq. 8 and 9, then $2-\alpha$ for Eq. 8 and $2-\delta$ for Eq. 9.

A numerical exercise was carried out to verify agreement with the latest Codata listing (2006). The adopted values are $c = 299792458$, $h = 6.62606837306\times10^{-34}$ and $G = 6.67291773245\times10^{-11}$. All results are within one or two standard deviations.

MODELING THE ENERGY OF STRINGY BLACK HOLE SOLUTIONS

Irina Radinschi* and Brindusa Ciobanu†

*Department of Physics "Gh. Asachi" Technical University Iasi, 700050, Romania
radinschi@yahoo.com
†Department of Physics "Gh. Asachi" Technical University Iasi, 700050, Romania
bciobanu2003@yahoo.com

Abstract. The localization of energy has attracted attention as one of the most intricate issues and is the subject of many studies since the outset of general relativity. In this paper we compute the energy distribution of a dual dilatonic solution in low-energy string theory using the Landau-Lifshitz, Weinberg and Møller prescriptions. We also point out the connections between the expressions for energy computed in these prescriptions, study some particular cases, and model the energetic properties of this solution using the Maple and Mathematica programs.

Keywords: energy,stringy black hole, dilaton gravity, string theory, energy-momentum complex
PACS: 04. 20.-q, 04. 20 Dw, 04. 70. Bw

INTRODUCTION

Recent studies [1] support the usefulness of energy-momentum complexes for energy-momentum localization in general relativity. Our paper is focused on the evaluation of the energy distribution of a dual pure scalar black hole solution in dilaton gravity [2] using the Landau-Lifshitz [3], Weinberg [4] and Møller [5] prescriptions.

ENERGY DISTRIBUTION OF THE DUAL PURE STRINGY BLACK HOLE SOLUTION

The solution under study in our work is expressed in terms of the string metric and is given by

$$ds^2 = \left(1-\frac{2\eta}{r}\right)^{\frac{-m-\sigma}{\eta}} dt^2 - \left(1-\frac{2\eta}{r}\right)^{\frac{\sigma-m}{\eta}} dr^2 - \left(1-\frac{2\eta}{r}\right)^{1+\frac{\sigma-m}{\eta}} r^2\left(d\theta^2+\sin^2\theta d\varphi^2\right) \quad (1)$$

where m is the mass, σ is the scalar charge and $\eta^2 = m^2+\sigma^2$. This solution was obtained from the spherically symmetric static solution of the heterotic string theory [3] using the Buscher formulae.

The Buscher transformations interchange the roles of m and σ. The results are given in the table below.

CP1018, *Frontiers of Fundamental and Computational Physics*,
edited by B. G. Sidharth, F. Honsell, O. Mansutti, K. Sreenivasan, and A. De Angelis

TABLE 1.

Black hole	**Energy - Landau-Lifshitz prescription**	**Energy - Weinberg prescription**	**Energy - Møller prescription**
dual Buscher solution	$(m-\sigma)\left(1-\frac{2\eta}{r}\right)\left(1-\frac{2\eta}{r}\right)^{\frac{2(\sigma-m)}{\eta}}$	$(m-\sigma)\left(1-\frac{2\eta}{r}\right)^{\frac{\sigma-m}{\eta}}$	$(-m-\sigma)\left(1-\frac{2\eta}{r}\right)^{-\frac{m}{\eta}}$
	limit $r\to\infty$ $m-\sigma$	limit $r\to\infty$ $m-\sigma$	limit $r\to\infty$ $-m-\sigma$
case $m=0$ $\sigma=\pm\eta$	$-\sigma\left(1-\frac{2\eta}{r}\right)\left(1-\frac{2\eta}{r}\right)^{\frac{2\sigma}{\eta}}$ $\sigma=-\eta$ Schwarzschild	$-\sigma\left(1-\frac{2\eta}{r}\right)^{\frac{\sigma}{\eta}}$ $\sigma=-\eta$ Schwarzschild	$-\sigma$
	limit $r\to\infty$ $-\sigma$	limit $r\to\infty$ $-\sigma$	
case $m=\sigma$	0	0	$-2m\left(1-\frac{2\eta}{r}\right)^{-\frac{m}{\eta}}$
			limit $r\to\infty$ $-m$
case $\sigma=0$	$m\left(1-\frac{2\eta}{r}\right)\left(1-\frac{2\eta}{r}\right)^{\frac{-2m}{\eta}}$	$m\left(1-\frac{2\eta}{r}\right)^{\frac{-m}{\eta}}$	$-m\left(1-\frac{2\eta}{r}\right)^{\frac{-m}{\eta}}$
	limit $r\to\infty$ m	limit $r\to\infty$ m	limit $r\to\infty$ $-m$

DISCUSSION

All these results describe the energetic properties of the dual pure scalar black hole solution. For performing the calculations we used Maple, which has attached the GrTensor platform.

ACKNOWLEDGMENTS

This work is supported by grant 51/2007 PN II, Resurse Umane, MC.

REFERENCES

1. K. S. Virbhadra, *Phys. Rev.* **D60**, 104041 (1999); E. C. Vagenas, *Mod. Phys. Lett.* **A21**, 1947 (2006); I-Ching Yang, and I. Radinschi, *Chin. J. Phys.* **42**, 40 (2004); I. Radinschi, and Th. Grammenos, *Int. J. Mod. Phys.* **A21**, 2853-2861 (2006).
2. S. Kar, *Class.Quant.Grav.* **16**, 101 (1999)
3. L. D. Landau and E. M. Lifshitz, *The Classical Theory of Fields*, Pergamon Press, 1987, p. 280.
4. S. Weinberg, *Gravitation and Cosmology: Principles and Applications of General Theory of Relativity*, John Wiley and Sons, Inc., New York, 1972, p. 165.
5. C. Møller, *Ann. Phys. (NY)* **4**, 347 (1958).

PART IV

EXPERIMENTS AND PHENOMENOLOGY AT THE FRONTIER OF FUNDAMENTAL PHYSICS

Weak Nuclear Forces cause the Strong Nuclear Force

E.L. Koschmieder

Center for Statistical Mechanics
The University of Texas at Austin, Austin TX 78712. USA
e-mail: koschmieder@mail.utexas.edu

Abstract. We determine the strength of the weak nuclear force which holds the lattices of the elementary particles together. We also determine the strength of the strong nuclear force which emanates from the sides of the nuclear lattices. The strong force is the sum of the unsaturated weak forces at the surface of the lattices.

Keywords: strong force, weak force, nuclear lattice, muon neutrino mass.
PACS: 14.60.Lm, 12.10.Dm, 12.90.+b.

INTRODUCTION

After have explained the masses of the electron and muon and of the stable mesons and baryons with cubic lattices consisting of either photons or of neutrinos in [1], we can now determine the strength of the weak and the strength of the strong nuclear forces. Both are 70 years old puzzles. We will, in the following, use lattice theory to determine the strength of the weak force which holds the lattices of the elementary particles together. We will then show that the strong force between two elementary particles is nothing but the sum of the unsaturated weak forces emanating from the lattice points at the surface of the lattice. Consequently the strong force between two elementary particles is on the order of 10^6 times stronger than the weak force in the interior of the particles. The strong force is caused by the weak force.

CP1018, *Frontiers of Fundamental and Computational Physics*,
edited by B. G. Sidharth, F. Honsell, O. Mansutti, K. Sreenivasan, and A. De Angelis

THE FORCE IN THE INTERIOR OF IONIC LATTICES

In order to determine the force in the interior of the cubic lattices with which we have explained the particle masses we will, as we have done before [2], use a classical paper by Born and Landé [3], (B&L), dealing with the potential and compressibility of regular ionic crystals. It is essential to realize that

- *for the existence of a cubic lattice it is necessary that the force between the lattice points has an attractive part and a repulsive part.*
- Otherwise the lattice would not be stable and collapse.

For the ionic crystals considered by B&L the Coulomb force between the ions is the attractive force, whereas the repulsive force originates from the electron clouds surrounding the ions. When the electron clouds of the ions approach each other during the lattice oscillations they repel each other. The magnitude of the repulsive force is not known per se and has to be determined from the properties of the crystal.

We will follow exactly the procedure in B&L in order to see whether their theory is also applicable to a cubic lattice made of neutrinos. In this case the Coulomb force is, of course, irrelevant. As B&L do, we say that the potential of a cell of the neutrino lattice has an attractive part $-\,\mathrm{a}/\delta$ and a repulsive part $+\,\mathrm{b}/\delta^{\mathrm{n}}$ with the unknown exponent n. δ is the distance in the direction between two neutrinos of the same type, either muon neutrinos and anti-muon neutrinos or electron neutrinos and anti-electron neutrinos.

The potential of a cell of an ionic cubic lattice is of the form

$$\phi = -\,\mathrm{a}/\delta + \mathrm{b}/\delta^{\mathrm{n}}\,, \tag{1}$$

Eq.(1) of B&L. The constant b is eliminated with the equilibrium condition $\mathrm{d}\phi/\mathrm{d}\delta = 0$. Consequently

$$\mathrm{b} = \frac{\mathrm{a}}{\mathrm{n}}\,\delta_0^{\mathrm{n}-1}\,, \tag{2}$$

where δ_0 is the lattice constant. The unknown exponent n of δ in Eq.(1) was determined by B&L with the help of the compression modulus κ which is known for ionic crystals. κ is defined by

$$\kappa = -\,1/V \cdot dV/dp\,. \tag{3}$$

The compression modulus of an ionic lattice is given by

$$\kappa = 9\,\delta_0^4/\mathrm{a}\,(\mathrm{n}-1)\,, \tag{4}$$

Eq.(4) in B&L. The interaction constant a of the Coulomb force in cubic ionic crystals resulting from the contributions of all ions on a single ion is given by Eq.(5) of B&L, it is

$$\mathrm{a} = 13.94\,\mathrm{e}^2 = 3.2161 \cdot 10^{-18}\,\mathrm{erg}\cdot\mathrm{cm}\,, \tag{5}$$

where e is the elementary electric charge. This equation is fundamental for the theory of ionic lattices and is based on an earlier paper by Madelung [4]. Consequently we find that

$$(\mathrm{n}-1) = 10.33\,\mathrm{r}_0^4/\mathrm{e}^2\kappa\,, \tag{6}$$

where $\mathrm{r}_0 = \delta_0/2$ is the distance between a pair of neighboring Na and Cl ions. For the alkali-halogenids, such as NaCl or KCl, B&L found that $\mathrm{n} \approx 9$. If $\mathrm{n} = 9$ is used in Eq.(4) to determine theoretically the compression modulus κ, then the theoretical values of κ agree, in a first approximation, with the experimental values of κ, thus confirming the validity of the ansatz for the potential in Eq.(1).

THE FORCE IN THE INTERIOR OF A NUCLEAR LATTICE

We now apply $\mathrm{n}-1 = 10.33\,\mathrm{r}_0^4/\mathrm{e}^2\kappa$, Eq.(6), to the neutrino lattice of the elementary particles in order to determine the potential in the interior of the particles. We must use for r_0 the distance between two neighboring neutrinos in the lattice, which is equal to the range of the weak nuclear force. The range of the weak force is $1{\cdot}10^{-16}$ cm, according to Perkins [5], and so we have

$$\mathrm{r}_0 = 1 \cdot 10^{-16}\,\mathrm{cm}\,. \tag{7}$$

We have used this value of r_0 throughout our explanation of the masses of the elementary particles in [1]. We must, furthermore, replace e^2 by the interaction constant g_w^2 of the weak force which holds the nuclear lattice together. According to Perkins [5]

$$\mathrm{g}_w^2 = 4\pi\hbar\,\mathrm{c}\cdot 1.05 \cdot 10^{-2}\,(\mathrm{M}_W/\mathrm{M}_p)^2\,, \tag{8}$$

where M_W is the mass of the W boson, $M_W = 80.403\,\mathrm{GeV/c^2}$, and M_p is the mass of the proton, $M_p = 0.938\,272\,\mathrm{GeV/c^2}$. That means that

$$g_w^2 = 2.9976 \cdot 10^{-17}\,\mathrm{erg \cdot cm}. \tag{9}$$

We must also use the compression modulus κ of the nucleon. The value of the compression modulus of the nucleon has been determined theoretically by Bhaduri et al. [6], following earlier theoretical and experimental investigations of the compression moduli of nuclei. Bhaduri et al. found that the compression modulus $K_A(N)$ of the nucleon ranges from 900 to 1200 MeV, or is 940 MeV or 900 MeV for particular sets of parameters. We determine κ of the nucleon with

$$\kappa = 9/\rho_{\#} K_{nm}\,, \tag{10}$$

from Eq.(1) in [6]. Bhaduri et al. write that "the compression modulus K_{nm} of nuclear matter is calculated by considering the nucleons as point particles", which they are not. Other assumptions are also sometimes made such as infinite nuclear matter, periodic boundary conditions, etc. Recent theoretical studies of the compressibility of "nuclear matter" [7,8,9] place the compressibility K_{nm} at values from between 250 to 270 MeV, not much different from what is was twenty years earlier in [6]. Considering the uncertainty of K_{nm} it seems to be justified to set, in the case of the nucleon, $K_{nm} = K_A$, where K_A is defined as the compression modulus for a finite system with A nucleons. It then follows from Eq.(10) with the number density $\rho_{\#}$ being the number density per fm^3, and with the radius of the nucleon $R_0 = 0.88 \cdot 10^{-13}$ cm, and with 1 MeV $= 1.6022 \cdot 10^{-6}$ erg that the compression modulus of the nucleon is

$$\kappa_n = 1.603 \cdot 10^{-35}\ \mathrm{cm^2/dyn}\,, \tag{11}$$

if we use for $K_A(N)$ the value 1000 MeV. We will keep in mind that κ_n is not very accurate because $K_A(N)$ is not very accurate.

If we insert (7), (9), and (11) into $n - 1 = 10.33 r_0^4/e^2\kappa$ (Eq.6) we find an equation for the exponent n in the term b/δ^n of the repulsive part of the potential in a nuclear lattice

•

$$n = 1 + 2.164 \cdot 10^{-12} = 1 + \epsilon\,, \tag{12}$$

with $r_0^4 = O(10^{-64})$, $g_w^2 = O(10^{-17})$ and $\kappa = O(10^{-35})$.

That means that

the potential ϕ in the interior of an elementary particle is given by

$$\phi = -\frac{a}{\delta} + \frac{b}{\delta^{1+\epsilon}} = \frac{a}{\delta}\left[\frac{(\delta_0/\delta)^{\epsilon}}{n} - 1\right], \tag{13}$$

which we can reduce with $n - 1 = \epsilon$, neglecting a term multiplied by $\epsilon^2 = O(10^{-24})$, using also $a = 13.94\,e^2$ and $1/n \cong (1-\epsilon)$, to

$$\phi \cong -\frac{a\,\epsilon}{\delta}\left[1 - \ln(\delta_0/\delta)\right] \cong -\frac{13.94\,g_w^2\,\epsilon}{\delta}\left[1 - \ln(\delta_0/\delta)\right]. \tag{14}$$

In equilibrium the value of ϕ in the nuclear lattice is about $g_w^2 \cdot \epsilon/e^2 \approx 2.7{\cdot}10^{-10}$ times smaller than the corresponding electrostatic potential in an ionic lattice. A graph of the potential in Eq.(14) versus δ is shown in Fig. 1.

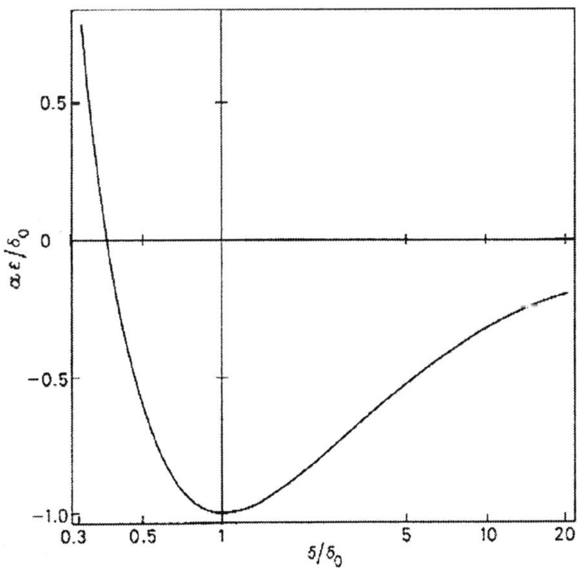

Fig.1: The potential ϕ of the weak force as a function of δ. After [2].

The minimum of the curve marks the equilibrium. From Eq.(14) follows with $F_w = d\phi/d\delta$ and $\delta = 2r$ that *the weak force in the interior of the nuclear lattice is*

$$F_w \cong \frac{3.485 \cdot g_w^2 \,\epsilon}{r^2} \cdot \ln(\frac{\delta}{\delta_0}) \,. \qquad \text{(dyn)} \tag{15}$$

For all distances $\delta > \delta_0$ the force F_w is attractive, for all distances $\delta < \delta_0$ the force is repulsive. The small value of $\epsilon \approx 10^{-12}$ means that small displacements $\delta/\delta_0 < 1$ of the neutrinos from their equilibrium position, which carry the neutrinos into the domain of their neighboring neutrino, cause a very strong repulsive force between both neutrinos.

We have thus determined the potential of the weak force in the interior of the lattice of the elementary particles with lattice theory. Let us consider how this was done.

- *Following exactly the procedure used by B&L to determine the potential in the interior of an ionic crystal,*

- *we have determined the potential in the interior of the lattice of an elementary particle by using the parameters of the nuclear lattice.*

As in an ionic lattice the potential in a nuclear lattice has an attractive and a repulsive part, as is necessary for the stability of the lattice. We have found that in equilibrium in a nuclear lattice the absolute values of the attractive and repulsive terms of the potential are very nearly the same, because ϵ is on the order of 10^{-12}.

THE MUON NEUTRINO MASS

One wonders, in view of the extraordinaryly small value of ϵ, whether the potential in Eq.(14) is not an academic result. It has to be shown that Eq.(14) is indeed relevant for elementary particles. We can show that the mass of the muon neutrino $m(\nu_\mu)$ depends on $\epsilon = n - 1 = 2.164 \cdot 10^{-12}$ and yields a correct mass $m(\nu_\mu)$ if only ϵ is on the order of 10^{-12}.

From the explanation of the $\pi^\pm$ mesons in [1] follows that the mass of the muon neutrinos is

$$m(\nu_\mu) = m(\bar{\nu}_\mu) \cong 50\,\text{milli}\,\text{eV}/c^2 \,, \tag{16}$$

Eq.(40) in [1]. One can, on the other hand, determine the mass of the muon neutrinos from the group velocity of the neutrino oscillations in their nuclear lattice. It follows from Eq.(26) of [1] that

$$m(\nu_\mu) \cong r_0^3\, c_{44}\,/c^2\,, \tag{17}$$

where c_{44} is one of the elastic constants of continuum mechanics, which is valid when $r_0 \to 0$. With the proper value of c_{44}, for details see http://arXiv: 0712.1849, we find that

$$m(\nu_\mu) = 1.284 \cdot 10^{-34}\,\mathrm{gr} = 72.1\,\mathrm{milli\,eV/c^2} = m(\bar{\nu}_\mu)\,. \tag{18}$$

The anti-muon neutrinos in the lattice move with the same group velocity as the muon neutrinos, so $m(\nu_\mu) = m(\bar{\nu}_\mu)$. The value of $m(\nu_\mu)$ in Eq.(18) differs by 50% from $m(\nu_\mu) \cong 50$ milli-eV/c^2 given by Eq.(16). We have thus learned that ϵ *must be on the order of* 10^{-12} to end up with a correct value of the muon neutrino mass $m(\nu_\mu)$.

To summarize: We have found that it follows from lattice theory that the value of the exponent $n = 1 + \epsilon$ in the repulsive term of the potential in the nuclear lattice leads to an, in the first approximation, correct value of the mass of the muon neutrino, albeit that ϵ is so extraordinarily small. We have thus confirmed the validity of the potential $\phi \cong -\,a/\delta + b/\delta^n = -\,a/\delta + b/\delta^{1+\epsilon}$. That means *we have found the potential of the weak force which holds the nuclear lattice together.*

THE STRONG FORCE BETWEEN TWO ELEMENTARY PARTICLES

Crucial for the understanding of the existence of a strong force between the sides of two cubic lattices is the observation that

- *the sides of two halves of a cubic lattice cleaved in vacuum exert a strong, attractive force on each other.*

It is an automatic consequence of lattice theory that

- *the weak force, which holds the lattice together, is accompanied by a strong force which emanates from the sides of the lattice.*

This seems to contradict the simple observation that two salt crystals stacked upon each other can be separated without any effort. This is only so because the surface of a cubic crystal cleaved in air oxidizes immediately, and then the sides do not attract each other any longer. The origin of the force emanating from the sides of a cubic ionic lattice is the Coulomb force between ions of opposite polarity, i.e. the force which holds the lattice together. The attractive force emanating from the side of a crystal cleaved in vacuum has a macroscopic value. The existence of this force has tangible consequences in space technology. The force between the sides of two cubic lattices was first studied by Born and Stern [11] (B&S).

If U_{12} is the potential between two halves of a crystal with the equal surfaces A, or the work that is necessary to separate the two halves of a cleaved crystal, then the capillary constant σ is given by Eq.(2) of B&S

$$\sigma = -\, U_{12}/A\,. \tag{19}$$

The capillary constant is, in the following, taken at zero degree absolute and against vacuum for the square outside area A of a cubic crystal. σ has been explained by B&S, but their formula cannot be used here because they use the value $n = 9$ of the alkali-halegonids. We will instead use Eq.(463) from Born [10] for the capillary constant σ_{100} of the (100) front surface of an ionic cubic crystal

$$\sigma_{100} = -\,\frac{e^2}{r_0^3}\,\frac{s_0(1)}{2}\cdot\left[1 - \frac{1}{n}\,\frac{s_0(n)}{s_0(1)}\cdot\frac{S_0(1)}{S_0(n)}\right]. \tag{20}$$

The sums $s_0(n)$ and $S_0(n)$ originate from the contributions of the different lattice points to the repulsive part of the potential. The sign of the second term on the right hand side in Eq.(22) comes from the repulsive part of the potential in Eq.(1). For the capillary constant in a nuclear lattice we set $e^2 = g_w^2$, $n = 1 + \epsilon$, (Eq.12), and $s_0(1) = -0.0650$ from [10] p.743. We find that $s_0(n) \cong s_0(1)$ since $n = 1 + \epsilon$ and $\epsilon \cong 10^{-12}$. Similarly we have $S_0(n) \cong S_0(1)$. Then we have

$$\sigma_{100} \cong \frac{0.065}{2}\,\frac{g_w^2}{r_0^3}\,\epsilon\,. \qquad (\text{dyn/cm}) \tag{21}$$

The work required to separate one half of a nuclear lattice from the other half is according to Eq.(19) given by

$$U_{12} \cong -\,\frac{0.065}{2}\,\frac{g_w^2\,\epsilon}{r_0^3}\cdot A\,. \tag{22}$$

We determine the area A with the number of the lattice points N in the cubic nuclear lattice

$$N = 2.854 \cdot 10^9 , \tag{23}$$

from Eq.(15) in [1]. It follows that $A = (\sqrt[3]{N} \cdot r_0)^2$. The strong attractive force between the sides of two nuclear lattices is

$$F_s = \frac{dU_{12}}{dr} = -\frac{d\sigma}{dr} \cdot A = \frac{3 \cdot 0.065\, g_w^2\, \epsilon}{2\, r^4} \cdot A\, . \tag{24}$$

- *The force emanating from the front surface of a cubic nuclear lattice, the strong force, is*

$$F_s = \frac{0.0975\; g_w^2\; \epsilon}{r^4} \cdot (\sqrt[3]{N} \cdot r_0)^2 . \qquad \text{(dyn)} \tag{25}$$

The strong force depends, first of all, on the *weak interaction constant* g_w^2, and secondly on the number of lattice points on a side of the lattice, via $(\sqrt[3]{N})^2 = 2.012 \cdot 10^6$. The strong force is also inversely proportional to the *fourth* power of the distance between the particles.

The ratio of the entire force emanating from the surface of the lattice, which is six times Eq.(25), to the weak force between the lattice points, Eq.(15), is then

$$\frac{F_s}{F_w} = \frac{0.33 \cdot 10^6}{r^2/r_0^2 \cdot \ln(r/r_0)} , \tag{26}$$

or the strong force is on the order of 10^6 times larger than the weak force but the ratio is sensitive to the value of r/r_0.

To summarize: According to lattice theory

- *the existence of the strong nuclear force between two elementary particles is an automatic consequence of our explanation of the masses of the elementary particles with cubic nuclear lattices,*

held together by the weak nuclear force. The lattices we have used for the explanation of the masses of the particles consist of photons or neutrinos. That means:

We do not use hypothetical particles.

CONCLUSIONS

We have found long sought after answers to the questions what is the weak nuclear force and the strong nuclear force, and why is the strong force so much stronger than the weak? The strong force is nothing but the sum of the large number of unsaturated weak forces at the surface of the lattice.

In order to understand the origin of the strong nuclear force one has to understand the structure of the elementary particles, which we have explained in [1] with nuclear lattices. With this concept we can explain the masses of the particles and the spin of the particles. We have now also understood the strength of the weak force which holds the nuclear lattice together and the cause of the strong force between two nuclear lattices.

REFERENCES

[1] Koschmieder, E.L. http://arXiv.org:physics/0602037. (2006)

[2] Koschmieder, E.L. Nuovo Cim. **101A**,1017. (1989)

[3] Born, M. and Landé, A. Verh.Dtsch.Phys.Ges. **20**,210. (1918)

[4] Madelung, E. Phys.Z. **19**,524. (1918)

[5] Perkins, D.H. *Introduction to High-Energy Physics* Addison-Wesley. (1982)

[6] Badhuri, R.K. et al. Phys.Lett.B **136**,189. (1984)

[7] Müller, H-M. et al. http://arXiv.org:nucl-th/9910038. (1999)

[8] Vretenar, D. et al. http://arXiv.org:nucl-th/0302070. (2003)

[9] Dexheimer, V.A. et al. http://arXiv.org:0708.0131. (2007)

[10] Born, M. *Atomtheorie des festen Zustandes* Teubner. (1923)

[11] Born, M. and Stern, O. Sitzungsber.Preuss.Akad.Wiss. **33**,901. (1919)

Supercritical fields and short living giant atoms

Walter Greiner

Frankfurt Institute for Advanced Studies, Johann Wolfgang Goethe University, Ruth-Moufang-Str. 1, 60438 Frankfurt am Main, Germany

Abstract. The decay of the vacuum in supercritical Coulomb fields is discussed. Connected with that is the time delay in giant nuclear systems like U+Cm etc. A perspective for future research is given.

Keywords: Quantumelectrodynamics, strong fields, superheavy nuclei, giant quasi-atoms
PACS: 12.20.-m, 25.70Jj, 25.70.L

THE VACUUM IN QUANTUM ELECTRODYNAMICS

It is generally accepted that the physical vacuum has a nontrivial structure. This conclusion was first made by Dirac on the basis of his famous equation for a fermion field which describes simultaneously particles and antiparticles. The Dirac equation in the vacuum has a simple form

$$(i\gamma^{\mu}\partial_{\mu} - m)\Psi(x) = 0\ , \tag{1}$$

where $\gamma^{\mu} = (\gamma^0, \boldsymbol{\gamma})$ are Dirac matrices, m is the fermion mass and $\Psi(x)$ is a 4-component spinor field. For a plane wave solution $\Psi(x) = e^{-ipx}u_p$ this equation is written as

$$(\widehat{p} - m)u_p = 0\ , \tag{2}$$

where $\widehat{p} = \gamma^0 E - \boldsymbol{\gamma}\mathbf{p}$. Multiplying by $(\widehat{p} + m)$ and requiring that $u_p \neq 0$ one obtains the equation $E^2 - \mathbf{p}^2 - m^2 = 0$ which has two solutions

$$E^{\pm}(\mathbf{p}) = \pm\sqrt{\mathbf{p}^2 + m^2}\ . \tag{3}$$

Here the $+$ sign corresponds to particles with positive energy $E_N(\mathbf{p}) = E^{+}(\mathbf{p})$, while the $-$ sign corresponds to solutions with negative energy. To ensure stability of the physical vacuum Dirac has assumed that these negative-energy states are occupied forming what is called now the Dirac sea. Then the second solution of eq. (3) receives natural interpretation: it describes holes in the Dirac sea. These holes are identified with antiparticles. Their energies are obviously given by $E_{\overline{N}}(\mathbf{p}) = -E^{-}(-\mathbf{p}) = \sqrt{\mathbf{p}^2 + m^2}$. Unfortunately, the Dirac sea brings divergent contributions to physical quantities such as energy density, and one should introduce a proper regularization scheme to get rid off these divergences. This picture has received numerous confirmations in quantum electrodynamics and other fields.

CP1018, *Frontiers of Fundamental and Computational Physics*,
edited by B. G. Sidharth, F. Honsell, O. Mansutti, K. Sreenivasan, and A. De Angelis

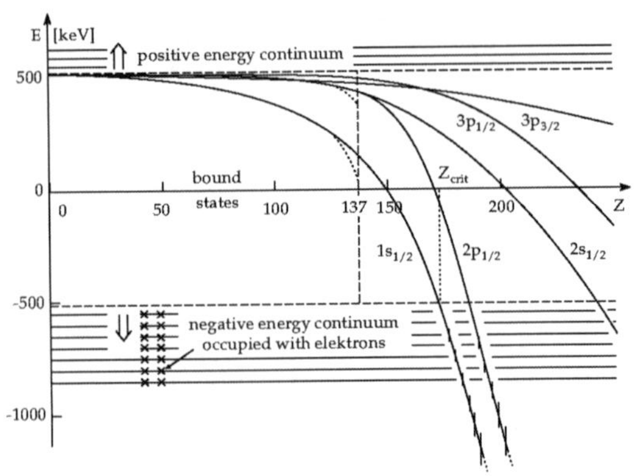

FIGURE 1. Lowest bound states of the Dirac equation for nuclei with charge Z. While the Sommerfeld fine-structure energies (dashed line) for $\xi = 1$ (s states) end at $Z = 137$, the solutions for extended Coulomb potentials (full line) can be traced down to the negative-energy continuum reached at the critical charge Z_{cr} for the $1s$ state. The bound states entering the continuum obtain a spreading witdth as indicated.

Adiabatic dynamics of heavy nuclear system

One of the most fascinating aspects is the structure of the vacuum in QED and its change into charged vacuum states under the influence of strong (supercritical) electric fields [1]. We shortly remind of this phenomenon.

Fig. 1 shows the diving of the deeply bound states into the lower energy continuum of the Dirac equation.

In the supercritical case the dived state is degenerate with the (occupied) negative electron states. Hence spontaneous e^+e^- *paircreation* becomes possible, where an electron from the Dirac sea occupies the additional state, leaving a hole in the sea which escapes as a positron while the electron's charge remains near the source. This is a fundamentally new process, whereby the neutral vacuum of QED becomes unstable in supercritical electrical fields. It decays within about $10^{-19}s$ into a charged vacuum. The charged vacuum is now stable due to the Pauli principle, that is the number of emitted particles remains finite. The vacuum is first charged twice because two electrons with opposite spins can occupy the $1s$ shell. After the $2p_{1/2}$ shell has dived beyond $Z_{cr} = 185$, the vacuum is charged four times, etc. This change of the vacuum structure is not a perturbative effect, as are the radiative QED effects (vacuum polarization, self-energy, etc.).

The time-dependence of the energy levels in a supercritical heavy-ion collision is depicted in Fig. 2. An electron (or hole) which was in a certain molecular eigenstate at the beginning of the collision can be transfered with a certain probability into different states by the dynamics of the collision. This can lead to the hole production in an inner shell by excitation of an electron to a higher state and/or hole production by ionization of an electron to the continuum. Further possibilities are induced positron production by excitation of an electron from the lower continuum to an empty bound level and direct pair production [2].

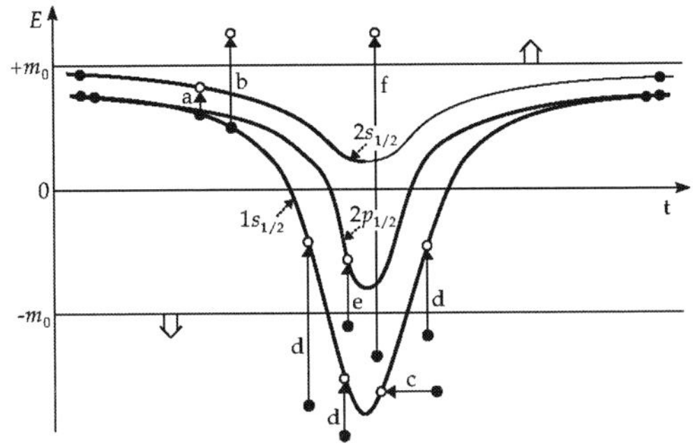

FIGURE 2. Time dependence.of the quasi-molecular energy levels in a supercritical heavy ion collision. The arrows denote various excitation processes which lead to the production of holes and positrons.

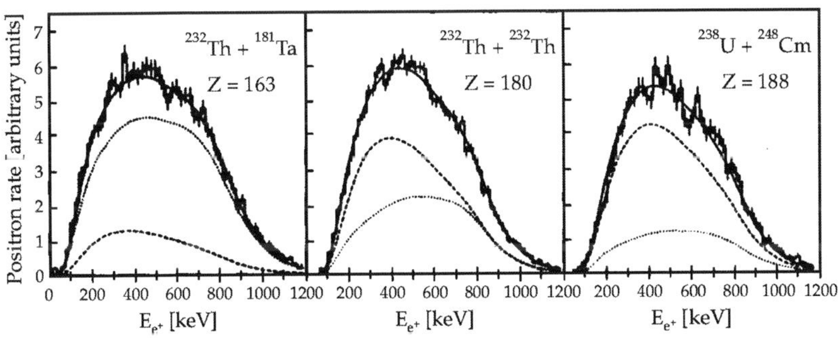

FIGURE 3. Positron energy spectra measured in collisions of Th+Ta, Th+Th, and U+Cm at energies of about 6 MeV per nucleon. The QED predictions (dashed lines) and the experimentally determined background from nuclear pair conversion (dotted lines) add up to the full lines which are in close agreement with experiment.

A comparison of the theoretical predictions and expectations and experimental data is shown in figure 3. Sharp positron peeks can be expected if there were a mechanism in the heavy ion collision leading to a time delay. This may be caused by a pocket in the potential between the two ions. Spontaneous pair production should then be enhanced in supercritical systems. Until now, however, the situation remains inconclusive [2].

ON SUPERHEAVY ELEMENT FORMATION AND BEYOND

Superheavy (SH) nuclei obtained in "cold" fusion reactions with Pb or Bi target [3] are along the proton drip line and very neutron-deficient with a short half-life. In fusion of actinides with ^{48}Ca more neutron-rich SH nuclei are produced [4] with much longer half-life. But they are still far from the center of the predicted "island of stability" formed by the neutron shell around N=184 (see the nuclear map in Fig. 4). In the "cold" fusion, the cross sections of SH nuclei formation decrease very fast with increasing charge of

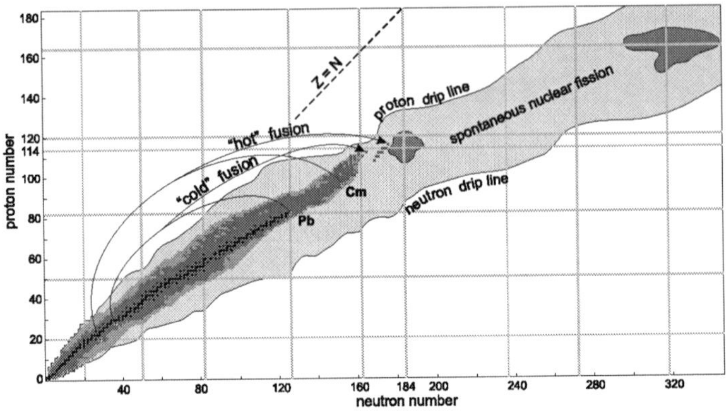

FIGURE 4. Nuclear map. Predicted islands of stability are shown around Z=114, N=184 and Z=164, N=318.

the projectile and become less than 1 pb for Z>112. Heaviest transactinide, Cf, which can be used as a target in the second method, leads to the SH nucleus with Z=118 being fused with ^{48}Ca. Using the next nearest elements instead of ^{48}Ca (e.g., ^{50}Ti, ^{54}Cr, etc.) in fusion reactions with actinides is expected less encouraging, though experiments of such kind are planned to be performed. In this connection other ways to the production of SH elements in the region of the "island of stability" should be searched for.

About twenty years ago transfer reactions of heavy ions with ^{248}Cm target have been evaluated for their usefulness in producing unknown neutron-rich actinide nuclides [5, 6]. The cross sections were found to decrease very rapidly with increasing atomic number of surviving target-like fragments. However, Fm and Md neutron-rich isotopes have been produced at the level of 0.1 μb. Theoretical estimations for production of primary superheavy fragments in the damped U+U collision have been also performed at this time within the semiphenomenological diffusion model [7]. In spite of obtained high probabilities for the yields of superheavy primary fragments (more than 10^{-2} mb for Z=120), the cross sections for production of heavy nuclei with low excitation energies were estimated to be rather small: $\sigma_{CN}(Z=114, E^*=30\,\mathrm{MeV}) \sim 10^{-6}$ mb for U+Cm collision at 7.5 Mev/nucleon beam energy. The authors concluded, however, that "fluctuations and shell effects not taken into account may conciderably increase the formation probabilities".

Here we study the processes of low energy collisions of very heavy nuclei (such as U+Cm) within the recently proposed model of fusion-fission dynamics [8]. The purpose is to find an influence of the nearest closed shells Z=82 and N=126 (formation of ^{208}Pb as one of the primary fragments) on nucleon rearrangement between primary fragments. We try to estimate how large the charge and mass transfer can be in these reactions and what the cross sections for production of surviving neutron-rich SH nuclei are. Direct time analysis of the collision process allows us to estimate also the lifetime of the composite system consisting of two touching heavy nuclei with total charge Z>180. Such "long-lived" configurations may lead to spontaneous positron emission from super-strong electric fields of quasi-atoms by a fundamental QED process (transition from

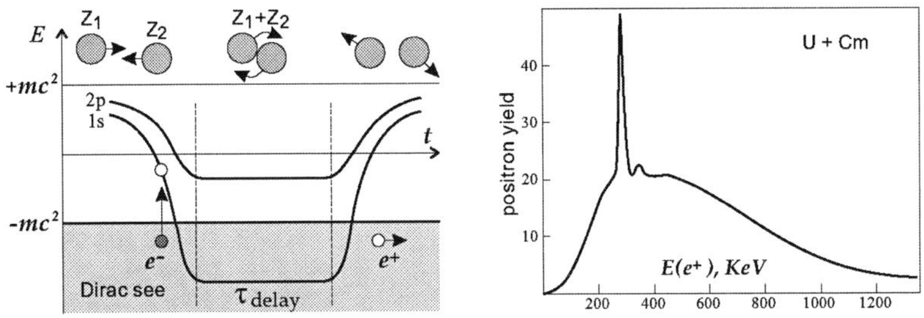

FIGURE 5. Schematic figure of spontaneous decay of the vacuum and spectrum of the positrons formed in supercritical electric field ($Z_1 + Z_2 > 173$).

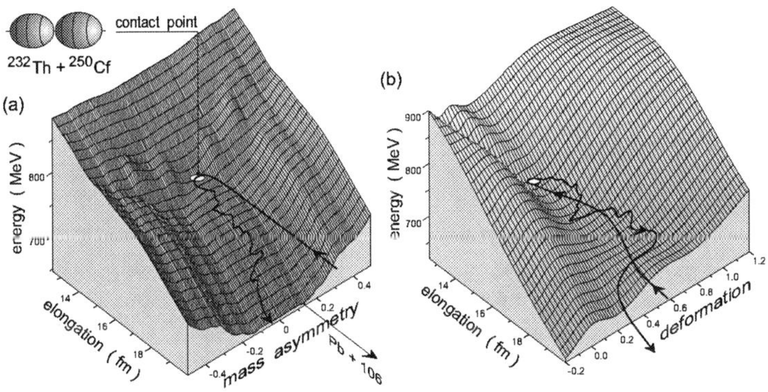

FIGURE 6. Potential energy surface for the nuclear system formed by ^{232}Th+^{250}Cf as a function of R and α ($\beta = 0.22$) (a), and R and β ($\alpha = 0.037$) (b). Typical trajectory is shown by the thick curves with arrows.

neutral to charged QED vacuum) [9, 10], see schematic Fig. 5.

Besides the distance between the nuclear centers, R, the mass transfer, $\alpha = (A_1 - A_2)/(A_1 + A_2)$ and dynamic deformations of nuclear surfaces, β, play a most important role in fusion-fission and deep inelastic processes of low energy heavy ion collisions. The corresponding multi-dimensional adiabatic potential energy surface was calculated here within the semi-empirical two-core approach [11] based on the two-center shell model idea [12]. Projections onto the $R - \alpha$ and $R - \beta$ planes of the three-dimensional potential energy are shown in Fig. 6 for the nuclear system formed in collision of ^{232}Th+^{250}Cf. There is no potential pocket typical for lighter systems, the potential is repulsive everywhere. However, the potential energy is not so much steep in the region of contact point and two nuclei may keep in contact for a long time increasing their deformations and transfering nucleons to each other (see below).

We studied the whole dynamics of such a nuclear system by numerical solution of the coupled Langevin-type equations of motion, inertialess motion along the mass-asymmetry coordinate was derived just from the corresponding master equation for nucleon transfer [8]. The inertia parameters μ_R and μ_β were calculated within the Werner-Wheeler approach [13]. Parameters of the friction forces and nucleon transfer rate were taken from Ref. [8], where they have been estimated from successful description of experimental regularities of heavy ion deep inelastic scattering and fusion-fission reactions. Damping of the shell effects due to the excitation energy has been taken into account in the level density parameter both on the dynamic stage of the reaction and in statistical treatment of decay of the primary fragments.

The cross sections are calculated in a natural way. A large number of events (trajectories) are tested for a given impact parameter. Each event in studied reactions ends in the exit channel with two excited primary fragments. The corresponding double differential cross-section is calculated as follows

$$\frac{d^2\sigma_\alpha}{d\Omega dE}(E,\theta) = \int_0^\infty bdb \frac{\Delta N_\alpha(b,E,\theta)}{N_{\rm tot}(b)} \frac{1}{\sin(\theta)\Delta\theta\Delta E}. \tag{4}$$

Here $\Delta N_\alpha(b,E,\theta)$ is the number of events at a given impact parameter b in which the system enters into the channel α with kinetic energy in the region $(E, E+\Delta E)$ and center-of-mass outgoing angle in the region $(\theta, \theta+\Delta\theta)$, $N_{\rm tot}(b)$ is the total number of simulated events for a given value of impact parameter.

The expression (4) describes the energy, angular, charge and mass distributions of the *primary* fragments formed in the reaction. Subsequent de-excitation of these fragments via fission or emission of light particles and gamma-rays was taken into account within the statistical model leading to the *final* fragment distributions. The sharing of the excitation energy between the primary fragments was assumed to be proportional to their masses. These fragments possess also some definite deformations at scission configuration. We assumed that the dynamic deformation relaxes rather fast to the ground state deformation (faster than neutron evaporation) and simply added the deformation energy to the total excitation energy of a given nucleus. Neutron emission during an evolution of the system was also taken into account. However, it was found that the pre-separation neutron evaporation does not influence significantly the final distributions.

Damped collisions of transuranium nuclei

Found in such a way mass and charge distributions of the primary and survived fragments formed in the ^{232}Th+^{250}Cf collision at 800 MeV center-of mass energy are shown in Fig. 7. The pronounced shoulder can be seen in the mass distribution of the primary fragments near the mass number A=208. It could be explained by existence of noticeable valley on the potential energy surface [see Fig. 6 (a)], which corresponds to formation of doubly magic nucleus ^{208}Pb ($\alpha = 0.137$). The emerging of the nuclear system into this valley resembles the well-known quasi-fission process and may be called "inverse (or anti-symmetrizing) quasi-fission" (the final mass asymmetry is larger than initial one). For $\alpha > 0.137$ (one fragment becomes lighter than lead) the potential energy sharply increases and the mass distribution of the primary fragments falls down rapidly at A<208 (A>274). The same is for the charge distribution at Z<82 (Z>106). As a result, in the charge distribution of survived heavy fragments, Fig. 7 (b), there is also a shoulder at Z=106 and the yield of nuclei with Z>107 was found in this reaction at the level of less than 1 pb. This result differs sharply from those obtained within the

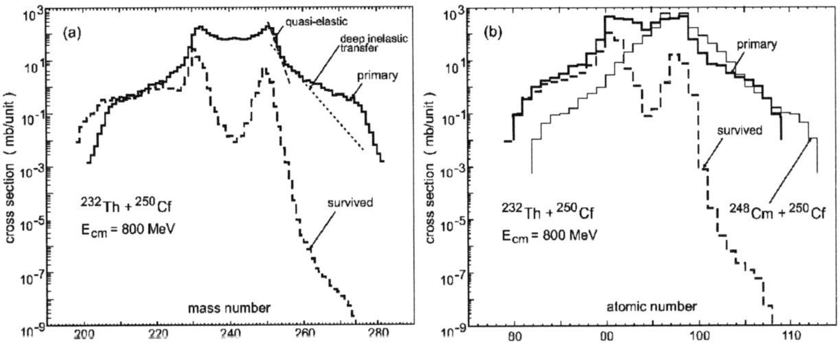

FIGURE 7. Mass (a) and charge (b) distributions of primary (solid histograms) and survived (dashed histograms) fragments in the ^{232}Th+^{250}Cf collision at 800 MeV center-of-mass energy. Thin solid histogram in (b) shows the primary fragment distribution in the hypothetical reaction ^{248}Cm+^{250}Cf.

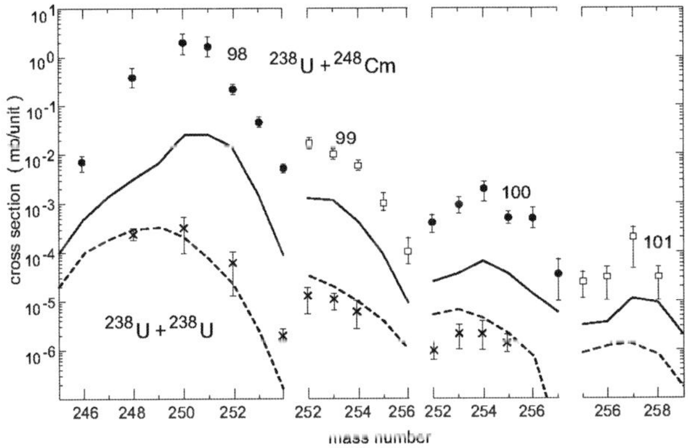

FIGURE 8. Isotopic yield of the elements 98÷101 in the reactions ^{238}U+^{238}U (crosses) [14] and ^{238}U+^{248}Cm (circles and squares) [5]. The curves show the result of our calculations.

diffusion model [7], where the yield of heavy primary fragments decreases slowly and monotonically with increasing charge number.

In Fig. 8 the available experimental data on the yield of SH nuclei in collisions of ^{238}U+^{238}U [14] and ^{238}U+^{248}Cm [5] are compared with our calculations. In these experiments thick targets have been used, which means that the experimental data were, in fact, integrated over the energy in the region of about 750÷850 MeV [14, 5]. The estimated excitation functions for the yield of heavy surviving nuclei, shown in Fig. 9, demonstrate not so sharp as in fusion reactions but still rather strong dependence on beam energy. In spite of that, the agreement of our calculations with experimental data is quite acceptable (worse for few-nucleon transfer and better for massive transfer processes).

The estimated isotopic yields of SH nuclei in the ^{232}Th+^{250}Cf, ^{238}U+^{238}U and ^{238}U+^{248}Cm collisions at 800 MeV center-of-mass energy are shown in Fig. 10. Thus there is a real chance for production and chemical study of the long-lived neutron-rich

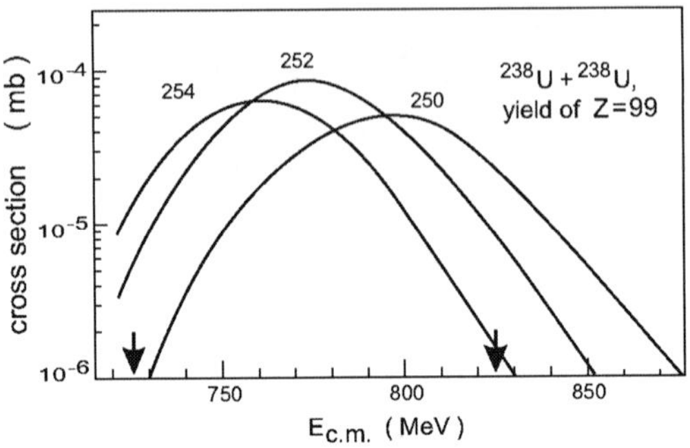

FIGURE 9. Excitation functions for the yields of Es isotopes (numbers near the curves) for the $^{238}U+^{238}U$ reactions. The arrows indicate potential energy of two nuclei in contact ($r_0 = 1.16$ fm) for tip or nose-to-nose (left) and side or belly-to-belly (right) configurations.

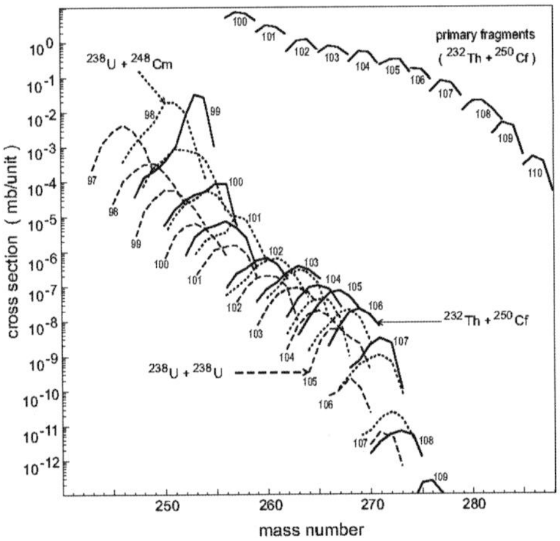

FIGURE 10. Yield of superheavy nuclei in collisions of $^{238}U+^{238}U$ (dashed), $^{238}U+^{248}Cm$ (dotted) and $^{232}Th+^{250}Cf$ (solid lines) at 800 MeV center-of-mass energy. Solid curves in upper part show isotopic distribution of primary fragments in the Th+Cf reaction. In the case of U+Cm the upper curve only is marked by Z-number (Z=98), the others are one by one up to Z=107.

SH nuclei up to $^{274}_{107}Bh$ produced in the reaction $^{232}Th+^{250}Cf$. As can be seen from Figs. 9 and 10 the yield of SH elements in damped reactions depends strongly on beam energy and on nuclear combination which should be chosen carefully. In particular, in Fig. 7 the estimated yield of the primary fragments obtained in the hypothetical reaction $^{248}Cm+^{250}Cf$ (both constituents are radioactive) is shown, demonstrating a possibility for production of SH neutron-rich nuclei up to the element 112 (complementary to lead fragments in this reaction).

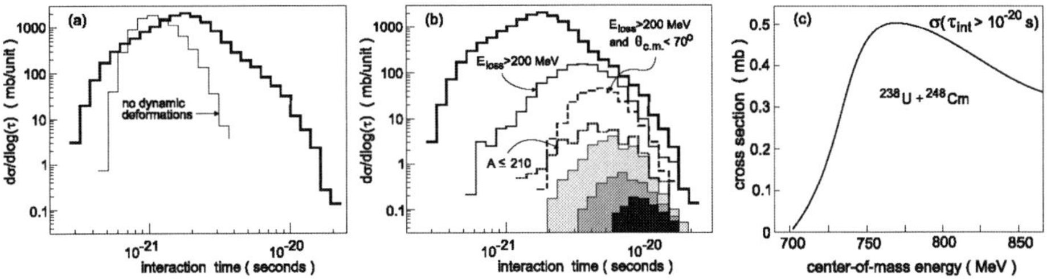

FIGURE 11. Reaction time distributions for the ^{238}U+^{248}Cm collision at 800 MeV center-of-mass energy. Thick solid histograms correspond to all events with energy loss more than 30 MeV. (a) Thin solid histogram shows the effect of switching-off dynamic deformations. (b) Thin solid, dashed and dotted histograms show reaction time distributions in the channels with formation of primary fragments with $E_{loss} > 200$ MeV, $E_{loss} > 200$ MeV and $\theta_{c.m.} < 70^o$ and $A \leq 210$, correspondingly. Hatched areas show time distributions of events with formation of the primary fragments with $A \leq 220$ (light gray), $A \leq 210$ (gray), $A \leq 204$ (dark) having $E_{loss} > 200$ MeV and $\theta_{c.m.} < 70^o$. (c) Cross section for events with interaction time longer than 10^{-20} s depending on beam energy.

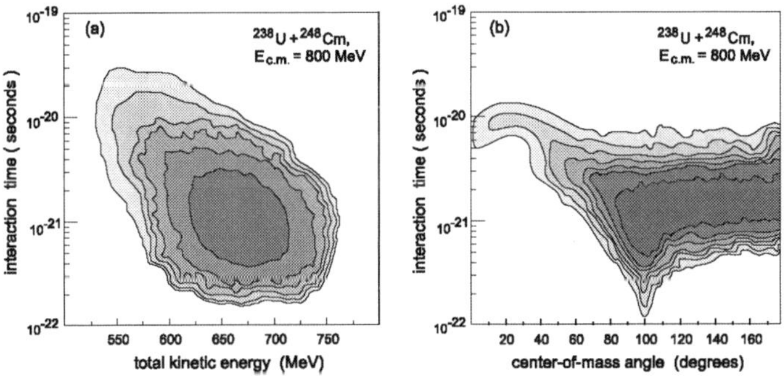

FIGURE 12. Energy-time (a) and angular-time (b) distributions of primary fragments in the ^{238}U $+^{248}$Cm collision at 800 MeV ($E_{loss} > 15$ MeV). The landscape is shown in logarithmic scale - lines are drawn over one order of magnitude. The quasi-elastic peak is removed.

Giant quasi-atoms and spontaneous positron formation

The time analysis of the reactions studied shows that in spite of non-existing attractive potential pocket the system consisting of two very heavy nuclei may hold in contact rather long in some cases. During this time it moves over multidimensional potential energy surface with almost zero kinetic energy (result of large nuclear viscosity), a typical trajectory is shown in Fig. 6. The total reaction time distribution, $\frac{d\sigma}{dlog(\tau)}$ (τ denotes the time after the contact of two nuclei), is shown in Figs. 11 and 12 for the ^{238}U+^{248}Cm collision. We found that the dynamic deformations are mainly responsible here for the time delay of the nucleus-nucleus collision. Ignoring the dynamic deformations in the equations of motion significantly decreases the reaction time, whereas the nucleon transfer influences the time distribution not so strong, see Fig. 11(a).

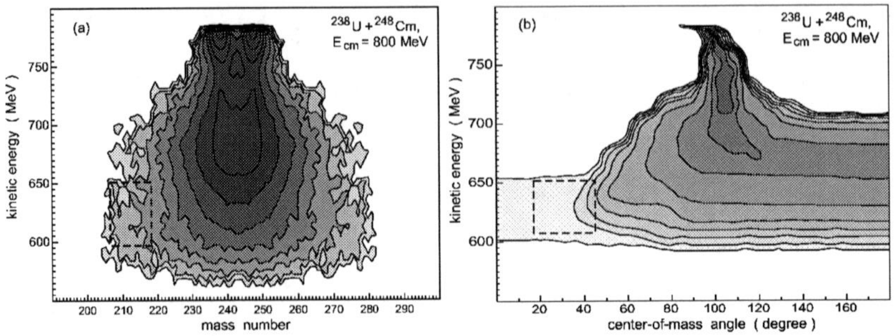

FIGURE 13. Energy-mass (a) and energy-angular (b) distributions of primary fragments in the $^{238}U+^{248}Cm$ collision at 800 MeV ($E_{loss} > 15$ MeV). Logarithmic landscape of the corresponding double differential cross sections. Lines are drawn over each half order of magnitude. The dashed rectangles show the regions of the longest events.

As mentioned above, the lifetime of a giant composite system more than 10^{-20} s is quite enough to expect for spontaneous e^+e^- production from the strong electric field as a fundamental QED process ("decay of the vacuun") [9, 10]. The absolute cross section for long events ($\tau > 10^{-20}$ s) was found to be maximal just at the beam energy ensuring two nuclei to be in contact, see Fig. 11(c). Note that the same energy is also optimal for production of the most neutron-rich SH nuclei (Fig. 9). Of course, there are some uncertainties in the used parameters, mostly in the value of nuclear viscosity. However we found only a linear dependence of the reaction time on the strength of nuclear viscosity, which means that the obtained reaction time distribution is rather reliable, see logarithmic scale on both axes in Fig. 11(a).

Formation of the background positrons in these reactions forces one to find some additional trigger for the longest events. Such long events correspond to the most damped collisions with formation of mostly excited primary fragments decaying by fission. However there is also a chance for production of the primary fragments in the region of doubly magic nucleus ^{208}Pb, which could survive against fission due to nucleon evaporation, see Figs. 13(a) and 7. The number of the longest events depends weakly on impact parameter up to some critical value. On the other hand, in the angular distribution of all the excited primary fragments (strongly peaked at the center-of-mass angle slightly larger than 90^0) there is the rapidly decreasing tail at small angles, see Fig. 13(b). Thus the detection of the surviving nuclei in the lead region at the center-of-mass angles less than 60^0 could be a definite witness for a long reaction time.

The production of long-lived neutron-rich SH nuclei in collisions of transuranium ions seems to be quite possible due to a large mass and charge rearrangement in the "inverse quasi-fission" process caused by the Z=82 and N=126 nuclear shells. Radiochemical identification of $^{267,268}Db$ isotopes, produced in the U+Cm or Th+Cf reactions, could be performed, for example, to test this conclusion. If the found cross section will be higher than 10 pb, then the subsequent experiments with such reactions could be planned aimed to the production of SH nuclei just in the region of the "island of stability". Parallel search for spontaneous positron emission from a supercritical electric field of long-lived quasi-atoms formed in these reactions is also quite promising.

CONCLUDING REMARKS - OUTLOOK

The extension of the periodic system into the sectors hypermatter (strangeness) and antimatter is of general and astrophysical importance. Indeed, microseconds after the big bang the new dimensions of the periodic system, we have touched upon, certainly have been populated in the course of the baryo- and nucleo-genesis. Of course, for the creation of the universe, even higher dimensional extensions (charm, bottom, top) come into play, which we did not pursue here. It is an open question, how the depopulation (the decay) of these sectors influences the distribution of elements of our world today. Our conception of the world will certainly gain a lot through the clarification of these questions.

The experimental verification of the decay of the vacuum is most important. A fundamental process like this must be experimentally clarified and soundly based. I hope that the proposed triggers for long time delay in the formation of giant nuclear systems may help to bring this important fundamental physics forward.

One of our goals has also been to demonstrate that energetic antiproton beams can be used to study new interesting phenomena in nuclear physics. We discuss the possible existence of a completely new kind of strongly interacting systems where both the nucleons and the antinucleons coexist within the same volume and where annihilation is suppressed due to the reduction of the available phase space. Such systems are characterized by large binding energy and high nucleon density. Certainly, antinucleons can be replaced by antihyperons or even by antiquarks. We have presented the first self–consistent calculation of a finite nuclear system containing one antiproton in a deeply bound state. For this study we have used several versions of the RMF model which give excellent description of ordinary nuclei. The presence of an antiproton in a light nucleus like ^{8}Be or ^{16}O changes drastically the whole structure of the nucleus leading to a much more dense and bound state. In heavy systems the presence of a few antinucleons distorts and deforms the nuclear system leading to a localized central region of highly increased density. This is a mechanism for cold compression. We also found that that nuclear systems with total baryon number zero show extremely deep and symmetric states.

It is clear however that these structural changes can occur only if the life time of the antibaryons in the nuclear interior is long enough.

One should bear in mind that originally the RMF model was formulated within the Hartree and no–sea approximations. Implementing the Dirac sea may require serious revision of the model and inclusion of additional terms. Hartree calculations including the Dirac sea and Hartree–Fock calculations including exchange terms lead to smaller nucleon potentials in normal nuclei. Shallower potentials will produce smaller attraction for antinucleons, but the qualitative effect that the presence of antiprotons reduces repulsion and enhances attraction for nucleons will remain valid. We expect that the additional binding and compression of the nucleus will appear even for an antinucleon potential as low as 200 MeV.

In summary, on the basis of the RMF model we have studied the structure of nuclear systems containing a few real antibaryons. We have demonstrated that the antibaryons act as strong attractors for the nucleons leading to enhanced binding and compression of the recipient nucleus. As our estimates show the life times of antibaryons in the nuclear environment could be significantly enhanced due to the reduction of the phase space available for annihilation. Narrow peaks in the pion or kaon spectra at the energy around

1 GeV are proposed as the most clear signature of deeply-bound antibaryon states in nuclei.

For the Gesellschaft für Schwerionenforschung (GSI), which I helped initiating in the sixties, the questions raised here could point to the way ahead. The FAIR project, meanwhile approved by the funding agencies, can – at least to some extent – follow this path.

Indeed, what is needed is a **vision on a long term basis**. The ideas proposed here, the verification of which will need the **commitment for 2–4 decades of research**, could be such a vision with considerable attraction for the best young physicists. The new dimensions of the periodic system made of hyper- and antimatter cannot be examined in the "stand-by" mode at CERN (Geneva); a dedicated facility is necessary also for this field of research, which can in future serve as a home for the universities. The GSI — which has unfortunately become much too self-sufficient — could be such a home for new generations of physicists, who are interested in the **structure of elementary matter**. GSI would then not develop just into a detector laboratory for CERN, and as such become obsolete. I can already see the enthusiasm in the eyes of young scientists, when I unfold these ideas to them — similarly as it was more than 35 years ago, when the nuclear physicists in the state of Hessen initiated the construction of GSI.

REFERENCES

1. W. Greiner, B. Müller, J. Rafelksi, *Quantum electrodynamics of strong fields*, Springer Verlag, 2nd edition, December 1985
2. J. Reinhardt and W. Greiner, *Quantum electrodynamics*, Springer Verlag, 3rd edition, February 2003
3. S. Hofmann and G. Münzenberg, Rev. Mod. Phys. **72**, 733 (2000).
4. Yu.Ts. Oganessian, V.K. Utyonkov, Yu.V. Lobanov, F.Sh. Abdullin, A.N. Polyakov, I.V. Shirokovsky, Yu.S. Tsyganov, G.G. Gulbekian, S.L. Bogomolov, B.N. Gikal, A.N. Mezentsev, S. Iliev, V.G. Subbotin, A.M. Sukhov, A.A. Voinov, G.V. Buklanov, K. Subotic, V.I. Zagrebaev, M.G. Itkis, J.B. Patin, K.J. Moody, J.F. Wild, M.A. Stoyer, N.J. Stoyer, D.A. Shaughnessy, J.M. Kenneally, P.A. Wilk, R.W. Lougheed, R.I. Il'kaev, and S.P. Vesnovskii, Phys. Rev. C **70**, 064609 (2004).
5. M. Schädel, W. Brüchle, H. Gäggeler, J.V. Kratz, K. Sümmerer, G. Wirth, G. Herrmann, R. Stakemann, G. Tittel, N. Trautmann, J.M. Nitschke, E.K. Hulet, R.W. Lougheed, R.L. Hahn, and R.L. Ferguson, Phys. Rev. Lett. **48**, 852 (1982).
6. K.J. Moody, D. Lee, R.B. Welch, K.E. Gregorich, G.T. Seaborg, R.W. Lougheed, and E.K. Hulet, Phys. Rev. C **33**, 1315 (1986).
7. C. Riedel, W. Nörenberg, Z. Phys. A **290**, 385 (1979).
8. V. Zagrebaev and W. Greiner, J. Phys. G **31**, 825 (2005).
9. J. Reinhard, U. Müller and W. Greiner, Z. Phys. A **303**, 173 (1981).
10. W. Greiner (Editor), *Quantum Electrodynamics of Strong Fields*, (Plenum Press, New York and London, 1983); W. Greiner, B. Müller and J. Rafelski, *QED of Strong Fields* (Springer, Berlin and New York, 2nd edition, 1985)
11. V.I. Zagrebaev, Phys. Rev. C **64**, 034606 (2001); in *Tours Symposium on Nucl. Phys. V*, AIP Conf. Proc., 2004, **704**, p. 31., see also invited talk at the IX International Conference on Nucleus Nucleus Collisions, Rio de Janeiro, Brazil, August 28 - September 1, 2006.
12. U. Mosel, J. Maruhn, and W. Greiner, Phys. Lett. B **34**, 587 (1971); J. Maruhn and W. Greiner, Z. Phys. **251**, 431 (1972).
13. F.G. Werner and J.A. Wheeler, unpublished; K.T.R. Davies, A.J. Sierk, and J.R. Nix, Phys. Rev. C **13**, 2385 (1976).
14. M. Schädel, J.V. Kratz, H. Ahrens, W.Brüchle, G. Franz, H. Gäggeler, I. Warnecke, G. Wirth, G. Herrmann, N. Trautmann, and M. Weis, Phys. Rev. Lett. **41**, 469 (1978).

Evidence for a New Light Boson from Cosmological Gamma-Ray Propagation?

Marco Roncadelli*, Alessandro De Angelis†,** and Oriana Mansutti‡

*INFN, Sezione di Pavia, via A. Bassi 6, I – 27100 Pavia, Italy
†Dipartimento di Fisica, Università di Udine, Via delle Scienze 208, I – 33100 Udine, and INAF and INFN, Sezioni di Trieste, Italy
**also at IST, Lisboa, Portugal
‡Dipartimento di Fisica, Università di Udine, Via delle Scienze 208, I – 33100 Udine, and INFN, Sezione di Trieste, Italy

Abstract. An anomalously large transparency of the Universe to gamma rays has recently been discovered by the Imaging Atmospheric Cherenkov Telescopes (IACTs) H.E.S.S. and MAGIC. We show that observations can be reconciled with standard blazar emission models provided photon oscillations to a very light Axion-Like Particle occur in extragalactic magnetic fields. A quantitative estimate of this effect is successfully applied to blazar 3C279. Our predictions can be tested in the near future by new observations performed both with IACTs and with the ARGO-YBJ air shower detector. Our result also offers an important observational test for models of dark energy wherein quintessence is coupled to the photon through an effective dimension-five operator.

Keywords: cosmic rays, gamma rays, axion, photon propagation
PACS: 14.80.Mz, 95.30.-k, 95.85.Pw, 95.85.Ry, 98.70Rz, 98.70Vc, 98.70.Sa

INTRODUCTION

An impressive amount of informations about the Universe in the very-high-energy (VHE) band has been collected over the last few years by the Imaging Atmospheric Cherenkov Telescopes (IACTs) H.E.S.S., MAGIC, CANGAROO III and VERITAS. Observations carried out by these IACTs concern gamma-ray sources over an extremely wide interval of distances, ranging from the parsec scale for Galactic objects up to the Gigaparsec scale for the fartest detected blazar 3C279. This circumstance allows not only to infer the intrinsec properties of the sources, but also to probe the nature of photon propagation throughout cosmic distances.

The latter isssue becomes particularly important in the energy range above 100 GeV, where the horizon of the observable Universe rapidly shrinks as the energy further increases. This comes about because photons from distant sources scatter off background photons permeating the Universe, thereby disappearing into electron-positron pairs [1]. The corresponding cross section $\sigma(\gamma\gamma \to e^+e^-)$ turns out to peak when the VHE photon energy E and the background photon energy ε are related by $\varepsilon \simeq (500\,\mathrm{GeV}/E)\,\mathrm{eV}$, so that the resulting cosmic opacity is dominated by the interaction with optical/infrared photons of the diffuse extragalactic background – usually called *extragalactic background light* (EBL) – for observations performed by IACTs.

Owing to the absorption process in question, photon propagation is controlled by the optical depth $\tau(E,D)$, with D denoting the source distance. Therefore, the observed

CP1018, *Frontiers of Fundamental and Computational Physics*,
edited by B. G. Sidharth, F. Honsell, O. Mansutti, K. Sreenivasan, and A. De Angelis

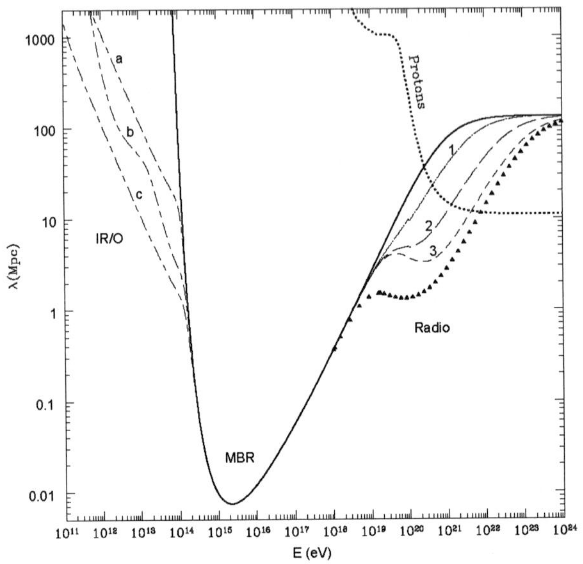

FIGURE 1. Pair-production photon mean free path as a function of beam photon energy (from ref. [3]).

photon spectrum $\Phi_{\rm obs}(E,D)$ is related to the emitted one $\Phi_{\rm em}(E)$ by

$$\Phi_{\rm obs}(E,D) = e^{-\tau(E,D)}\,\Phi_{\rm em}(E)\ . \tag{1}$$

Unlike the CMB, the EBL is produced by galaxies during the whole age of the Universe and possibly by a first generation of stars formed before galaxies were assembled. Based on stellar evolution models in galaxies as well as on deep galaxy counts, several groups have attempted a determination of the spectral energy distribution of the EBL and ultimaly of the optical depth $\tau(E,D)$ for 100 GeV $< E <$ 100 TeV [2]. Because galaxies were brigther in the past than they are now, evolutionary effects should in principle be included in the evaluation of $\tau(E,D)$ but they become unimportant at sufficiently low redshift and will therefore be neglected throughout. Correspondingly we have $\tau(E,D) \simeq D/\lambda_\gamma(E)$, with $\lambda_\gamma(E)$ denoting the photon mean free path for $\gamma\gamma \to e^+e^-$. As a consequence, eq. (1) becomes

$$\Phi_{\rm obs}(E,D) \simeq e^{-D/\lambda_\gamma(E)}\,\Phi_{\rm em}(E)\ . \tag{2}$$

The function $\lambda_\gamma(E)$ has been computed within realistic models of the EBL and is reported in Fig. 1 (from ref. [3]).

We see that $\lambda_\gamma(E)$ decreases roughly like a power law from the Hubble radius 4.2 Gpc slightly below 100 GeV to about 2 Mpc at 100 TeV. Thus, eq. (2) entails that the observed flux is *exponentially* suppressed both at high energy and at large distances, so that sufficiently far-away sources become *hardly visible* in the VHE range. Moreover, the observed spectrum of distant sources gets *much steeper* than the emitted one.

Yet, observations have *not* detected the behaviour predicted by eq. (2). A first indication in this direction was reported by the H.E.S.S. collaboration in connection with the discovery of the two blazars H2356-309 ($z = 0.165$) and 1ES1101-232 ($z = 0.186$) at $E \sim 1\,\text{TeV}$ [4]. Stronger evidence comes from the observation of the blazar 3C279 ($z = 0.538$) at $E \sim 0.5\,\text{TeV}$ by the MAGIC collaboration [5]. In particular, the signal from 3C279 collected by MAGIC in the region $E < 220$ GeV has more or less the same statistical significance as the one in the range 220 GeV $< E <$ 600 GeV (6.1σ in the former case, 5.1σ in the latter) [5].

Turning the argument around and assuming *standard* photon propagation, the observed spectrum can only be reproduced by an emission spectrum *much harder* than for any other observed blazar. Such a hard spectrum is also largely inconsistent with the predictions of current blazar models [6]. So, we are led to believe that the effect under consideration ought to be due to an *anomalous* photon propagation over cosmological distances rather than to new phenomena occurring inside the blazars themselves. From this point of view, the Universe appears to be *more transparent* to VHE gamma rays than previously thought, up to the point that a conflict with the standard scenario can be envisaged.

A cautionary remark is in order. Available observational informations are insufficient to fully settle the issue and in fact it has been pointed out that an anomalously hard emission spectrum can be produced within a highly unconventional blazar model [7]. However, according to this option it looks difficult to explain why the physical mechanisms occurring in the most distant blazars are so different from those described by standard blazar models.

A way out of the considered difficulty has recently been proposed by the present authors and rests upon an oscillation mechanism occurring in extragalactic magnetic fields, whereby a photon can become a new very light spin-zero particle, named *Axion-Like Particle* (ALP) [8]. Once produced, the ALP travels unimpeded throughout the Universe and can convert back to a photon before reaching the Earth, thereby acting as if the observed photons had travelled a distance *largely in excess* of their mean free path. Thanks to the exponential dependence of the observed flux on $\lambda_\gamma(E)$, even a small increase in $\lambda_\gamma(E)$ gives rise to a large effect. More specifically, this mechanism yields an observed spectrum *much harder* than the one predicted by eq. (2) for distant blazars, thereby leading to agreement with observations for *standard* emission spectra.

Our aim is to review the main features of our proposal as well as its application to blazar 3C279.

PHOTON-ALP OSCILLATIONS

Motivations for ALPs

In spite of the enormous success scored by the Standard Model of strong, weak and electromagnetic interactions in describing physics at energies up to the Fermi scale $G_F^{-1/2} \simeq 250\,\text{GeV}$, nobody would seriously regard it as the final theory. Instead, on the basis of phenomenological as well as conceptual reasons, the Standard Model is currently viewed as the low-energy manifestation of some more fundamental and richer theory of all elementary-particle interactions including gravity. Therefore, the lagrangian of the Standard Model is expected to be modified by small non-renormalizable terms describing interactions among known and new particles.

Attempts to go beyond the Standard Model are the central research topic in high-energy physics since almost three decades and each specific proposal is characterized by a set of new particles along with their mass spectrum and interactions with the standard world. ALPs turn out to be a generic prediction of many extensions of the Standard Model and have attracted considerable interest over the last few years. Besides than in four-dimensional models [9], they naturally arise in the context of compactified Kaluza-Klein theories [10] as well as in superstring theories [11]. Moreover, it has been argued that ALPs with mass $m \sim 10^{-33}$ eV are a good candidate for quintessential dark energy [12] which presumably triggers the present accelerated expansion of the Universe.

Specifically, ALPs are spin-zero light bosons defined by the following low-energy effective lagrangian

$$\mathscr{L}_{\mathrm{ALP}} = \frac{1}{2}\partial^{\mu} a \partial_{\mu} a - \frac{1}{2} m^2 a^2 - \frac{1}{4M} F^{\mu\nu} \tilde{F}_{\mu\nu} a \,, \tag{3}$$

where $F^{\mu\nu}$ is the electromagnetic field strength, $\tilde{F}_{\mu\nu}$ is its dual and a denotes the ALP field. According to the above view, it is assumed that the inverse two-photon coupling M is much larger than $G_F^{-1/2} \simeq 250\,\mathrm{GeV}$. On the other hand, the ALP mass m is supposed to be much smaller than $G_F^{-1/2} \simeq 250\,\mathrm{GeV}$ and for definiteness we take $m < 1\,\mathrm{eV}$. As far as generic ALPs are concerned, the parameters M and m are regarded as *independent*. The situation is at variance with the case of the standard *Axion* [13], for which the relationship

$$m = 0.7k \cdot \left(\frac{10^{10}\,\mathrm{GeV}}{M}\right)\,\mathrm{eV} \tag{4}$$

holds, with k a model-dependent constant of order one [14].

General considerations

A characteristic feature of ALPs is the trilinear vertex γ-γ-a described by the last term in $\mathscr{L}_{\mathrm{ALP}}$, whereby one ALP couples to two photons. This interaction gives rise to photon-ALP conversion, which leads in turn to a very interesting variety of physical processes, including the one to which the present paper is devoted.

One of these processes consists in ALP photo-production through the *Primakoff process*, which takes place when an incoming photon scatters on a charged particle thereby becoming an ALP upon the exchange of a virtual photon. Hot, dense plasmas in stellar cores are ideal environments where the Primakoff process involving thermal photons can occur. Once produced, the ALPs escape because their mean free path is much larger than the stellar radius, thereby carrying off energy. Owing to the virial equilibrium, the core has a negative specific heat. Therefore it reacts to such an energy loss by getting hotter. As a result, the rate of nuclear reactions sharply increases, bringing about a substantial change in the observed properties of stars. Still, current models of stellar evolution are in fairly good agreement with observations. Hence, M has to be large enough to provide a sufficient suppression of unwanted ALP effects. This argument has been applied in a quantitative fashion to the Sun, to main-sequence stars and to red-giants stars, with the result [15]

$$M > 10^{10}\,\mathrm{GeV}\,. \tag{5}$$

Remarkably enough, for $m < 0.02\,\mathrm{eV}$ almost the same conclusion, namely

$$M > 0.86 \cdot 10^{10}\,\mathrm{GeV}\,, \tag{6}$$

has been reached by the negative result of the CAST experiment designed to detect ALPs emitted by the Sun [16].

Another process implied by the γ-γ-a vertex is *direct* photon-ALP conversion without the need of charged particles playing the role of catalysts, provided an *external* magnetic (electric) field is present. Whenever the external field extends over a large region and the momentum transfer is small, photon-ALP conversion becomes coherent and can be regarded as an *oscillation* phemomenon. Indeed, the γ-γ-a vertex produces an *off-diagonal* element in the mass matrix for the photon-ALP system in the presence of an external field. Therefore, the interaction eigenstate *differ* from the propagation eigenstates and photon-ALP oscillations show up [17]. The situation is analogous to what happens in the case of massive neutrinos with different flavours, apart from an important difference. All neutrinos have equal spin, and so neutrino oscillations can freely occur. Instead, ALPs are supposed to have spin zero whereas the photon has spin one, and so one of them can transform into the other only if the spin mismatch is compensated for by an external field.

One consequence of photon-ALP oscillations is that a lower bound on M stronger than condition (5) can be derived for ALPs with mass $m < 10^{-10}\,\mathrm{eV}$. In this connections, two methods have been put forward. One concerns the energetics of the supernova 1987a. Because the emitted neutrinos have been observed and the whole energy budget is rather well known, an upper bound on the ALP flux can be derived [18]. Alternatively, observation of time-lag between opposite-polarization modes in pulsar radio emission similarly yields an upper bound on the two-photon coupling of an ALP [19]. In either case, one gets

$$M > 3 \cdot 10^{11}\,\mathrm{GeV}\,. \tag{7}$$

Application to VHE gamma-ray observations

Our proposal now starts to become clear. We imagine that photons emitted by a distant blazar can oscillate to ALPs in the presence of cosmic magnetic fields. So, the emitted flux gets reduced along the line-of-sight because some photons become ALPs. If this were the whole story, the observable prediction would merely be a *dimming* [20]. Things can be totally *different* when photon absorption becomes important. To see how this comes about, suppose that a sizeable fraction of the emitted photons convert to ALPs close enough to the source and that a nonnegligible fraction of the ALPs in queston are in turn converted back to photons close enough to the Earth. Because ALPs propagate unimpeded, in such a situation the observed photon flux can be considerably *larger* than the one predicted by eq. (2). Below, we will turn this qualitative picture into a quantitative estimate.

In principle, we have to evaluate the probability $P_{\gamma\to\gamma}(D)$ that a photon remains a photon after propagation from the source to us when allowance is made for photon-ALP oscillation as well as for photon absorption.

However, an exact treatment would be exceedingly difficult, mainly because of the uncertainty concerning the configuration of the magnetic field responsible for photon-ALP oscillations. Actually, the line-of-sight to a distant blazar is expected to cross magnetic fields on different scales. A magnetic field is expected to be present in the source itself. In addition, the Galactic magnetic field can give a nontrivial contribution to the effect under consideration. Finally, extragalactic magnetic fields can play an important role. Throughout, we restrict our attention to extragalactic magnetic fields, whose existence has been demonstrated very recently by AUGER observations [21]. A complementary picture involving only magnetic fields in the source and in the Milky Way has been considered in ref. [22].

Unfortunately, almost nothing is known about the morphology of extragalactic magnetic fields, which reflects both their cosmic origin and the evolutionary history of baryonic matter. While it is evident that their coherence length cannot be arbitrarily large, no reliable estimate of its value is presently available. As far as our analysis is concerned, this means that we cannot suppose that extragalactic magnetic fields are constants over the whole distance to the source – yet, their spatial dependence is unknown. The usual way out of this difficulty amounts to suppose that extragalactic magnetic fields **B** have a domain-like structure [23]. That is, **B** is supposed to be constant over a domain of size L_{dom} equal to its coherence length, with **B** randomly changing its direction from one domain to another but keeping approximately the same strength. Reference values to be used throughout are $B \simeq 10^{-9}\,\text{G}$ and $L_{\text{dom}} \simeq 1\,\text{Mpc}$, which are close to existing upper limits but consistent with them [24]. Remarkably enough, such a picture of cosmic magnetic fields turns out to be consistent with AUGER observations [25].

Thus, the whole propagation process of the photon beam can be recovered by iterating the propagation over a single domain as many times as the number of domains crossed by the beam, taking each time a random value for the angle θ between **B** and a fixed fiducial direction. In this way, we are effectively led to the much easier problem of photon-ALP oscillation in a *constant* magnetic field.

Another simplification is brought about by the fact that photon absorption is insensitive to the specific propeties of the photon-ALP oscillation – it only depends on the photon mean free path λ_γ. So, we can first identify the regime in which $P_{\gamma\to\gamma}$ is maximal over a single magnetic domain and next we can work out how much this probability is suppressed by photon absorption.

Consider in the first place the propagation of a photon beam over one domain without photon absorption. Since now **B** is constant, the probability that a photon converts to an

ALP after a distance y can be computed exactly and reads [17]

$$P^{(0)}_{\gamma\to a}(y) = \sin^2 2\alpha \, \sin^2\left(\frac{\Delta_{\rm osc} y}{2}\right) , \qquad (8)$$

where the photon-ALP mixing angle α is

$$\alpha = \frac{1}{2}\arcsin\left(\frac{B}{M\Delta_{\rm osc}}\right) \qquad (9)$$

and the oscillation wave number reads

$$\Delta_{\rm osc} = \left[\left(\frac{m^2-\omega_{\rm pl}^2}{2E}\right)^2 + \left(\frac{B}{M}\right)^2\right]^{1/2} , \qquad (10)$$

so that the oscillation length is $L_{\rm osc} = 2\pi/\Delta_{\rm osc}$. Actually, eq. (10) pertains to the situation in which the beam propagates in a cold intergalactic plasma with plasma frequency

$$\omega_{\rm pl} = \left(\frac{4\pi\alpha n_e}{m_e}\right)^{1/2} \simeq 3.69 \cdot 10^{-11}\left(\frac{n_e}{\rm cm^{-3}}\right) {\rm eV} , \qquad (11)$$

where n_e denotes the electron density. Because we are dealing with weak magnetic fields, their contribution to the vacuum refractive index is totally negligible.

A look back at eqs. (8), (9) and (10) shows that the photon-ALP transition probability is maximal in the strong-mixing regime, namely for $E \gg |m^2 - \omega_{\rm pl}^2|M/2B$. A conservative estimate of the density of intergalactic plasma yields $n_e \simeq 10^{-7}\,{\rm cm}^{-3}$ [26], resulting in the plasma frequency $\omega_{\rm pl} \simeq 1.17 \;\; 10^{-14}\,{\rm eV}$. Therefore, the strong-mixing condition takes the explicit form $|(m/10^{-10}\,{\rm eV})^2 - 1.37 \cdot 10^{-8}| \ll 0.38(E/{\rm GeV})(B/10^{-9}\,{\rm G})$ $(10^{10}\,{\rm GeV}/M)$. Since we are interested in the energy range $E > 10^2\,{\rm GeV}$, we find that for e.g. $M > 4 \cdot 10^{11}\,{\rm GeV}$ the bound $m \ll 10^{-10}\,{\rm eV}$ has to be satisfied. Recalling eq. (4), we see that the standard axion is excluded. We also remark that the present mechanism works for arbitrarily small values of m, provided M happens to be considerably smaller than the Planck mass $M_P \simeq 1.22 \cdot 10^{19}\,{\rm GeV}$. As a consequence, our result also applies to models of dark energy wherein quintessence enjoys a photon coupling described by $\mathscr{L}_{\rm ALP}$ [12], thus ultimately providing an important observational test for these models.

ENERGY SPECTRUM FOR BLAZAR 3C279

It is straightforward to solve the beam propagation equation once photon absorption by the EBL is taken into account and produces a finite photon mean free path λ_γ [23]. In the strong-mixing regime, the probability for a photon to become an ALP after a distance $y \leq L_{\rm dom}$ becomes

$$P_{\gamma\to a}(y) \simeq \frac{1}{2} e^{-y/(2\lambda_\gamma)} \sin^2\left(\frac{y\delta}{2\lambda_\gamma}\right) , \qquad (12)$$

whereas the probability that a photon remains a photon now reads

$$P_{\gamma\to\gamma}(y) \simeq \frac{1}{2} e^{-y/\lambda_\gamma} \left[1+\cos^2\left(\frac{y\delta}{2\lambda_\gamma}\right)\right] , \tag{13}$$

where we have introduced the dimensionless parameter

$$\delta \equiv \frac{B\lambda_\gamma}{M} \simeq 0.11 \left(\frac{B}{10^{-9}\,\mathrm{G}}\right)\left(\frac{10^{11}\,\mathrm{GeV}}{M}\right)\left(\frac{\lambda_\gamma}{\mathrm{Mpc}}\right) . \tag{14}$$

The relation between the photon energy E and the source redshift z yields for 3C279 $\lambda_\gamma \simeq 450$ Mpc at $E = 500\,\mathrm{GeV}$ [27], so that we have $\delta \simeq 12.4$ in this case study.

Over distances $y \gg L_{\mathrm{dom}}$, the transition probabilities $P_{\gamma\to a}(y)$ and $P_{\gamma\to\gamma}(y)$ arise as the incoherent average of Eqs. (12) and (13) over $N \simeq (y/L_{\mathrm{dom}})$ domains crossed by the beam, respectively. Assuming (as before) that the beam propagates along the y direction and choosing the x and z directions arbitrarily in the orthogonal plane, the problem becomes truly three-dimensional, because of the random orientation of the magnetic field. Consequently, the beam state is described by the vector (γ_x, γ_z, a).

We have derived the propagation equations describing the absorption of photons due to the interaction with the EBL and their oscillations to ALPs (and vice-versa). As in Ref. [23], we are led to the transfer equation

$$\begin{pmatrix} \gamma_x \\ \gamma_z \\ a \end{pmatrix} = \mathrm{e}^{iEy}\left[T_0\,\mathrm{e}^{\lambda_0 y} + T_1\,\mathrm{e}^{\lambda_1 y} + T_2\,\mathrm{e}^{\lambda_2 y}\right]\begin{pmatrix} \gamma_x \\ \gamma_z \\ a \end{pmatrix}_0 \tag{15}$$

where

$$\lambda_0 \equiv -\frac{1}{2\lambda_\gamma} , \tag{16}$$

$$\lambda_1 \equiv -\frac{1}{4\lambda_\gamma}\left[1+\sqrt{1-4\delta^2}\right] , \tag{17}$$

$$\lambda_2 \equiv -\frac{1}{4\lambda_\gamma}\left[1-\sqrt{1-4\delta^2}\right] , \tag{18}$$

$$T_0 \equiv \begin{pmatrix} \sin^2\theta & -\cos\theta\sin\theta & 0 \\ -\cos\theta\sin\theta & \cos^2\theta & 0 \\ 0 & 0 & 0 \end{pmatrix} , \tag{19}$$

$$T_1 \equiv \begin{pmatrix} \frac{1+\sqrt{1-4\delta^2}}{2\sqrt{1-4\delta^2}}\cos^2\theta & \frac{1+\sqrt{1-4\delta^2}}{2\sqrt{1-4\delta^2}}\cos\theta\sin\theta & -\frac{\delta}{\sqrt{1-4\delta^2}}\cos\theta \\ \frac{1+\sqrt{1-4\delta^2}}{2\sqrt{1-4\delta^2}}\cos\theta\sin\theta & \frac{1+\sqrt{1-4\delta^2}}{2\sqrt{1-4\delta^2}}\sin^2\theta & -\frac{\delta}{\sqrt{1-4\delta^2}}\sin\theta \\ \frac{\delta}{\sqrt{1-4\delta^2}}\cos\theta & \frac{\delta}{\sqrt{1-4\delta^2}}\sin\theta & -\frac{1-\sqrt{1-4\delta^2}}{2\sqrt{1-4\delta^2}} \end{pmatrix} , \tag{20}$$

$$T_2 \equiv \begin{pmatrix} -\frac{1-\sqrt{1-4\delta^2}}{2\sqrt{1-4\delta^2}}\cos^2\theta & -\frac{1-\sqrt{1-4\delta^2}}{2\sqrt{1-4\delta^2}}\cos\theta\sin\theta & \frac{\delta}{\sqrt{1-4\delta^2}}\cos\theta \\ -\frac{1-\sqrt{1-4\delta^2}}{2\sqrt{1-4\delta^2}}\cos\theta\sin\theta & -\frac{1-\sqrt{1-4\delta^2}}{2\sqrt{1-4\delta^2}}\sin^2\theta & \frac{\delta}{\sqrt{1-4\delta^2}}\sin\theta \\ -\frac{\delta}{\sqrt{1-4\delta^2}}\cos\theta & -\frac{\delta}{\sqrt{1-4\delta^2}}\sin\theta & \frac{1+\sqrt{1-4\delta^2}}{2\sqrt{1-4\delta^2}} \end{pmatrix}. \quad (21)$$

Here θ denotes the angle between the x axis and the extragalactic **B** in a single domain. Starting with an unpolarized photon beam, we propagate it by iterating eq. (15) as many times as the number of domains crossed by the beam, taking each time a random value for the angle θ (this reflects the random orientation of **B**). We repeat such a procedure 10 000 times and finally we average over all these realizations of the propagation process. The resulting spectrum is exhibited in Fig. 2. We find that about 13% of the photons arrive to the Earth for $E = 500$ GeV, resulting in an enhancement by a factor of about 20 with respect to the flux expected in the absence of the proposed oscillation mechanism; the comparison is made with the "best-fit model" described in Kneiske *et al.* (2004) [2]. The same calculation gives a fraction of 76% for $E = 100$ GeV (to be compared to 67% without the oscillation mechanism) and a fraction of 3.4% for $E = 1$ TeV (to be compared to 0.0045% without the oscillation mechanism). In addition, we have checked the stability of our result against independent variations of **B** and $L_{\rm dom}$ within a factor of 10 about their preferred values. The corresponding spectrum is represented by the gray band in Fig. 2. We remark that the standard deviation of the above averaging procedure lies well inside the gray band. Our predictions can be tested in the near future by new observations performed both with IACTs and with the ARGO-YBJ air shower detector.

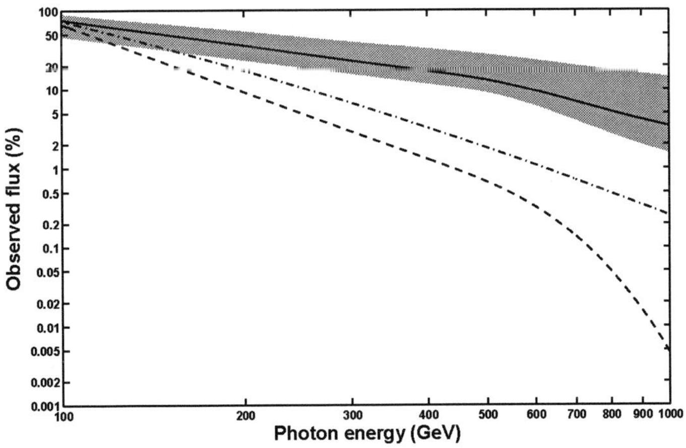

FIGURE 2. The two lowest lines give the fraction of photons surviving from a source at the same distance of 3C279 without the oscillation mechanism, for the "best-fit model" of EBL (dashed line) and for the minimum EBL density compatible with cosmology [2]. The solid line represents the prediction of the oscillation mechanism for $B \simeq 10^{-9}$ G and $L_{\rm dom} \simeq 1$ Mpc within the "best-fit model" of EBL. The gray band is the envelope of the results obtained by independently changing **B** and $L_{\rm dom}$ within a factor of 10 about their preferred values.

ACKNOWLEDGMENTS

We thank Nicola Cabibbo, Tanja Kneiske, Luciano Maiani and Massimo Persic for suggestions and comments.

REFERENCES

1. G. G. Fazio, F. W. Stecker, Nature **226** (1970) 135. F. Aharonian, *Very High Energy Cosmic Gamma Radiation* (World Scientific, Singapore, 2004).
2. J. R. Primack, R. S. Sommerville, J. S. Bullock, J. E. G. Devriendt, AIP Conference Proceedings **558** (2001) 463. T. M. Kneiske, K. Mannheim, D. H. Hartmann, Astron. Astrophys. **386** (2002) 1. T. M. Kneiske, T. Bretz, K. Mannheim, D. H. Hartmann, Astron. Astrophys. **413** (2004) 807. F. Aharonian *et al.*, Astron. Astrophys. **448** (2006) L19. F. W. Stecker, M. A. Malkan, S. T. Scully, Astrophys. J. **648** (2006) 774. D. Mazin, F. Goebel, Astrophys. J. **655 2** (2007) L13.
3. P. Coppi, F. Aharonian, Astrophys. J. **487** (1997) L9.
4. F. Aharonian *et al.* (H.E.S.S. Collaboration), Nature **440** (2006) 1018.
5. M. Teshima *et al.*, on behalf of the MAGIC Collaboration, arXiv:0709.1475v1 (2007) (to appear in the proceedings of the 30th International Cosmic Ray Conference, Merida, Yucatan, Mexico, July 2007). A. De Angelis, on behalf of the MAGIC Collaboration, arXiv:0708.3571 (2007) (to appear in the proceedings of the Second Multiwavelength Workshop for Next Generation Gamma-Ray Experiments, Chicago, Illinois, USA, August 2007).
6. G. Fossati, L. Maraschi, A. Celotti, A. Comastri, G. Ghisellini, Mon. Not. R. Astron. Soc. **299** (1998) 433.
7. F. W. Stecker, S.T. Scully, Astron. Astrophys **478** (2008) L1. F. W. Stecker, M. G. Baring, E. J. Summerlin, Astrophys. J. **667** (2007) L29.
8. A. De Angelis, M. Roncadelli, O. Mansutti, Phys. Rev. D **76** (2007) 121301.
9. E. Masso, R. Toldra, Phys. Rev. D **52** (1995) 1755. E. Masso, R. Toldra, Phys. Rev. D **55** (1997) 7967. C. Coriano, N. Irges, Phys. Lett. B **651** (2007) 298. C. Coriano, N. Irges, S. Morelli, JHEP 07.08 (2007) 008.
10. S. Chang, S. Tazawa, M. Yamaguchi, Phys. Rev. D **61** (2000) 084005. K. R. Dienes, E. Dudas, T. Gherghetta, Phys. Rev. D **62** (2000) 105023.
11. N. Turok, Phys. Rev. Lett. **76** (1996) 1015. P. Svrcek, E. Witten, JHEP 0606 (2006) 051.
12. S. M. Carroll, Phys. Rev. Lett. **81** (1998) 3067.
13. J. H. Kim, Phys. Rep. **150** (1987) 1. H. Y. Cheng, Phys. Rep. **158** (1988) 1.
14. S. L. Cheng, C. Q. Geng, W. T. Ni, Phys. Rev. D **94** (1995) 3132.
15. G. G. Raffelt, Phys. Rep. **198** (1990) 1. G. G. Raffelt, *Stars as Laboratories for Fundamental Physics* (University of Chicago Press, Chicago, 1996). M. Yu. Khlopov, S. G. Rubin, *Cosmological pattern of microphysics in inflationary Universe* (Kluwer Academic Publishers, Dordrecht, 2004).
16. K. Zioutas *et al.*, Phys. Rev. Lett. **94** (2005) 121301. S. Andriamoje *et al.*, JCAP 0704 (2007) 010.
17. P. Sikivie, Phys. Rev. Lett. **51** (1983) 1415; (E) ibid. **52** (1984) 695. L. Maiani, R. Petronzio, E. Zavattini, Phys. Lett. B **175** (1986) 359. G. G. Raffelt, L. Stodolsky, Phys. Rev. D **37** (1988) 1237.
18. J. W. Brockway, E. D. Carlson, G. G. Raffelt, Phys. Lett. B **383** (1996) 439. J. A. Grifols, E. Masso, R. Toldra, Phys. Rev. Lett. **77** (1996) 2372.
19. S. Mohanty, S. N. Nayak, Phys. Rev. Lett. **70** (1993) 4038; (E) ibid. **71** (1993) 1117.
20. A. De Angelis, O. Mansutti, M. Roncadelli, Phys. Lett. B **659** (2008) 847. D. Hooper, P. D. Serpico, Phys. Rev. Lett. **99** (2007) 231102. K. A. Hochmuth, G. Sigl, arXiv:0708.1144 (2007).
21. J. Abraham *et al.* [Pierre Auger Collaboration], Science **318** (2007) 939.
22. M. Simet, D. Hooper, P. D. Serpico, Phys. Rev. D **77** (2008) 063001.
23. C. Csáki, N. Kaloper, M. Peloso, J. Terning, JCAP **05** (2003) 005.
24. P. P. Kronberg, Rept. Prog. Phys. **57** (1994) 325. D. Grasso, H. Rubinstein, Phys. Rep. **348** (2001) 163. S. R. Furlanetto, A. Loeb, Astrophys. J. **556** (2001) 619.
25. A. De Angelis, M. Persic, M. Roncadelli, Mod. Phys. Lett. A **23** (2008) 315.
26. P. J. E. Peebles, *Principles of Physical Cosmology* (Princeton University Press, Princeton, 1993).
27. T. M. Kneiske, T. Bretz, K. Mannheim, D. H. Hartmann, Astron. Astrophys. **413** (2004) 807.

Very high energy photons from distant extragalactic sources observed by MAGIC: AGNs and GRBs

V. Scapin and the MAGIC Collaboration [1]

University of Udine and INFN Trieste, via delle Scienze 208, 33100 Udine, Italy

Abstract. Very High Energy (VHE) photons are a powerful probe of fundamental physics under extreme conditions. During the recent years, the observation of γ–rays throws light on the high energy Universe thanks to the current and future satellite-borne detectors and ground–based telescopes. Ground-based gamma-ray astronomy is part of a new field of fundamental research of Astroparticle Physics and recently made spectacular discoveries, studying sources that, already known in several other energy bands (i.e. gamma-ray, X-ray, radio and/or optical frequencies), are expected to be VHE gamma-rays emitters. In this contribution I present some recent results obtained with the MAGIC telescope, the world's largest single-dish Imaging Atmospheric Cherenkov Telescope (IACT). In particular I will discuss the VHE photons originated in extragalactic gamma-rays emitters: Active Galactic Nuclei (AGN) and Gamma Ray Bursts (GRBs).

INTRODUCTION

The most violent phenomena in the Universe can be seen as cosmic particle accelerators. The gamma rays they emit, not deflected by magnetic fields, can carry also useful information about the production of hadronic cosmic rays. Very High Energy (VHE) γ are a most interesting part of the spectrum of photon emission from non–thermal astrophysical sources[1]. Detection of VHE emission will require clarification of the underlying emission processes: in the case of Gamma–Ray Bursts (GRBS), the detection of VHE–γ's is a strong challenge, because even the possibility of producing VHE–γ's is still debated. Moreover the study of photons that are produced in AGNs and GRBs and travel long distances on their way to the Earth, open the possibility to discuss propagation physics and the transparency of the Universe.

VHE-physics scenario

VHE-γ astrophysics developed only in the last few years, because detecting VHE photons is a complex and difficult experimental problem. The atmosphere shields us from photons of UV, and higher frequencies: in particular photons in the X–ray and soft γ–ray ranges can be directly detected from satellite-borne instruments. VHE–γ's, on the contrary, can be detected only by ground-based detectors placed at high altitudes. With

[1] http://wwwmagic.mppmu.mpg.de/collaboration/

CP1018, *Frontiers of Fundamental and Computational Physics*,
edited by B. G. Sidharth, F. Honsell, O. Mansutti, K. Sreenivasan, and A. De Angelis

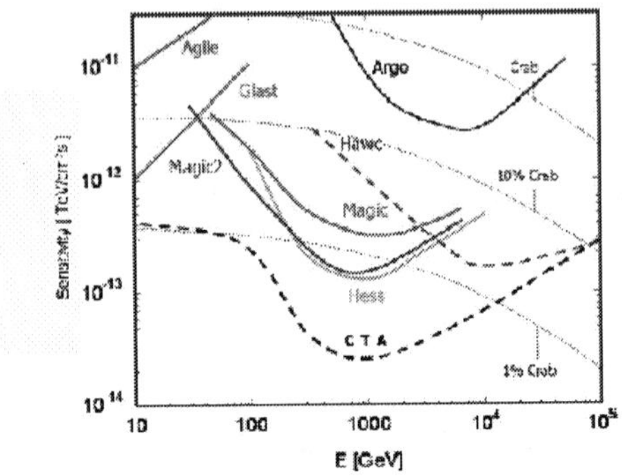

FIGURE 1. Sensitivities of some present and future HE gamma detectors [2], measured as the minimum intensity source detectable at 5 σ. Data for EAS and satellite detector are based on one year of data taking; data for Cherenkov telescopes are based on 50 hours of data.

large effective areas and high sensitivities this class of detectors is sensitive to VHE and UHE photon fluxes (the photon fluxes decrease steeply with the photon's energy). The sensitivities of current and incoming gamma detectors are shown in figure 1. Ground-based VHE Extensive Air Shower (EAS) Arrays such as Milagro [3] and ARGO [4] and the new generation of Cherenkov telescopes like CANGAROO, H.E.S.S. [5], MAGIC and VERITAS [6] detect the secondary particles of the atmospheric showers produced by the primary photons and the cosmic rays of energy higher than the primaries observed by satellites. In the last years, mostly thanks to Cherenkov detectors, 71 VHE sources have been detected: 7 supernova remnants, 18 pulsar wind nebulae, 3 diffuse emission, 4 binary system, 19 AGNs and 21 unidentified sources.

The MAGIC Telescope is a new generation Imaging Atmospheric Cherenkov Telescope (IACT) at La Palma, Canary Islands, Spain (28.3°N, 17.8°W, 2240 m asl), and currently is the world's largest single-dish IACT (with a 17 m diameter). Operative since fall 2003 recently concluded the third year of full scientific activity.

THE MAGIC TELESCOPE

The MAGIC telescope scientific activity is based on the detection of the Cherenkov photons produced in the Earth's atmosphere by electromagnetic showers initiated by VHE–γ–rays.

A VHE γ–ray entering domain impacting the Earth's atmosphere initiates a shower by means of the pair production and the bremsstrahlung mechanisms. This atmospheric shower has a maximum development at around 10–12 km altitude depending upon the primary particle's energy. The charged particles of an electromagnetic shower emit Cherenkov radiation, in the form of a UV–blue light cone of the duration of few nanoseconds and an opening, dominated form the Cherenkov angle, of about one degree. At the ground level, a circle of around 100 m radius is illuminated for the same time interval. The parabolic shaped reflector, resulting in a total mirror area of 236 m^2 tasselleted

FIGURE 2. Left: the MAGIC telescope [7]; right: MAGIC II structure [8].

in 974 Al-mirrors, allows MAGIC to collects part of this Cherenkov light and focus it onto a multi-pixel camera, composed of 576 photomultipliers, with a typical efficiency of about 30%. The total field of view of the camera is 3.5°. The incident light pulses are converted into optical signals and transmitted, via optical fiber, to a 2-level trigger system. The selected events are digitized by fast Flash ADCs, with a data acquisition rate of some hundreds Hertz. The primary particle energy and incoming direction reconstruction is done studying the intensity (and area) of the elliptical image and from its orientation in the camera. In particular, studying the shape of the image is possible to distinguish between photon's and hadron's induced shower and to reject the background due to charged particles atmospheric showers. Due to its Carbon-Fibre-Reinforced-Plastic structure, that provides stiffness and lightness, MAGIC is the only IACT with fast repositionning capabilities, designed especially for the observation of fast transients like GRBs. In its current configuration, when triggered from the GRB Coordinate Network (GCN), the telescope can slew to the source position on average in 42 s. An Active Mirror Control system has been implemented to adjust every mirror in the exact position on the paraboloid under varying gravitational load, camera sagging and other similar effects. In its current configuration MAGIC has an angular resolution on a point-like source of about 0.1°, and a trigger energy threshold of 50 GeV. A second telescope, MAGIC II, similar to the first one, will be inaugurated on September 2008. Stereoscopic observations will allow an increase in the sensitivity by at least a factor 2 and improvements in the energy and direction reconstruction of the primary γs.

POSSIBLE VHE-γ EMISSION MECHANISMS

The origin of photons from astrophysical sources can be traced back to the release of gravitational energy by the collapse of matter towards a central massive object. The rotational dynamics of such a collapse gives rise to an accretion disk and, possibly, emission of plasma in relativistic jets. Even if the low energy part of the spectrum can be well explained as electron synchrotron radiation, the origin of the HE-VHE component is still under debate.

For the classes of sources of our main interest here, Blazars and GRBs, the most popular interpretation is the self-synchrotron Compton (SSC) mechanism. Synchrotron

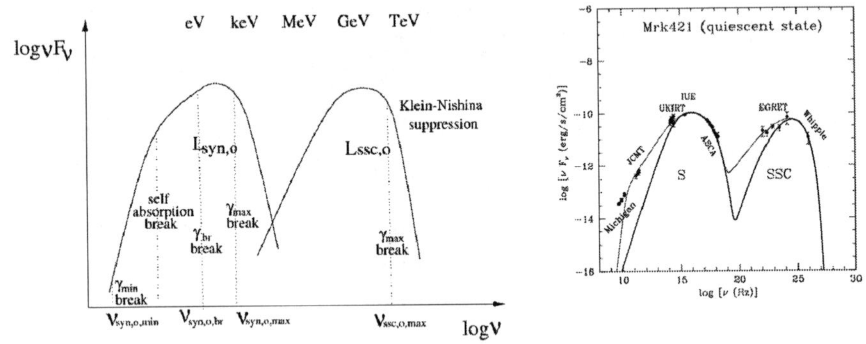

FIGURE 3. Left: Schematic picture of the multi-frequency spectrum of a typical TeV blazar. Here $L_{syn,o}$ and $L_{ssc,o}$ are the observed bolometric luminosities of synchrotron and SSC components, respectively. Corresponding to the break in the relativistic electron energy spectrum in the emission region, a break feature appears in the observed synchrotron spectrum. Around the TeV energy region, the Klein-Nishina effect suppresses the observed flux compared to the Thomson regime. We exclude the information of $\nu_{ssc,o,br}$, because of sparse observational data points and complication from the Klein-Nishina effect [9]. Right: One-zone SSC model spectra for the steady state emission of Mrk 421. The thick solid line shows the spectrum for the model in [9], the thin solid line shows the best fit spectrum of Mrk 421

photons, emitted from ultra-relativistic electrons, are Compton scattered by the parent electron population. The synchrotron and Inverse Compton components peak at, respectively, UV/X-ray and GeV–TeV energies. Two different particle populations are involved in alternative and/or complementary hadronic VHE emission models. A first electron population is accelerated within the jet while a second one is generated by showers initiated by primary protons (or nuclei) previously accelerated in the jet or is a population of extremely energetic photons. In this case the energy of the primary protons is expected to be one or two orders of magnitude larger than the energy of gammas, since the leading mechanism for photon production is the π^0 decay.

AGN AND VHE EMISSIONS

Around 1% of galaxies host an AGN. Central engines of AGNs are supermassive black holes of 10^6–10^9 solar m$\odot$. In $\sim$10% of AGNs, revistivically moving, collimated outflows are emitted. If a relativistic jet is viewed at small angle to its axis the source is called a blazar ($\sim$0.3% of all AGN). The spectral energy distributions (SEDs) of blazars are generally characterized by two broad humps, peaking at, respectively, IR/X-ray and GeV-TeV frequencies.

The ordinary interpretation of blazars VHE emission is due to SSC (Figure 3).

A pure SSC model predicts a definite correlation between the synchrotron and Compton emission during a flare. Evidence of neutrinos in a flare would be a smoking gun for the presence of a hadronic component. Of course, the accurate knowledge of emission mechanisms requires simultaneous broadband γ-ray and X-raydata.

GRB AND VHE EMISSION

Over 40 years after their discovery GRBs are still among the great mysteries of modern astrophysics. The nature of their progenitors and the primary mechanisms driving GRB's explosions are still largely unknown. The (isotropic) luminosity makes them the most luminous events occurring in the Universe since the Big Bang. Their sky distribution is isotropic. GRBs are short duration events from milliseconds to minutes. The prompt emission (peaked at typical energies $\sim$100 keV), is followed by a fading longer-lived "afterglow" (in the X–ray, radio, and optical bands).

The progenitors of long—duration (> 2 s) GRBsare likely hypernovae, the ones of short–durations (<2 s) GRBs are thought to be merging system of two compact objects (black–holes or neutron-stars)(NS). In both types of GRBs the central compact object is a black–hole of 10-100 m$\odot$. The only significant detections so far of GRBs in the γ–ray band were made by EGRET which could detect seven GRBs emitting gamma-rays with energies between 100 MeV and 18 GeV. The spectrum of at least one burst (GRB941017) contained a very hard long-duration component with a differential photon flux spectral index $\alpha = -1$ with no cut-off up to 200MeV.

The detection of VHE emission from the GRB can allow to discriminate among different and competing theoretical models. The mainstream model for GRB emission is an expanding relativistic fireball [11, 12, 13], with the beamed radiation due to internal/external shocks (prompt/afterglow phase, respectively). In the framework of the fireball model, gamma-ray emission in the GeV-TeV range in the prompt and afterglow phase is predicted by several authors. Possible emission mechanisms appear in leptonic and hadronic models. In the first case, inverse-Compton (IC) scattering by electrons in internal or external shocks [15, 16], IC in the afterglow shocks [17, 18] and IC by electrons responsible of optical flashes [19, 20] have been proposed. Even considering pure electron synchrotron radiation, measurable VHE gamma-ray emission for a significant fraction of GRBs is predicted [18]. Hadronic models comprise proton synchrotron emission, photon-pion production and neutron cascades [21, 22]. Recent observations of X-ray flares in the early afterglow phase by the SWIFT [23] satellite [24] suggested the possibility of correlated Gev–TeV emission occurring with a delay of 10^3 s–10^4 s respect to the prompt where strong GeV-TeV flares from IC scattered photons in the forward shock [25]. If the emission processes are synchrotron and IC, then a blazar-like SED is predicted, with a double-peak shape extending into the VHE band.

PROPAGATION OF γ-RAYS

AGN and GRB are high–z sources, so their photons reach us after travelling long distances. Electron-positron pair production in the interaction of VHE photons with extragalactic IR background photons causes opacity of the Universe to gamma rays: the photon mean free path (figure 4) is smaller than the source distance (for photons of 70 GeV the cosmological γ–ray horizon is $z \sim 1$). The relevant background component for VHE-γ-rays is the optical and infrared background radiation: the Extragalactic Background Light (EBL). The attenuation of the measured spectra in the TeV region can thus be used to derive constraints on the EBL density [27]. First limits on the EBL were ob-

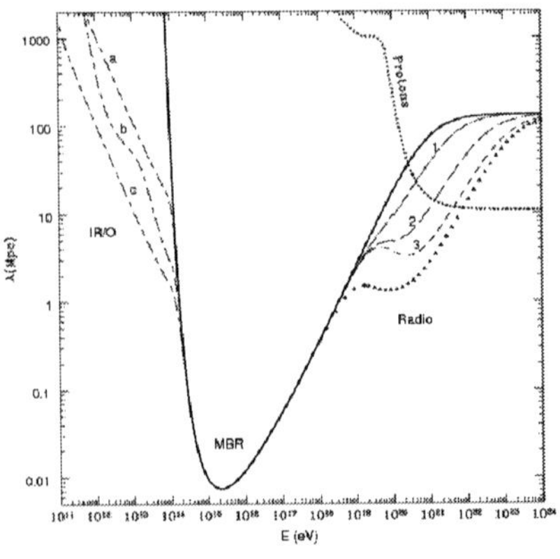

FIGURE 4. Mean free path as a function of the photon energy [26].

tained in [28], while recent determinations from the detection of distant VHE sources are reported in [29, 30].

Present data on the absorption seem not clearly compatible in the pure QED and other kinds of interaction can be speculated. Photons can interact with (possibly quintessential) very light axion-like particles, which might change the absorption length [31, 32]. In particular, in the DARMA model [32], such effect might enhance the photon flux via a regeneration mechanism. Such an interaction would be mediated by the intergalactic magnetic fields. Mechanisms in which the absorption is changed through violation of the Lorentz invariance are also to be proved: such models are particularly fascinating within quantum gravity (QG) scenarios [33].

The AGNs detected at VHE (see Table 1) are almost uniformly distributed in the sky and they are very far from us. The spectral index measured is plotted in figure 5 and - withih the small redshift range probed - show a substantial lack of correlation between measured TeV slopes and redshift. Also the observed spectrumof 3C 279 must be not so steep: MAGIC detected It will be important to increase the statistic in order to increase our knowledge. The trasparency of the Universe at γ–rays seems to be larger that the one expected from attenuation by EBL. This might also indicate the existence of new propagation physics.

As for GRBs the mean redshift of Swift's GRB is $z = 2.8$ so the VHE emission can be totally attenuated by cosmological $\gamma\gamma$ absorption. However, detecting VHE-emission from a GRB is an exiting challenge hence the search goes on. A relatively nearby (z<1) GRB would be the winner!

MAGIC EXTRAGALACTIC RESULTS

In this final part I present some highlights from recent MAGIC observations of extragalactic γ-ray emitters.

TABLE 1. TeV blazar data [10].
Col.1: source name.
Col.2: source redshift.
Col.3: observed 0.2-2 TeV photon spectral index, and associated statistical uncertainty. The corresponding systematic uncertainties are typically ~ 0.1 for H.E.S.S. and ~ 0.2 for MAGIC.
Col.4: observed >0.2 TeV flux (in erg cm^{-2} s^{-1}), and associated statistical uncertainty (from observed spectral normalization only).
Col.5: Cherenkov telescope (CT) or array with which the data in col.3 have been collected: symbols stand for H=H.E.S.S., M=MAGIC, W=Whipple.

Source	z	α_γ	F_γ	CT
(1)	(2)	(3)	(4)	(5)
Mrk 421	0.031	2.33 ± 0.08	1.03($\pm$0.03)E-10	M
Mrk 501	0.034	2.28 ± 0.05	1.71($\pm$0.11)E-11	M
		2.45 ± 0.07	3.84($\pm$1.00)E-12	M
PKS 2344+514	0.044	2.95 ± 0.12	1.21($\pm$0.10)E-11	M
Mrk 180	0.045	3.30 ± 0.70	8.46($\pm$3.38)E-12	M
1ES 1959+650	0.047	2.72 ± 0.14	3.04($\pm$0.35)E-11	M
BL Lacertae	0.069	3.60 ± 0.50	3.28($\pm$0.26)E-12	M
PKS 2005-489	0.071	4.00 ± 0.40	3.32($\pm$0.48)E-12	H
PKS 2155-304	0.116	3.37 ± 0.07	2.89($\pm$0.18)E-11	H
1ES 1426+428	0.129	3.55 ± 0.46	2.53($\pm$0.43)E-11	W
1ES 0229+200	0.139	2.50 ± 0.19	4.46($\pm$0.71)E-12	H
H 2356-309	0.165	3.09 ± 0.24	2.55($\pm$0.68)E-12	H
1ES 1218+304	0.182	3.00 ± 0.40	1.01($\pm$0.26)E-11	M
1ES 1101-232	0.186	2.94 ± 0.20	4.35($\pm$0.69)E-12	H
1ES 0347-121	0.188	3.10 ± 0.23	3.86($\pm$0.73)E-12	H
1ES 1011+496	0.212	4.00 ± 0.50	6.40($\pm$0.32)E-12	M
PG 1553+113	>0.25	4.20 ± 0.30	5.24($\pm$0.87)E-12	M,H
3C 279	0.536			M

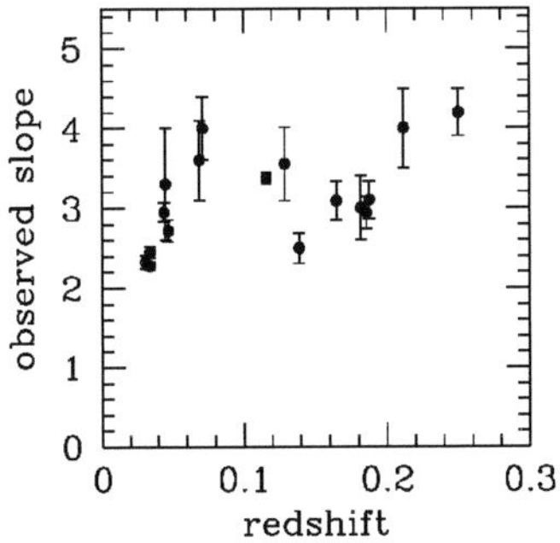

FIGURE 5. Observed spectral indexes for AGN in the VHE region [10].

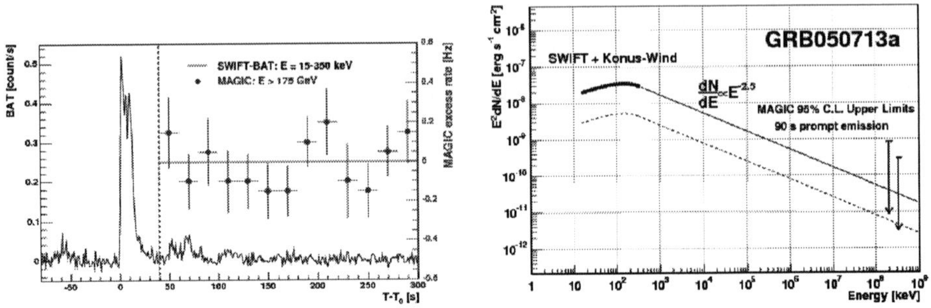

FIGURE 6. Plots of the MAGIC data about the GRB 050713a [35]: left, excess event rate compared with SWIFT-BAT observations; right: upper limits on the spectrum.

Observation of GRBs with MAGIC

MAGIC is the best suited IACT for GRB observations. Thanks to its fast repositioning and low energy threshold, MAGIC can react promptly when triggered by an alert from the GRB Coordinates Network to search for prompt and early afterglow emission from GRBs. In the first two years of operation, after the alert system implementation, MAGIC reacted to 24 Swift BAT alerts (suitable by sky location and zenith angle) and carried out nightly observations that correspond to an 8.3% duty cycle and a mean observation frequency of about one GRB per month [34]. Unfortunately, until now no evidence for VHE signals was found. The threshold energy for the observations was between 80 and 200 GeV, depending on the zenith angle of the burst. In two cases (GRB 050713a and GRB 050904) the observation took place during the the prompt emission phase. Figure 6 shows the case of the observation of GRB 050713a, in which MAGIC was able to start the data taking only 40 s after the onset of the prompt. However, no excess at > 175 GeV was detected, neither during the prompt emission phase nor during the afterglow. The upper limits to the MAGIC flux are compatible with simple extrapolations of the Swift $\Gamma \simeq 1.6$ power-law spectrum to hundreds of GeV [35]. Observation of GRBs in the VHE domain will provide important information on the physical conditions in GRB outflows. Thanks to improvements of MAGIC I and the incoming MAGIC II, it is only a question of time until MAGIC will observe a sufficiently close-by GRB that will allow us to make a strong statement on the underlying VHE emission mechanisms.

Discovery of VHE γ-ray emission form the radio Quasar 3C 279

The quasar 3C279 is one of the best-studied flat–spectrum radio quasars. It is located at a comparatively large redshift, z=0.536: E>200 GeV observations of such distant sources were until recently impossible, due to both the expected steep spectrum (ie. the declining side of the Compton component) and the attenuation of the VHE γ-rays by the EBL. The MAGIC Telescope observed 3C279 in early 2006. MAGIC detected the signif-

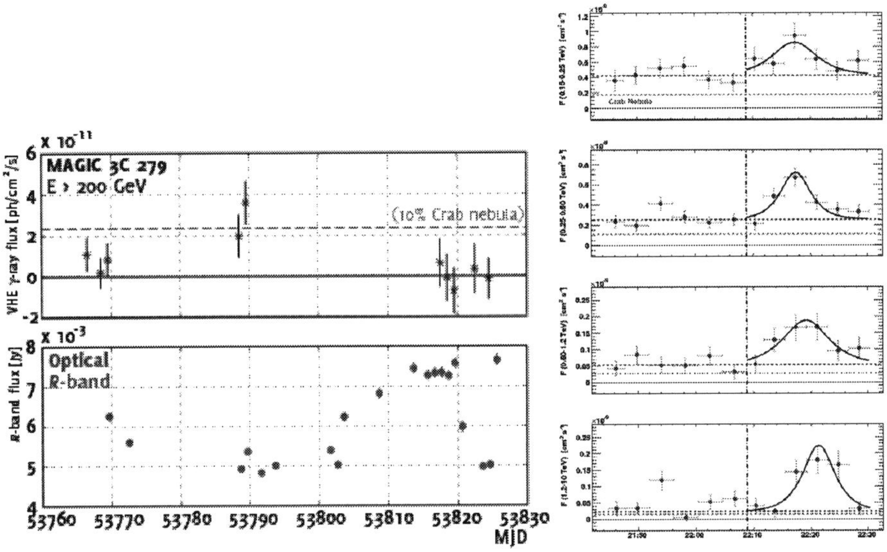

FIGURE 7. Left: preliminary lighcurve [36] of 3C 279 by MAGIC on February 23rd, 2006.; right: Integral flux of Mkn 501 detected by MAGIC in four different energy ranges [37].

icant VHE signature from this source on the observation night of 2006 February 23 [36]. As emphasized above, the VHE detection of 3C 279 is important on account that:

- it has enlarged the cosmologicaldistance over which sources have been detected at VHE frequencies;
- for the first time VHE γ-rays were detected from a flat-spectrum radio quasar; and
- a precise measurement of the source's energy spectrum may give strong constraints on the EBL energy density between z=0 and z=0.536.

Lorentz invariance tests?

It is widely speculated that space-time is a dynamical medium, subject to quantum-gravitational (QG) effects that cause the fabric of space-time to fluctuate on the Planck time and distance. In MAGIC observations the blazar Mkn 501 showed large scale flux variations by an order of magnitude, and short-term flux fluctuations with flux-doubling times down to even two minutes. Such fast changes had never been observed in BL Lacs, and may give rise to revisions of popular acceleration models. An energy–dependent time delay of the VHE photons during bursts was found in a detailed analysis [37]. This gave rise to speculations about a possible QG signature as some models predict an effect of exactly that kind. To really test the Lorentz invariance we need more observation, at different redshifts, of fast transient phenomena. It's clear that GRBs are, in principle, good candidates to test this kind of models.

CONCLUSIONS

Mostly thanks the current generation of Cherenkov telescopes, the number of discovered VHE-γ sources swelled in the recent years. The study of extragalactic sources like AGNs and GRBs gives us clues for studying the origin of the VHE emission of powerfull non thermal extragalactic sources, poses questions on the transparency of the Universe and might suggest new propagation physics.

REFERENCES

1. A. De Angelis and O. Mansutti and M. Persic, arXiv:0712.0315v3 [astro-ph]
2. Observations with the MAGIC Telescope, Institut für Theoretische Physik und Astrophysik, Fakultät für Physik und Astronomie, Universität Würzburg, http://www.astro.uni-wuerzburg.de/mphysics/
3. The Milagro Gamma-Ray Observatory, LANL, http://www.lanl.gov/milagro/
4. The Argo-YBJ Experiment, INFN Roma 3, http://193.204.162.110/ nucleare/argo/argo.html
5. The H.E.S.S. Telescopes, Max Plank Institute, http://www.mpi-hd.mpg.de/hfm/HESS/public/telescope/hn_telescopes.htm
6. VERITAS Gallery, http://veritas.sao.arizona.edu/component/option,com_gallery2/Itemid,34/
7. MAGIC Picture Gallery, Max-Plank-Institut für Physik, Munich, http://wwwmagic.mppmu.mpg.de/gallery/
8. MAGIC picture gallery, http://wwwmagic.mppmu.mpg.de/gallery/pictures/
9. M. Kino et al., Astrophys. J. **564** (2002) 97
10. M. Persic and A. De Angelis, arXiv:0711.2317v1 [astro-ph]
11. M.J. Rees and P. Meszaros, Mon. Not. R. Astron. Soc. **258** (1992) 41
12. P. Meszaros and M.J. Rees, Astrophys. J. **405** (1993) 278
13. R. Sari et al., Astrophys. J. **497** (1998) L17
14. P. Meszaros, Rep. Prog. Phys. **69** (2006) 2259
15. R.P. Pilla and A. Loeb, Astrophys. J. **494** (1998) L167
16. P. Meszaros, M. Ress, and H. Papathanassoiu, Astrophys. J. **432** (1994) 181
17. C.D. Dermer and J. Chiang and K. Mitman, Astrophys. J. **537** (2000) 785
18. B. Zhang and P. Meszaros, Astrophys. J. **559** (2001) 110
19. A. Beloborodov, Astrophys. J. **618** (2005) L13
20. A. Peer and E. Waxman, Astrophys. J. **603** (2004) 448
21. M. Boettcher and C.D. Dermer, Astrophys. J. **499** (1998) L131
22. E. Waxman, Phys. Rev. Lett. **75** (1995) 386
23. Official NASA Swift homepage, http://swift.gsfc.nasa.gov/docs/swift/swiftsc.html
24. D. Burrows, Science **309** (2005) 1833
25. X. Wang and Z. li and P. Meszaros, Astrophys. J. **641** (2006) L89
26. P. Coppi and F.A. Aharonian, Astrophys. J. **487** (1997) L9
27. F.W. Stecker, Int. Astron. Union Symp. **204** (2001) 135
28. F.W. Stecker, O.C. De Jager, and M.H. Salomon, Astrophys. J. **390** (1992) L49
29. F.A. Aharonian et al., Nature **440** (2006) 1018
30. D. Mazin and M. Raue, Astron. Astrophys. **471** (2007) 439
31. A. De Angelis, O. Mansutti, and M. Roncadelli, Phys. Lett. B **659** (2008) 847
32. A. De Angelis, M. Roncadelli, and O. Mansutti, Phys. Rev. D **76** (2007) 121301
33. G. Amelino-Camelia et al., Nature **393** (1998) 763
34. J. Albert et al., Astrophys. J. **667** (2007) 358
35. J. Albert et al., Astrophys. J. **641** (2006) L9
36. M. Teshima, E. Prandini et al., Procs. ICRC Conf., 2007, Merida, Mexico
37. J. Albert et al., Submitted to Phys. Rev. Lett., arXiv:0708.28897v1 [astro-ph]

From Tevatron to LHC

Marina Cobal-Grassmann

University of Udine and INFN Sezione di Trieste

Abstract. A short review of the main physics results and goals at the two hadron collider machines (Tevatron and LHC) is given, with particular attention to their discovery potential for New Physics beyond the Standard Model.

Keywords: High energy physics, Colliders, Standard Model, SUSY, Higgs, Exotics
PACS: 20.29.85.Fj

WHAT DO WE KNOW AT THE START OF LHC?

The Standard Model (SM) of particle physics is a theory that describes three (electromagnetic, weak and strong) of the four known fundamental interactions between the elementary particles that make up all matter. It unifies the Electroweak (EW) Theory and Quantum Chromodynamics (QCD) into a structure denoted by the gauge groups SU(3)×SU(2)×U(1), where forces are mediated by gluons, photons and W/Z bosons. This quantum field theory is consistent with both quantum mechanics and special relativity, and it includes three families of quarks and leptons, with a diverse mass spectrum going from fraction of a GeV (ν) up to 10^2 GeV (top, W/Z). Transitions between quarks generations, with violation of CP is allowed, while transitions between lepton generations have been observed so far only in the neutrino sector.

To date, almost all experimental tests of the three forces described by the SM have agreed with its predictions and SM turned out overall to be an extremely complete and successful description of known phenomena. However, it falls short of being a complete theory of fundamental interactions, primarily because of its lack of inclusion of gravity, the fourth known fundamental interaction, but also because of the eighteen numerical parameters (such as masses and coupling constants) that must be put "by hand" into the theory (rather than being derived from first principles).

In addition there are many other arguments suggesting that the SM is incomplete or additional structures are required, like the hierarchy problem, naturalness of the EW scale, the couplings' unification at the GUT scale, the fact that neutrino masses provide experimental indications that something is missing.

Regardless of our personal level of pragmatism and indifference towards theoretical speculations, as scientists we have to accept the existence of physics Beyond the Standard Model (BSM), and we have a duty to search for it!.

CP1018, *Frontiers of Fundamental and Computational Physics*,
edited by B. G. Sidharth, F. Honsell, O. Mansutti, K. Sreenivasan, and A. De Angelis

HADRON COLLISIONS

The right places to answer these still open questions are certainly the hadron colliders. Beams of protons collide with other beams of protons or anti-protons at nearly the speed of light in particle accelerators, and burst into showers of secondary particle. At high energies we go inside protons and (anti-)protons where internal quarks and gluons collide (see Figure 1).

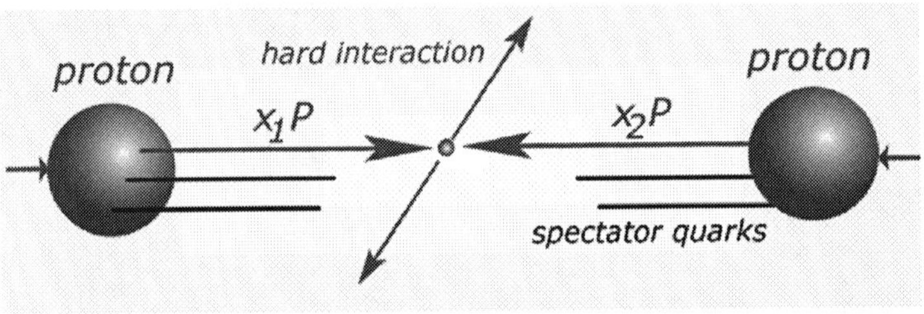

FIGURE 1. Sketch of a hard interaction between the proton constituents. Each one of the interacting partons brings a fraction x of the proton momentum.

The collisions take place inside each of the detectors placed on the accelerator ring. The detectors' job it is to observe as many collisions as possible, to recognize and record the particles that come flying out, and to preserve the information for later studies. The detector's first task is to decide which of the collisions that occur each second are worth recording. It would be impossible to save them all. Computers are instructed to weed out events which are not relevant, and to register only the most interesting collisions in detail for later analysis. Next, the detector must record as much information as possible about every particle in the events selected: the particle's path, its energy, and its charge. Because each kind of particle that flies off from a collision has its own distinctive behaviour, the detector can use this information to identify the particle's characteristic electronic signature and determine that it was present at the scene of the collision.

To create a complete picture of an event, a detector uses layers of different kinds of materials arranged around the central collision point. Each layer is designed to collect and transmit a different kind of information about the moving particles that interact with the material it contains. Electronic signals transmit signals of the interactions to computers, tracing the paths of the particles, along with their energy and charge. Putting all these signals together, the detector builds up a kind of electronic photograph, as though the detector functioned as a several thousands-ton camera, creating a portrait of each collision it records.

At the Fermilab laboratory (Chicago, USA), the Tevatron, with its two experiments CDF and D0, is a proton-anti-proton collider active since 1983. Nowadays it is producing collisions at a center of mass energy of 2 TeV. In 2007, the monthly integrated luminosity increased from 25 to 45 pb^{-1}. It is expected that, by October 2009, a total amount of data corresponding to 6-7 fb^{-1} will be available for analysis.

Starting from the end of 2008, the 27 km long LHC collider [1] at CERN (Geneva) should jump into the game, with 14 TeV of energy in the center of mass, and collisions every 25 ns. At LHC there will be four experiments (see Figure 2, left plot), two of them devoted to discover and understand New Physics beyond the Standard Model. ATLAS [2], about 45 meters long, more than 25 meters high, and with a weight of about 7,000 tons, is one of them (Figure 2, right plot).

STANDARD MODEL PHYSICS

The Tevatron has been an extremely powerful tool to test the SM. In 1995, CDF has seen the first evidence of top-anti-top pairs production [3]. More recently, Tevatron gave evidence for WZ, ZZ and for the electroweak production of single top [4]. In addition, Tevatron has brought several precision measurements and new discoveries - for example: the observation of the B_s mixing and the cascade b, the world's best single measurement of the W mass (with an uncertainty of 0.05%), and an uncertainty of 1.2% on the top mass, combining CDF and D0.

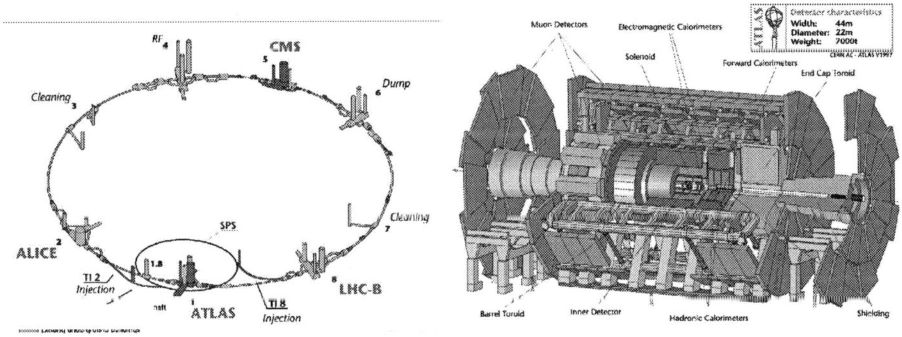

FIGURE 2. Left plot: Layout of the LHC tunnel including the four experiments and the infrastructures. Right plot: the ATLAS experiment, with the indication of the various sub-detectors.

The contribution of LHC in this field will be important as well. With 1 fb^{-1} of collected data, the W mass will be measured with a total error of ±40±25 MeV, and the triple gauge boson coupling will be studied looking at $W^{\pm}Z^0$ and $W^{\pm}\gamma$ events,

For what concern the top-quark sector, LHC will be a "top-factory", with millions of top-anti-top pairs produced every year. An uncertainty of less than 1 GeV on the top mass measurement seems to be reachable with 1 fb^{-1}. But already with only 100 pb^{-1} it will be possible to observe a clear top peak over the background from QCD W/Z+jets and single top events (even without identifying the two jets originated from the b-quarks produced in the two tops decay: $t \rightarrow Wb$), as shown in Figure 3. The top-anti-top production cross section can be measured in this case with a systematic uncertainty of ~15%.

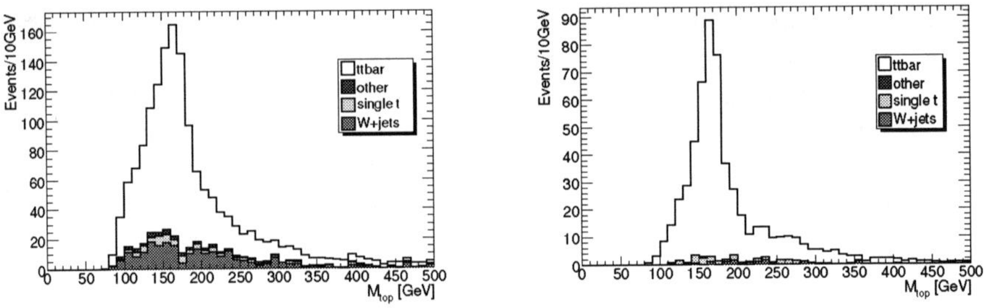

FIGURE 3. Left plot: Top mass distribution obtained with 100 pb^{-1} of collected data at the ATLAS experiment. The relevant backgrounds (apart from the QCD multi-jets) are shown as well. Righ plot: The same distribution if jets originated from a b-quark are identified (b-tagging).

A LIGHT HIGGS?

The fermion and gauge-boson masses of the SM are dynamically generated via the Higgs mechanism when spontaneously breaking electroweak symmetry. The Higgs mechanism, postulated by Peter Higgs in the 1960s [5], was incorporated into modern particle physics by Steven Weinberg and is an essential part of the SM.

In the SM, at temperatures high enough so that the symmetry is unbroken, all elementary particles except the scalar Higgs boson are massless. At a critical temperature, the Higgs field spontaneously slides from the point of maximum energy in a randomly chosen direction. Once the symmetry is broken, the particles get a mass. The mass can be interpreted to be a result of the interactions of the particles with the "Higgs ocean". However, any Higgs boson has been identified yet.

LHC will have the important task to discover the Higgs, namely the particle responsible for the breaking of the SU(2)xU(1) electroweak symmetry and for the generation of the particle masses. On top of that, it will explore the nature of the Higgs boson, and establish the detailed dynamical mechanism with which the EW symmetry is broken: pure SM? Supersymmetry? Extra dimensions? Something else?

Tevatron put already some limits on the Higgs, as shown in Figure 4.

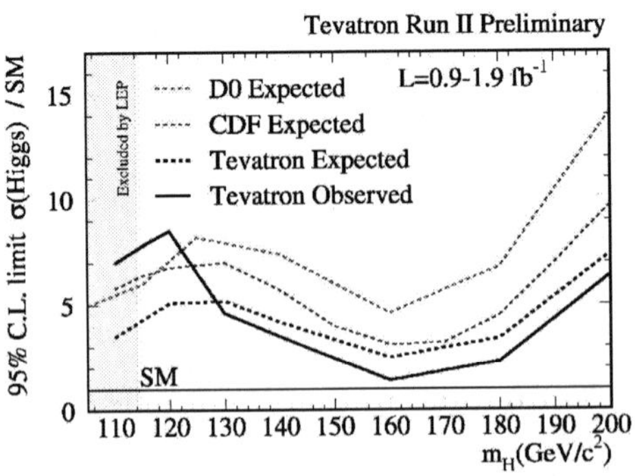

FIGURE 4. Left plot: 95% C.L. on the σ(Higgs) divided by the SM expectation for different Higgs masses at the Tevatron.

A fit of the electroweak parameters, using all the available data, suggests a Higgs mass of less than 182 GeV.

The LHC potential is quite strong. To give an example, at the LHC an Higgs with mass of 130 GeV will be produced via top loop with a cross section of about 35 pb. The same Higgs mass produced via vector boson fusion has a cross section of 4 pb. Many analyses have been performed to explore the LHC sensitivity to the various decay channels. At a given Higgs mass, several channels contribute to this sensitivity. The results indicate that a good discovery potential has to be expected for a SM Higgs, with 10 fb^{-1}, as shown in Figure 5.

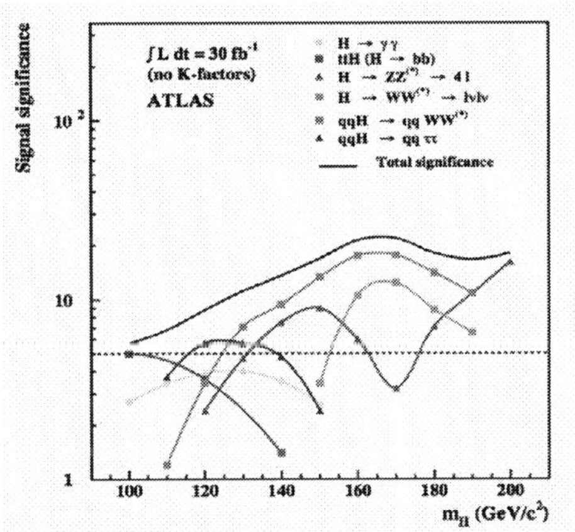

FIGURE 5. 95% C.L. on the σ(Higgs) divided by the SM expectation for different Higgs masses at the Tevatron. Righ plot: 5σ discovery curves for various channel vs. Higgs mass at LHC (ATLAS).

THE HIERARCHY PROBLEM: NEW PHYSICS?

If a Higgs with mass < 1 TeV is discovered, the SM will be complete. But there is an additional problem: when radiative corrections are applied to the bare Higgs mass, a quadratically divergent integral appears.

A cut-off sets the scale where new particles and physical laws must come in:

$$\delta m_H^2 \propto \int_0^\infty d^4k \frac{k^2 + m_f^2}{\left(k^2 + m_f^2\right)^2} + \ldots \xrightarrow{\text{cut-off}} \int_0^{\Lambda_{\text{cut-off}}} (\ldots) \propto \Lambda_{\text{cut-off}}^2$$

But, above the EW scale, we only know of two scales: the GUT one (~10^{16} GeV) and the Planck one (~10^{19} GeV). The question is why the Higgs boson is so much lighter than the Planck mass (or a GUT scale, or a heavy neutrino mass scale): one would expect that the large quantum contributions to the square of the Higgs boson mass would inevitably make the mass huge, comparable to the scale at which new physics appears, unless there is an incredible fine-tuning cancellation between the quadratic radiative corrections and the bare mass.

It should be remarked that the problem cannot be even formulated in the strict context of the Standard Model, for the Higgs mass cannot be calculated. In a sense, the problem amounts to the worry that a future theory of fundamental particles, in which the Higgs boson mass will be calculable, should not have excessive fine-tunings. Accepting the physical reality of the hierarchy problem, it is expected that new physics should make an appearance at energy scales not much higher than the scale of energy required to produce the Higgs boson, and thereby provide an explanation for the small mass which is experimentally favoured.

SUPERSYMMETRY

The light scalar Higgs boson is unprotected from divergencies at the GUT/Planck scales. On the contrary, all the other light particles of the SM are protected against large scales. This happens since, due to the chiral symmetry, mass corrections to fermions are logarithmic in the energy (not quadratic). On the other hand, the gauge symmetry protects the bosons (there are no corrections to photon or gluon masses).

Fermion and boson loops contribute with different signs to the Higgs radiative corrections: if there existed a symmetry relating these two, this could protect the mass of the scalar!. Indeed, Supersymmetry (SUSY) realises this by transforming i.e. a scalar boson into a spin-½ fermion, whose mass is protected. Hence, the scalar mass is also protected (precisely through SUSY). This solve the naturalness and the hierarchy problems of the SM (at least technically). In SUSY, the number of bosons is the same as the number of fermions. To create super-multiplets, one needs to add one super-partner to each SM particle: these have opposite spin statistics but otherwise equal quantum numbers. To describe the model, one has to add many more parameters.

The SUSY particles masses are unknown (to preserve a light Higgs, should be $M_{SUSY} \leq 1$ TeV), but the cross sections should be quite large. To avoid the proton decay, a discrete R-parity (which we will assume here as a conserved quantity) is introduced, which value is +1 for SM particles and -1 for SUSY partners. As a consequence, all interactions with an odd numbers of SUSY particles are forbidden (SUSY particles are produced in pairs). A conserved R-parity requires the existence of a lightest stable SUSY particle (LSP). Since no exotic strong or electromagnetic bound states have been observed, the LSP should be neutral and colourless. LSP is stable, therefore it shows up as missing energy in the detector. R-parity provides dark matter candidates sneutrino (ruled out?), gravitino and neutralino. It has to be added that imposing the local gauge invariance of SUSY, requires the existence of spin-3/2 and spin-2 particles: therefore, the spin-2 graviton, assumed to mediate the gravitational forces is naturally introduced.

At LHC we look at SUSY in different ways.. The signature is driven by the particle masses: we look at high P_T jets from squarks and gluinos, leptons from gauginos and sleptons, missing energy from LSPs. Figure 6 shows the result of an inclusive search at LHC for a SUSY signal of 1 TeV. A sensitive variable to detect SUSY decays is the "effective mass", defined as the sum of the missing transverse energy and the P_T of all jets and leptons. Requiring at least one high P_T lepton reduces the QCD background by a factor of 20-30, with a signal loss of only a factor of about 3.

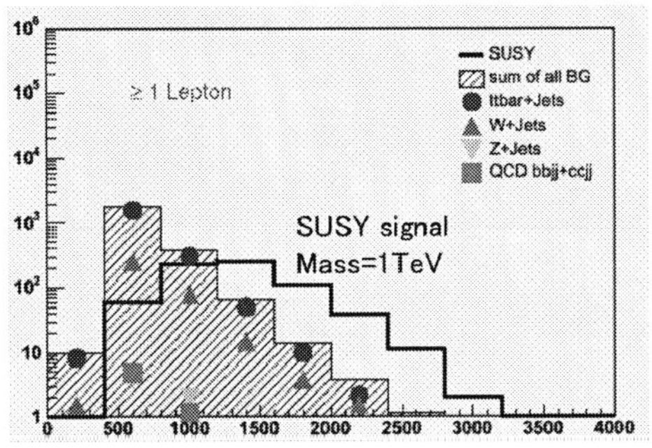

FIGURE 6. Effective mass distribution for a 1 TeV SUSY signal. The relevant backgrounds are shown as well (from the ATLAS experiment).

Another example is the search for squarks and gluinos. Figure 7 shows the 5σ discovery contours for squarks and gluinos with mass of 1 TeV. The cross sections in this case are of the order of about 1 pb, which means ~10^4 events/year at low luminosity. The region on the left is forbidden since the LSP would be a charged stau. The region on the right is the one in which the electroweak symmetry is not broken.

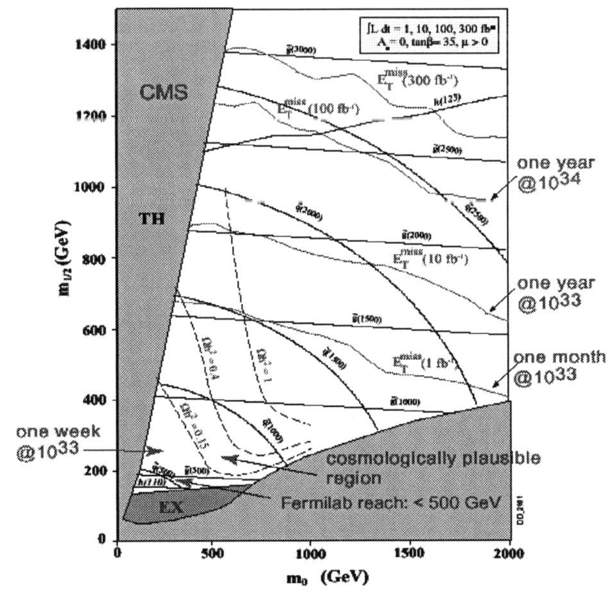

FIGURE 7. 5σ discovery contour for squarks and gluinos (from the CMS experiment).

Table 1 lists the performance of LHC on the reachable squarks and gluino masses, for different luminosities, together with the present limits from Tevatron.

TABLE 1. Performance of LHC for the mass limit on squark and gluino, for different luminosities, together with the present limit reached at Tevatron by the D0 and CDF experiments.

Time period	Luminosity ($cm^{-2}s^{-1}$)	Squark/gluino masses
1 month	10^{33}	~1.3 TeV
1 year	10^{33}	~1.8 TeV
1 year	10^{34}	~2.5 TeV
Ultimate	$\int = 300\ fb^{-1}$	~2.5 -3 TeV
D0 & CDF	$\int = 0.3\ fb^{-1}$	$>_{2\sigma}$ 0.35 TeV

LITTLE HIGGS

Similar to SUSY are the "little Higgs" theories [6], so called because they generate a Higgs particle with a relatively small mass. They also predict new particles with masses near 1 TeV, characterised by quantum numbers different from those of SUSY. The cancellation of the quantum corrections occurs between fields of the same spin: fermions cancel fermions and bosons cancel bosons (this is a so-called non-linearly realised symmetry). The breaking of a new SU(5) symmetry leads to new O(TeV) particles, among which are a new top quark (T), new gauge fields ($W^{\pm}_H$, Z_H, A_H) and five new Higgs bosons.

The precise masses and properties of these new particles, depend from the model, but predictions can be made. To give an example a ϕ^{++} Higgs boson, can be produced at the LHC via quark-antiquark collision ($\overline{q}q \rightarrow \phi^{++}\phi^{-} \rightarrow 4l$), but it has a too small cross section for discovery. However, production can also go through vector-boson fusion ($uu \rightarrow dd\phi^{++} \rightarrow ddW^{+}W^{-}$). In this case the signature of the boson is given by two leptons with the same charge. Figure 8 (left plot) shows the transverse mass distribution for a 1 TeV boson and for 200 fb^{-1} of integrated luminosity. The main backgrounds are also shown.

EXOTICS

Many models (Sequential Standard Model, Grand Unified Theories, "left right" models) predict extra gauge bosons (Z'). The present experiments explore to about 1 TeV, but early LHC running will go beyond this. Figure 8 (right plot) shows the mass peak of a reconstructed 3 TeV mass Z' in GUT model, together with the SM background.

Extradimensions

There is a strong theoretical interest in models with extra-dimensions (ED) [7]. Relatively large EDs in which gravitons propagate are not excluded: the SM particles could be confined in a smaller sub-space, a *brane*. This sub-space would be our 4-dimensional world, and only gravity would "see" the other dimensions. These models solve the so-called hierarchy problem. The EDs influence the cross sections of standard accelerator processes, and allow the production of gravitons which can be coupled to ordinary particles, and might be visible at the LHC.

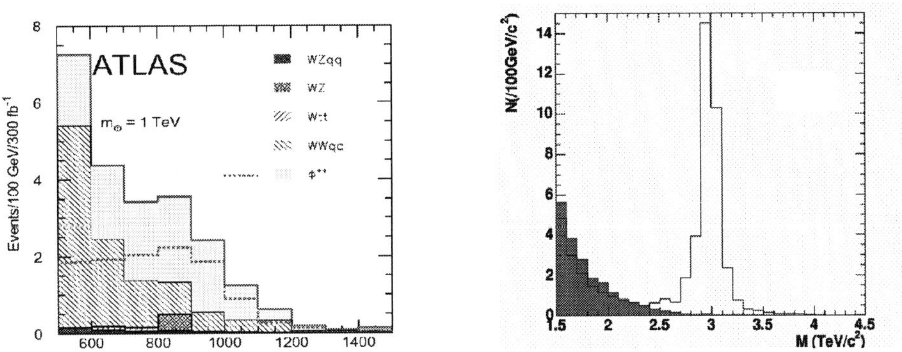

FIGURE 8. Left plot: Transverse mass distribution from a 1 TeV ϕ^{++} boson with 300 fb^{-1}. Right plot: Reconstructed mass distribution of a 3 TeV Z' in GUT model.

Gravitons

Once produced at LHC, gravitons escape in extra-dimensions. They can therefore be observed as missing energy, their signature being characterised by a jet (or a photon) + missing E_T. However, they can also appear as virtual particles. In this case their presence would be signalled as extra-production of standard particles or by the presence of resonances.

Black Holes

Strong gravity in extra-dimensions allows black hole production at colliders [8]. Extra dimensions have very small radius and imply that gravity is much stronger at shorter distances. For this reason, if one gets an energy of the order of ~TeV into a "small" enough region, a black hole (BH) will form spontaneously. These black holes would be characterised by a small radius (~ 10^{-18}-10^{-19} m) and by a high temperature (~ 10^{16} K). The cross-section for the formation of a black hole is of the order of σ_{BH} ~ πr^2, where r is the Schwarzschild radius in 4+d dimensions. With M_D (the true fundamental scale) ~ 2–3 TeV → σ_{BH} ~ O(pb) which implies a fast discovery for M_{BH} < 4 TeV, d = 2-6 . These black holes decays instantaneously on the LHC time scales, and evaporate because of the Hawking radiation into leptons, photons and jets. The final state multiplicity increases with M_{BH}. The thermal decay time scale is $\tau \sim 10^{-27}$ s. It may be possible to determine from the observed final state energy spectrum and the BH cross-section the characteristic Hawking temperature T_H of the BH: T_H ~ M_D · $(M_D/M_{BH})1/(d+1)$.

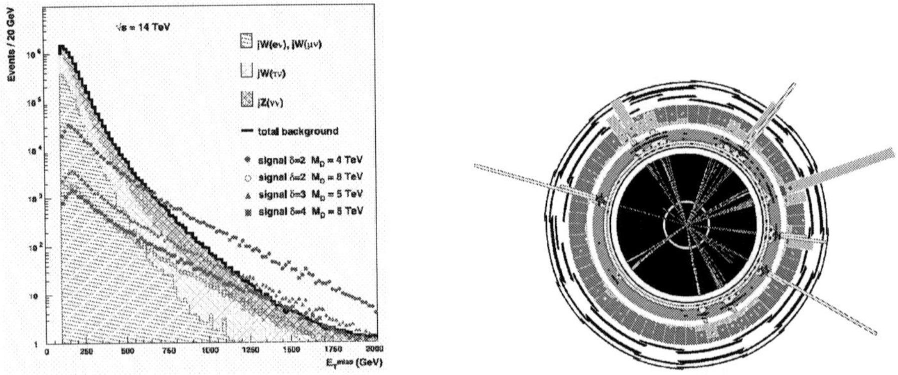

FIGURE 9. Left plot: Missing E_T distribution for an ADD graviton after 100 fb^{-1} of data collected (ATLAS experiment).The dominant backgrounds are shown as well. Right plot: Event display of a black hole event at the LHC, evaporating in leptons, photons and jets.

CONCLUSIONS

There are still (too) many things which we do not know and understand. For this reason, any good high energy physics cannot believe that the Standard Model is the end of the story. This field of study is asking some of the most fundamental questions in modern physics, and to answer to these questions, the tools to be used are of incredible complexity. Still, one of the most exciting intellectual adventures of mankind. The LHC should help the next step in the understanding of the infinitely small and the infinitely large. After it, our view of particle physics could change forever.

REFERENCES

1. L. Rossi, *IEEE Trans. Appl. Supercond.* 17 (2007) 1005-1014
2. ATLAS Detector and Physics Performance Technical Design Report LHCC 99-14/15.
3. CDF Collaboration, *Phys. Rev.* D**50**, 2966 (1994); CDF Collaboration, *Phys. Rev. Lett.* **73**, 225 (1994); *Phys. Rev.* D**52**, 2605 (1995); CDF Collaboration, *Phys. Rev.* D**51**, 4623 (1995).
4. D0 Collaboration, Physical Review Letters **98**, 181802 (2007).
5. P W Higgs, "Broken Symmetries, Massless Particles and Gauge Fields", *Phys. Lett.* **12** 132 (1964); P W Higgs, "Broken Symmetries and the Masses of Gauge Bosons", *Phys. Rev. Lett.* **13** 508 (1964).
6. N. Arkani-Hamed, A. G. Cohen, H. Georgi, "Electroweak symmetry breaking from dimensional deconstruction". *Phys. Lett. B.* **513**: 232-240 (2001); N. Arkani-Hamed, A. G. Cohen, T. Gregoire, J. G. Wacker , "Phenomenology of electroweak symmetry breaking from theory space", *JHEP* 0208: 020 (2002); N. Arkani-Hamed, A. G. Cohen, T. Gregoire,E. Katz, A. E. Nelson, J. G. Wacker , "The Minimal moose for a little Higgs", *JHEP* 0208: 021 (2002).
7. G. Nordström, „Über die Möglichkeit, das elektromagnetische Feld und das Gravitationsfeld zu vereinigen", *Physik. Zeitschr.* **15** 504-506 (1914); T. Kaluza, "On the problem of unity in physics", *Sitzungsber. Preuss. Akad. Wiss. Berlin.* (*Math. Phys.*) 966-972 (1921); Paul S. Wesson, *Space-Time-Matter, Modern Kaluza-Klein Theory*, (1999) World Scientific, Singapore ISBN 981-02-3588-7; Oskar Klein, Quantum theory and five dimensional theory of relativity, *Z. Phys.* **37** 895-906 (1926).
8. S. Dimopoulos, G. Landsberg, arXiv:hep-ph/0106295.

Top quark cross section measurement in ATLAS: Commissioning Analysis

Alhroob.M* and Belhouari.A †

*University of Udine, Via delle Scienze 208, 33100 Udine.
†INFN Trieste and ICTP, V.le Miramare, Trieste.

Abstract. An accurate determination of the top-quark pair production cross-section at the LHC provides a valuable check of the Standard Model. In this article we present a commissioning analysis to measure to quark cross section with the ATLAS detector. The cross section is determined in the semileptonic channel using a counting method, based solely on counting the numbers of top candidate events that pass selection, and subtracting all backgrounds in order to get the yield of $t\bar{t}$ events in the sample. Our cross section study is performed without relying on the tagging of b-quark initiated jets.

Keywords: Top quark,Counting method, cross section
PACS: 10

INTRODUCTION

The LHC will produce millions of top quark pairs. During the first period of data taking, these events will provide an essential calibration tool for ATLAS, playing a key role in the detector commissioning phase. The top cross section measurement, will be one of the first relevant one.

The determination of the top pair production cross-section is one of the standard measurements that will be carried out once the first data samples are available from the ATLAS experiment.

COUNTING METHOD

The cross-section for top quark pair production by using counting method is defined as:

$$\sigma_{t\bar{t}} = \frac{N_{data} - N_{back}}{\varepsilon \times \int \mathscr{L}} \qquad (1)$$

where: N_{data} is the total number of events which passed the cuts, N_{back} is the number of estimated background events which passed the cuts, ε is the total efficiency with which pure signal events pass the triggers and selection cuts, $\int \mathscr{L}$ is the integrated Luminosity. We tried to simulate what will happen when the real data will be available by using two Monte Carlo (MC) samples: sample M which represents the MC data and sample D which represents the real data. Sample M is used to estimate the background level and the efficiency. In our study both samples D and M are normalized to 100 pb^{-1}.

CP1018, *Frontiers of Fundamental and Computational Physics*,
edited by B. G. Sidharth, F. Honsell, O. Mansutti, K. Sreenivasan, and A. De Angelis

DATA AND SELECTION

The event generator used for the top signal is MC@NLO. The main expected backgrounds arise from QCD W and Z production together with jets, as well as from the production of single top quarks. The W+jets and Z+jets were generated with the Alpgen MC, while the single top samples were generated using AcerMC. Minor contributions which have been also evaluated come from WW, WZ, ZZ, Wbb and $Wc\bar{c}$ events. We start the identification of semileptonic $t\bar{t}$ events by requiring a lepton trigger to be fired. For our *default* selection, the events should have one and only one electron with $P_T >$ 20 GeV, Missing $\not{E}_T > 20$ GeV.At least four jets with $P_T > 20$ GeV, of which at least three jets must have $P_T > 40$ GeV.

M_{top} can be reconstructed by choosing the three jets whose total Lorentz vector has the highest P_T to be the candidate 3-jet combination arising from the hadronically decaying top quark. Additionally, since two of the 3-jets in a hadronic top decay are from a decaying W-boson, we can also define the "W-mass constraint" by requiring that a pair of jets from the 3-jet combination exists whose mass is within 10 GeV of the W-mass. Figure 1 (the right plot is after having applied the W mass constraint) and Figure 2 show the top mass and M_W distributions respectively. Other cuts can be applied to increase the purity such as $\cos\theta^*$ [1] cut, which means $|\cos\theta^*|$ of the three leading jets should be less than 0.7, and central region cut which requires the 3 jets to each have $|\eta| < 1$.

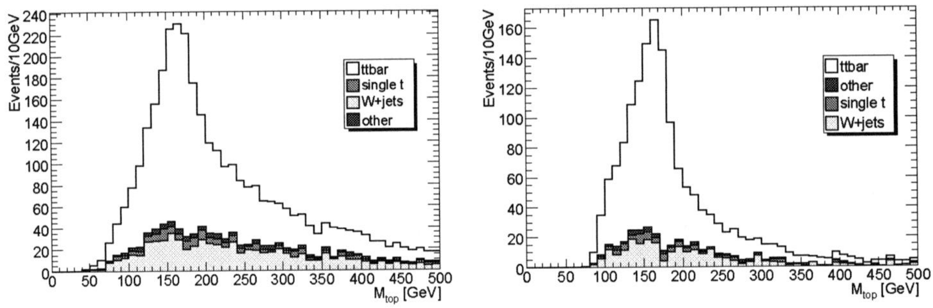

FIGURE 1. Expected distribution of the three-jet invariant mass before (Left plot), After(Right plot): the W-boson mass constraint, normalized to 100 pb^{-1}

CROSS SECTION

The estimated cross-sections are: $845.18pb^{-1} \pm 2.47\%$stat after the default cuts, and $868.18pb^{-1} \pm 3.25\%$stat after the W-mass constraint,

[1] $\cos\theta^*$ is defined as the angle between the jet and the beam direction in the center of mass frame of the event

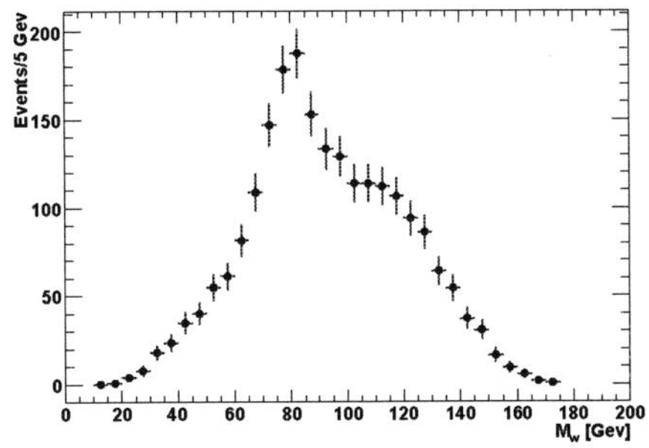

FIGURE 2. Expected distribution of the di-jet invariant mass, normalised to 100 pb^{-1}

TABLE 1. Systematic uncertainties on the electron commissioning counting cross-section measurement, in percent.

Source	Default analysis (%)	Default+W-boson constraint (%)
Lepton ID efficiency	1.0	1.0
Lepton trigger efficiency	1.0	1.0
Jet Energy Scale (+5%)	12.4	7.5
Jet Energy Scale (-5%)	13.3	9.7
CTEQ PDFs	2.3	2.5
MRST PDFs	0.9	1.0
(CTEQ-MRST)/CTEQ	1.9	2.1
ISR/FSR	10.0	3.5
MC uncertainty	12.0	5.0

Systematic errors

There are a number of sources of systematic uncertainties which affect the results. These include the lepton identification and trigger efficiency, the dependence on the jet energy scale, PDF uncertainties, uncertainties in the modeling of the Initial and Final state radiation (ISR/FSR) and in the Monte Carlo signal modelling itself. The main systematic uncertainties for the counting analysis are listed in Table 1.

Results

In the starting period of LHC when the collected data will be from an integrated luminosity of 100 pb^{-1}, we can observe a $t\bar{t}$ signal and determine its production cross-section using a counting method. We expect the following level of uncertainties (for the default selection + the W-boson mass constraint in the electron analysis)
$\frac{\Delta\sigma}{\sigma}(\%) = 3.5(\text{stat}) \pm 10.9(\text{syst}) \pm 5.0(\text{MC unc.}) \pm 5.0(\text{Lum})$

A novel background reduction strategy for high level triggers and processing in gamma-ray Cherenkov detectors

G. Cabras*, A. De Angelis*, B. De Lotto*, M.M. De Maria*, F. De Sabata*, O. Mansutti*, M. Frailis†, M. Persic†, C. Bigongiari**, M. Doro**, M. Mariotti**, L. Peruzzo**, A. Saggion**, V. Scalzotto**, R. Paoletti‡, A. Scribano‡, N. Turini‡, A. Moralejo§, D. Tescaro§ and the MAGIC Collaboration

*Dipartimento di Fisica dell'Università di Udine and INFN, Udine, Italy
†Osservatorio Astronomico di Trieste and Dipartimento di Fisica dell'Università di Udine, Italy
**Dipartimento di Fisica dell'Università di Padova and INFN, Padova, Italy
‡Dipartimento di Fisica dell'Università di Siena and INFN, Siena, Italy
§Institut de Fìsica d'Altes Energies, Edifici Cn., E-08193 Bellaterra (Barcelona), Spain

Abstract. Gamma ray astronomy is now at the leading edge for studies related both to fundamental physics and astrophysics. The sensitivity of gamma detectors is limited by the huge amount of background, constituted by hadronic cosmic rays (typically two to three orders of magnitude more than the signal) and by the accidental background in the detectors. By using the information on the temporal evolution of the Cherenkov light, the background can be reduced. We will present here the results obtained within the MAGIC experiment using a new technique for the reduction of the background. Particle showers produced by gamma rays show a different temporal distribution with respect to showers produced by hadrons; the background due to accidental counts shows no dependence on time. Such novel strategy can increase the sensitivity of present instruments.

Keywords: Gamma ray astrophysics; Cherenkov telescopes; Techniques for signal extraction.
PACS: 95.30.-k; 95.55.-n; 95.75.Pq; 95.85.Pw

During the recent years, there has been an impressive progress in gamma-ray astrophysics thanks to ground-based gamma detectors. Such a progress has impact both on fundamental physics and on astrophysics [1].

Ground-based telescopes cannot detect cosmic gamma rays directly, as such particles are absorbed when entering the atmosphere. They detect instead the radiation emitted by the ultra-relativistic particles of the showers produced by the interaction of photons with the atmosphere. In particular, IACTs *(Imaging Atmospheric Cherenkov Telescopes)* detect the Cherenkov light emitted by secondary charged particles of a shower, and study the properties of the primary particle from the image formed by the Cherenkov light on photodetectors placed in the focal plane. The physical and phenomenological characterization of these showers, and the morphological study of the shape of the Cherenkov image collected in the camera plane, give the possibility to select the showers induced by the gamma events among the huge amount of cosmics.

MAGIC is a IACT located in the Canary island of La Palma (28.75°N, 17.86°W, 2225 m a.s.l.); its reflecting surface, 17 m in diameter, has a parabolic shape (which

CP1018, *Frontiers of Fundamental and Computational Physics,*
edited by B. G. Sidharth, F. Honsell, O. Mansutti, K. Sreenivasan, and A. De Angelis

preserves the time structure of the Cherenkov light flash). The field of view of the focal PMT camera is about 3.5°. Fast PMT analog signals are routed via optical fibers to the DAQ-system electronics, where the signals are digitized and saved to disk. Further details can be found in [2].

Thanks to a major upgrade, since February 2007 the data acquisition works via ultra-fast FADCs, digitizing the signal at the speed of 2 GSamples/s [3]. Particle showers produced by gamma rays have a temporal distribution of $(2 \div 5)$ ns, narrower than the ones typical for hadronic showers, of around $(5 \div 10)$ ns. Using the time structure of the signal, we can obtain an enhancement of the telescope performance basically for two reasons: a reduction in the amount of NSB (Night Sky Background) light integrated with the real signal (due to a smaller integration window), and the possibility to reconstruct with a good time resolution the timing characteristics of the showers [4] [5, 6].

For this work, to find the maximum pulse within the useful range of the FADCs ($\sim$ 30 ns), we use a cubic spline function. The position of its maximum is used to estimate the arrival time of the signal in a pixel. The intensity of the signal is then obtained by integrating the spline in a range of 7.5 ns. In this way, the time resolution of each pixel has been estimated to be around 0.4 ns RMS for a signal of 40 photoelectrons (phe), through the study of the pixel time spread in calibration events. This value may improve with the use of a more sophisticated pulse reconstruction.

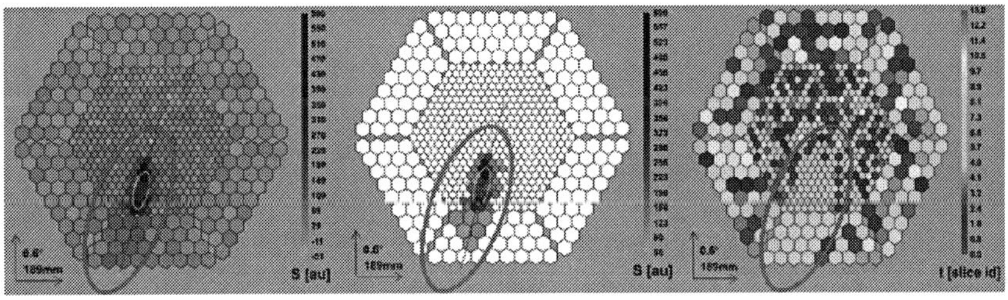

FIGURE 1. Shower image before (left) and after (center)image cleaning. The ellipse superimposed on the camera corresponds to the Hillas approximation. The third map (right) shows the arrival time for each pixel: it can be seen that there is a tight coincidence for pixels belonging to the gamma shower image.

After the calibration of the data, for each event, the shower image appears like in figure 1, on the left: it is surrounded by pixels containing only noise from the light of the diffuse background. The image cleaning procedure removes pixels which apparently do not form part of the shower image (figure 1, center).

Only low energy events have been selected, through an upper cut in the signal of the two highest pixels, since we want to concentrate our algorithm on the region of the order of 50 GeV, the most difficult to detect for Cherenkov telescopes. We require a time coincidence between core and boundary pixels within a window spanning from 0.5 to 3.5 ns. After the cleaning, the shape of the surviving image is studied by an approximation to an ellipse, as introduced by Hillas [7]. The information is therefore saved into different parameters related with this ellipse, like its width, its length and its orientation in the camera. To test the performance of the cleaning, we study the enhancement of the signal to noise ratio obtained after a cut in the Hillas ALPHA

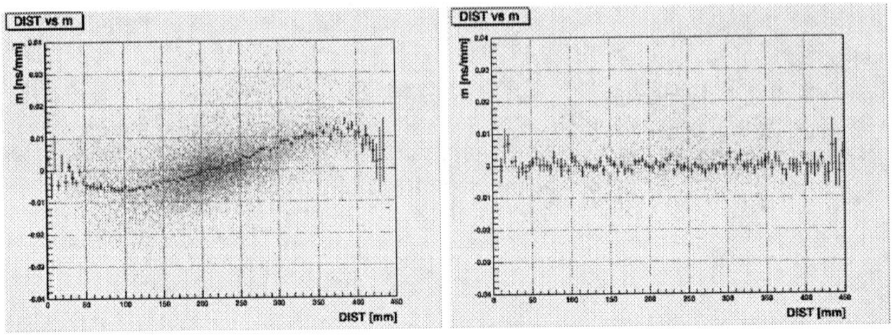

FIGURE 2. TIME GRADIENT (called m) vs DIST (297 mm= 1 deg). The correlation for simulated γ-rays is evident (left); this correlation does not appear for hadrons (right).

parameter[1].

The information given by the time evolution of the images, has moreover been used for developing new parameters in the gamma/hadron separation algorithm based on a particular neural network architecture called Random Forest [8]. We studied the additional background rejection power added to the classical Hillas parameters.

The TIME RMS parameter estimates the arrival time spread of the Cherenkov photons in the pixels belonging to the cleaned image (μ, γ-ray and hadron induced showers have a different characteristic time spread [9]). Previous studies [10] determined that along the major axis of gamma induced Cherenkov image a time structure is present (see figure 1, right). This can be well approximated as a linear TIME GRADIENT. The slope of this gradient is related to the angle between the telescope and shower axes; moreover, for showers of a given direction, it is well correlated with the Impact Parameter (IP) of the shower, and so with the DIST image parameter[2], as can be seen in figure 2.

A data sample of 5.6 h of Crab Nebula observation (February 2007; zenith angle between 5° and 30°) has been used to check the results on real data; the results are consistent with the ones previously described.

Three different analyses of the same previously described Crab Nebula data have been performed: the first one by a standard analysis, the second one by an image cleaning (IC) with time constraint, and the third one by a time-IC and the TIME RMS and TIME GRADIENT in the Random Forest parameter list. In figure 3 we compare two α-plots obtained from the first and the third analysis.

Time parameters allows $\sim$50% better background suppression keeping the same amount of excess events. Similar improvements are seen at all energies. Improvements (of 15%) have been found also in the event energy reconstruction. In fact the TIME GRADIENT gives information about the real IP of the shower and therefore it helps to distinguish distant high energy showers from closer, low energy ones.

---

[1] ALPHA indicates the orientation of the image ellipse with respect to the source position: for gammas it is expected to be close to 0.

[2] IP: distance between the telescope pointing axis and the axis of the shower. In case of real data DIST is a good estimator of the IP.

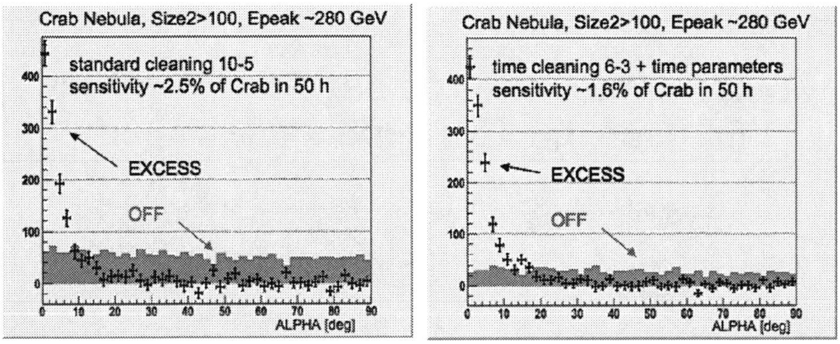

FIGURE 3. Comparison of two α-plots between analysis 1 (left) and 3 (right). Keeping the same number of excess events background is reduced by about 50%. Energy threshold: 280 GeV.

We can conclude that the use of timing parameters in the analysis of current MAGIC data results in a significant improvement in background rejection, and thus in sensitivity.

The integral sensitivity of the telescope obtained with the standard analysis for energies larger than 280 GeV was 2.5% of Crab in 50 hours, while through this novel analysis we obtain a value of 1.6% of Crab.

Moreover, at the lowest energies the time cleaning also contributes, besides reducing the background, to enlarge the event statistics.

The new scenario for temporal analysis of Cherenkov flashes has just been explored for the first time, and further developments are close to come. Within the year 2008, MAGIC will be upgraded to a two telescope system. A dedicated Monte Carlo chain has been developed in order to estimate its performance: observations in stereoscopic mode will increase the current sensitivity of the instrument by a factor 3 [11].

ACKNOWLEDGMENTS

This work was partly financed by the Ministero dell'Università e della Ricerca within the PRIN 2005.

REFERENCES

1. A. De Angelis, O. Mansutti and M. Persic, arXiv:0712.0315 [astro-ph]
2. http://wwwmagic.mppmu.mpg.de/introduction/factsheet/
3. F. Goebel et al., Proc. of the 30th Int. Cosmic Ray Conf. (ICRC 2007), arXiv:0709.2363 [astro-ph]
4. D. Tescaro et al., Proc. of the 30th Int. Cosmic Ray Conf. (ICRC 2007), arXiv:0709.2363 [astro-ph]
5. S. Raducci, Diploma Thesis, University of Udine (2004)
6. D. Tescaro, Diploma Thesis, University of Padova (2005)
7. A.M. Hillas, Proc. of the 19th Int. Cosmic Ray Conf. (ICRC 1985), Conf. Papers **3** (1985) 445
8. J. Albert et al, arXiv:0709.3719 [astro-ph]
9. R. Mirzoyan et al, Astropart. Phys. **25** (2006) 342
10. M. Heß et al., Astropart. Phys. **11** (1999) 363
11. E. Carmona et al., Proc. of the 30th Int. Cosmic Ray Conf. (ICRC 2007), arXiv:0709.2363 [astro-ph]

Representations and image classification methods for Cherenkov telescopes

C. Malagón*, J.A. Barrio†, D. S. Parcerisa* and D. Nieto†

*Dpto. Ingeniería Informática, Univ. Antonio de Nebrija, E-28040 Madrid, Spain
†Dpto. Física Atómica, Univ. Complutense de Madrid, E-28040 Madrid, Spain

Abstract. The problem of identifying gamma ray events out of charged cosmic ray background (so called hadrons) in Cherenkov telescopes is one of the key problems in VHE gamma ray astronomy. In this contribution, we present a novel approach to this problem by implementing different classifiers relying on the information of each pixel of the camera of a Cherenkov telescope.

Keywords: Gamma Ray Astronomy, Cherenkov telescopes, pattern recognition
PACS: 95.55.Ka, 95.75.Pq, 95.85.Pw

INTRODUCTION

Since Hillas parameter (second moments of image in telescope camera) analysis was developed back in 1985 to separate between gamma-like and hadron-like events as recorded by Cherenkov telescopes [1], many techniques have been used for gamma/hadron separation based on such parameters. Bock et al. [2] performed a case study for most of these techniques. We propose to apply usual machine learning techniques to the full image recorded by a Cherenkov telescope, on two different domain representations: a) a pixel-based representation; b) a so-called histogram-sector representation, relying on a transformation based on typical OCRs feature extraction.

Data sample

For this study, we have used gamma and proton events (as the latter represents the majority of hadronic cosmic rays) simulated with Corsika code [3], plus MAGIC Reflector and Camera reconstruction standard software [4].

Machine learning techniques

We have used three individual classifiers: (a) Decision trees, grown using Quinlan's C4.5 algorithm [5]; (b) Multilayer Perceptron, trained with the classical backpropagation algorithm [6]; and K Nearest Neighbour (K-NN) algorithm [7] (K=11, euclidean distance metric). Regarding ensemble classifiers, which will only be applied to decision trees classifier, we will deal with voting classification algorithms, namely Boosting (Adaboost algorithm [8]) and Bagging (ensemble method by Breiman [9]).

CP1018, *Frontiers of Fundamental and Computational Physics*,
edited by B. G. Sidharth, F. Honsell, O. Mansutti, K. Sreenivasan, and A. De Angelis

RESULTS

Supervised Machine Learning Methods Results

All the above-mentioned machine learning methods have been fed with above-described gamma and proton samples, both for the pixel-based and the histogram-sector representation and their results presented in terms of ROCc (Receiver Operating Characteristic curves). Figure 1 (left panel) shows the results when classifying the whole data set, regardless incident energies. Pixel-based representation has also been successfully applied to energies below 100 GeV [10], as shown in Figure 1 (right panel).

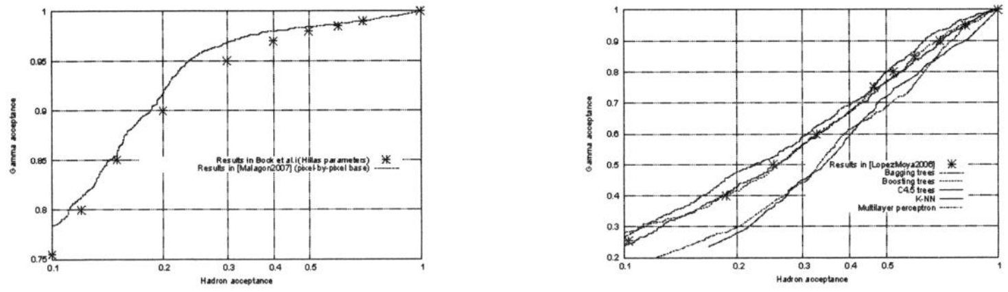

FIGURE 1. ROC curve the full data set (left) and for energies lower than 100 GeV (right)

SOM Results

We have applied the Self-Organizing Maps algorithm [12] as a feature extractor, reducing the dimensionality of the input space. After applying SOM algorithm to our data, two unlabelled clusters are obtained, corresponding to gamma and hadron events.

ACKNOWLEDGMENTS

We thank the MAGIC collaboration for providing us with the simulated gamma and proton sample that we have used to test the method and R. Bock for fruitful discussions.

REFERENCES

1. Hillas A. M., *Proc. 19th ICRC*, La Jolla, **4**, 445 ,1985.
2. Bock, R.K. et al., *Nucl. Instrum. Meth. A* **516**, 511-528 , 2004.
3. D. Heck et al., *Forschungszentrum Karlsruhe Report FZKA* **6019**, 1998.
4. Bretz, T. (MAGIC Coll.), Gaug, M., et al (MAGIC Coll.), *Proc. 29th ICRC*, Pune, India, 2005.
5. Quinlan, R., *Machine Learning* **1**, 81-106, 1986.
6. Rumelhart D., McClelland J. And the PDP Research Group. *MIT Press*, 1986.
7. Aha, D. and Kibler, D., *Machine Learning* **6**, 37-66, 1991.
8. Freund, Y., Schapire, R. E., *Proc. 13th Int. Conf. on Machine Learning*. San Francisc, 148-146, 1996.
9. Breiman, L., *Machine Learning* **24**, 123-140, 1996.
10. Malagón, C., Barrio, J.A., Nieto, D., Reyes, R. de los, *Proc. 30th ICRC*, Mérida, Mexico, 2007.
11. Lopez M., *PhD. thesis*, UCM, Madrid 2006
12. Kohonen, T. *Self-Organizing Maps*. Springer, 1995.

PART V

COMPUTATIONAL PHYSICS, COMPLEX SYSTEMS AND THE STRUCTURE OF MATTER

Tonelli Principle: finite reduction and fixed energy Molecular Dynamics trajectories

Alessandro Turco*, Daniele Passerone† and Franco Cardin**

*S.I.S.S.A., Trieste
†EMPA, Zurich
**Dip. Matematica Pura ed Applicata, Padova

Abstract. We propose a novel theoretical and practical alternative to the Maupertuis functional in the field of molecular dynamics: Tonelli functional [1]. Our aim is to adapt this technique to the study of rare events where the initial and the final state of a system are known and we look for transition paths. We reached it with a rigorous mathematical development of the functional and an efficient numerical algorithm. We couple the Tonelli functional with an exact finite dimensional reduction: this is not far from a multiscale approach. Indeed, the reduction will help the study of the high frequencies of the systems. It can be seen as a magnifying glass for atomic trajectories. A test on a 38-atom Lennard-Jones cluster shows the main features of our work.

Keywords: Molecular Dynamics, Rare Events, Computational Strategies, Lagrangian Mechanics
PACS: 02.70, 31.15

INTRODUCTION

The variational principles of classical mechanics govern the existence and the characterization of dynamical trajectories connecting two points in configuration or phase space. Such dynamical trajectories joining two fixed points can show different regimes according the the underlying potential energy profile. Let us think, for example, of a rare transition event overcoming a barrier between two stable states with a total energy just above the energy of the transition state. The trajectory describing such event will have regions with fast oscillations close to the reactant and the product basins, and a region close to the transition state with well known temporal instabilities.

We propose two techniques which can improve the understanding of the problem. On one hand we introduce the Tonelli principle [2] that allows us to describe properly situations in which the total energy is close to the potential energy. On the other hand we discuss a finite reduction scheme that leads to a computationally convenient strategy for treating the different timescales of the trajectory. If the process is not diffusive, the very transition process is rather short, and a sampling limited to the transition part of the path can give information on the transient timescale of the process. It is in this kind of processes that two-point boundary algorithms can play their role. The computation of a single trajectory can be interesting also from the point of view of path sampling. As pointed out by several authors (see, e.g., Bai and Elber [3] and references therein) two-point boundary algorithms applied to a description where all microscopic degrees of freedom are considered explicitly can become exceedingly expensive when the total time of the path becomes too large. Nevertheless, Bai and Elber point out that short time

CP1018, *Frontiers of Fundamental and Computational Physics*,
edited by B. G. Sidharth, F. Honsell, O. Mansutti, K. Sreenivasan, and A. De Angelis

trajectories can become statistically important if one focuses on the transitional part of a reactive path from a reactant to a product.

However, the introduction of a novel variational principle (although already well known in the field of differential geometry and classical mechanics of periodic orbits [4]) into the realm of complex multidimensional systems is useful *per se*. If, as we hope to have achieved, a correspondent computational strategy for obtaining accurate paths is developed, then we have a new tool for understanding the microscopic mechanisms underlying a physical or chemical process.

TONELLI PRINCIPLE AND ITS FINITE DIMENSIONAL REDUCTION

We present here the mathematical details of these techniques. We will focus on the construction of the tools we need, while we will omit the proofs (that can be found in our paper [1]). Let's consider a system of $\mathcal{N}$ particles interacting through a potential V at a fixed total energy E. The potential has to satisfy some important hypothesis (about its smoothness and its growth at infinity), but they are sufficiently weak to include the Lennard-Jones potential in the set of admissible choices. As announced, we are interested in trajectories with fixed extremal values, say $q(0)$ and $q(1)$. The total time of the transition is an unknown of the problem, but it can be computed starting from the geometric description of the dynamical path, like in the Maupertuis formulation. This is why, for the moment, we work in a reference time interval $[0,1]$. For each curve, sufficiently smooth, joining the two extremals and for every reasonable value of E, we can define

$$\mathcal{T}_E(q(\cdot)) = \int_0^1 K(\dot{q}(\tau))\,d\tau \cdot \int_0^1 (E - V(q(\tau))\,d\tau \qquad (1)$$

where K denotes the kinetic energy of the system. The stationary points of this functional are the traces of the mechanical curves: we mean that they describe the correct paths (at the desired energy E) with a fictitious parametrization. The interesting point is that only a simple liner reparametrization is needed to recover the physical trajectory (and so the exact total time). If we define

$$\omega^2 := \frac{\int_0^1 (h - V(q(\tau)))\,d\tau}{\int_0^1 K(\dot{q}(\tau))\,d\tau} \qquad (2)$$

the total time will be ω^{-1}.

The determination of a saddle point of a functional is however a difficult and demanding task. Increasing the number of degree of freedom, the computer time becomes easily out of reach. The finite dimensional reduction techniques can work in the direction of lowering the dimensionality of the problem, without loosing precision and accuracy. For that we need to adapt them to the Tonelli functional and to derive an efficient algorithm for the coupling. The first step is the introduction of the Fourier representation of the trajectories. We can write

$$q(t) = q_0 + (q_1 - q_0)t + \sum_{n \in \mathbf{N}} a_n \sin(\pi n t) \qquad (3)$$

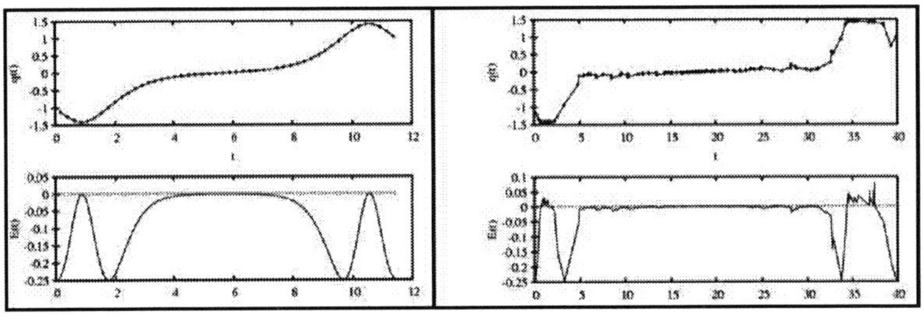

FIGURE 1. A comparison between the results of the Tonelli functional (on the left) and Maupertuis one (on the right) for a one dimensional double well potential with $E = 0.001$. the fact that potential energy is larger than total energy in the right panel (an absurd fact) is a signal that the algorithm had problems to converge in the Maupertuis' case. The panel is a snapshot of the situation during the failed optimization process for V E. Time units on the x-axis are consistent with the arbitrary units of the potential.

The coefficients a_n are now our unknowns. The finite dimensional reduction consists in dividing them in two sets $u = (a_1, \ldots, a_N)$ and $v = (u_{N+1}, \ldots)$ and in treating them differently. Also the value of ω is at this stage an unknown and we will add it to the finite vector u. We can prove that for every choice of the vector u, the following functional G_u is a contraction map over the space of the v (infinite dimensional) vectors. The only requirement is that N is sufficiently large.

$$[G_u(v)]_n := \frac{1}{(\pi n)^2} \left[\frac{\nabla V\left(q_0 + (q_1 - q_0)t + u + v\right)}{\omega^2} \right]_n \qquad \text{for } n > N \tag{4}$$

Moreover, if we are able to find $\bar{u}$, the vector corresponding to the dynamical path, the functional $G_{\bar{u}}$ will give the related $\bar{v}$ as its unique fixed point. Clearly, the search of the fixed point of a contractive map is an extremely convenient numerical procedure. For that, a second cut in the Fourier series is needed. We have to choose a value M that will be our precision level (the map G is still contractive with this cut and its rate of convergence is even increased). Exploiting the fact that G is contractive for every possible choice of u, we derived the following algorithm.

- set q_0 and q_1, the initial and final points of the trajectory and find an initial path π_0 connecting q_0 and q_1.
- *start iteration:* expand the path π_0 in Fourier series, and keep the first N harmonics. This generates the finite dimensional part of the problem, that can be solved using Tonelli principle and iterative methods like the conjugate residual. In the first step, set the harmonics $N+1$ to M to zero. This generates a new path $u_1 = \sum_1^N a_n \sin(\pi n t)$;
- u_1 is inserted into G, which is a contraction if N was carefully chosen. That leads to a set v_1, which can be referred to as the Fourier components $N+1$ to M of the trajectory;
- the new path π_1 is found as $\pi_1 = u_1 + v_1$ and we can compute the time reparametrization, recovering ω in (2) ;
- iterate the procedure until convergence.

RESULTS

The Tonelli functional presents some interesting features that can be appreciated also in very little and simple systems. We show the performance of a classical stationarization for a one dimensional double well potential. We took a total energy $E = 0.001$ very close to the maximum of the potential energy and we compare the results of the Tonelli and the classical Maupertuis functional. As you can see in Figure 1, there are two kind of advantages: the trajectory computed with Tonelli is more accurate and it does not presents problems in the regions where $V \simeq E$. Moreover the total computed time is exactly equal to the known real one.

The coupling between this new functional (new in the field of molecular dynamics) and the reduction technique has been tested on a 38-atom Lennard-Jones cluster. More precisely we look for the isomerization of this cluster. For this test we set $N = 32$ and $M = 320$. The map G is contractive with these choices, the stationarization of the finite part with Tonelli is not too demanding and the precision of the computed trajectory is assured by the high value of M. In Figure 2 you can see the improvements gained in passing from a 32-slice description and the complete 320-component algorithm.

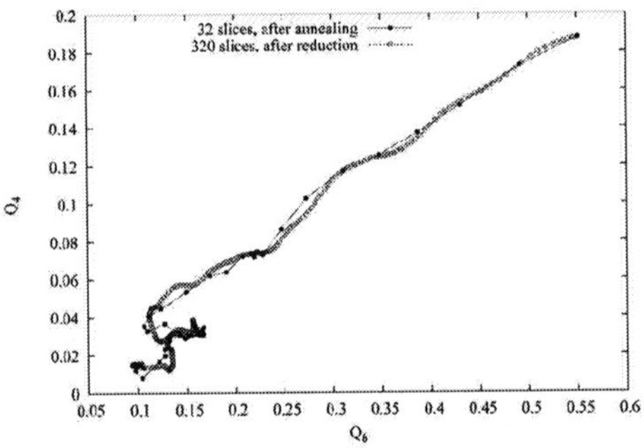

FIGURE 2. Comparison between the orientational order parameters for the coarse trajectory (32 slices) and for the complete one at the same energy.

REFERENCES

1. A. Turco, D. Passerone, F. Cardin, *Tonelli Principle: finite reduction and fixed energy molecular dynamics trajectories*, to appear (2008)
2. Tonelli L., *Sulle orbite periodiche*, Rend. R. Accad. dei Lincei, XXI-1 (1912) pp. 251-258 and 332-334.
3. D. Bai and R. Elber, *Calculation of Point-to-Point Short-Time and Rare Trajectories with Boundary Value Formulation*, J. Chem. Theory Comput. 2, 484 (2006).
4. Ambrosetti A., Coti Zelati V., *Periodic solutions of singular Lagrangian systems.* Progress in Nonlinear Differential Equations and their Applications, 10. Birkhäuser Boston, Inc., Boston, MA, 1993. xii+157 pp.

First-Principles Investigations of Intrinsic and Si-doped GaAs Nanowires: Structural Stability and Electronic Properties

Nahid Ghaderi[a,b], Maria Peressi[a,c] and Nadia Binggeli[a,d]

[a]Theory@Elettra Group, CNR-INFM DEMOCRITOS National Simulation Center, Trieste
[b]Department of Physics, Isfahan University of Technology, Iran
[c]Department of Theoretical Physics, University of Trieste, Strada Costiera 11, I-34014 Trieste
[d]The Abdus Salam International Centre for Theoretical Physics, Strada Costiera 11, I-34014 Trieste

Abstract. High quality GaAs nanowires (NWs) are nowadays experimentally grown on different substrates, regularly shaped and nicely oriented. At variance with the bulk phase, they show a prevalence of wurtzite structure with respect to zincblende and a p-type behavior with respect to the possible amphoteric behavior upon Si incorporation. Motivated by these recent experimental findings we investigate by first principles pseudopotential calculations the structural stability and electronic properties of GaAs NWs. The nature of Si dopants in GaAs NWs and the relative stability of donors and acceptors have been also studied. Our results show that wurtzite NWs are more stable than zincblende NWs for diameters of 50 Angstrom or less. On the basis of the formation energy only, Si dopants have amphoteric behaviour in GaAs NWs and segregates at the surface of the wire.

Keywords: Semiconductors, Nanowires; GaAs; zincblende; wurtzite; Density Functional Theory; First-principles Pseudopotentials; Donors; Acceptors
PACS: 61.46.Km; 62.23.Hi; 71.15.-m; 71.55.-i; 73.22.-f

INTRODUCTION

Semiconductor nanowires (NWs) offer nowadays new perspectives for nano-electronic devices. Recently, high quality GaAs NWs have been experimentally grown on different substrates: they are regularly shaped and nicely oriented, with wurzite (WZ) or zincblende (ZB) structure depending on the substrate and on the growth conditions [1]. It is well known that Si has amphoteric behaviour in bulk; in NWs, at variance, mainly p-type doping is observed upon Si incorporation.

The main issues of the present work are therefore to understand by means of accurate first-principles total energy and electronic structure calculations why NWs are observed also in WZ structure and why the behaviour of Si dopants in NWs is different with respect to the bulk.

CP1018, *Frontiers of Fundamental and Computational Physics*,
edited by B. G. Sidharth, F. Honsell, O. Mansutti, K. Sreenivasan, and A. De Angelis

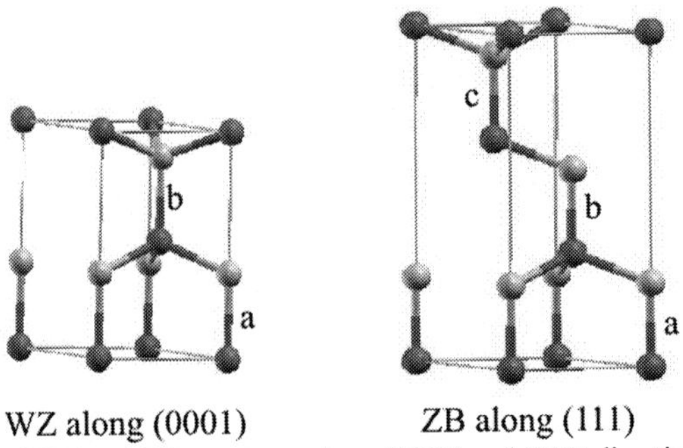

FIGURE 1. Stacking of WZ and ZB structures along (0001) and (111) directions respectively; a,b,c indicate the non equivalent anion-cation double layers.

NUMERICAL APPROACH

Calculations have been performed within the framework of Density Functional Theory in the local density approximation using first-principles pseudopotentials to describe the valence electrons and plane wave as a basis set to expand the electronic wavefunctions. Periodically repeated supercells are used for a convenient reciprocal space formulation of the problem. The Quantum-Espresso package has been used [2].

The relative stability of bulk ZB phase of GaAs is correctly described by our numerical simulations, that give a difference in the cohesive energy between ZB and WZ phase of ~11 meV/atom, in good agreement with the literature. It is meaningful to compare ZB and WZ structures along (111) and (0001) directions respectively, where they are similar but with a different stacking of anion-cation double layers: the stacking sequence for ZB is *abcabc*... and for WZ is *ababab*... (Fig. 1).

A NW is characterized by its structure (ZB or WZ), its shape around the growth axis (Triangular (T) or Hexagonal (H)) and the number of atoms in each double layer ((n_a,n_b,n_c) for ZB and (n_a,n_b) for WZ) (Fig. 2). In all cases the exposed facets are non – polar, containing the same number of Ga and As atoms.

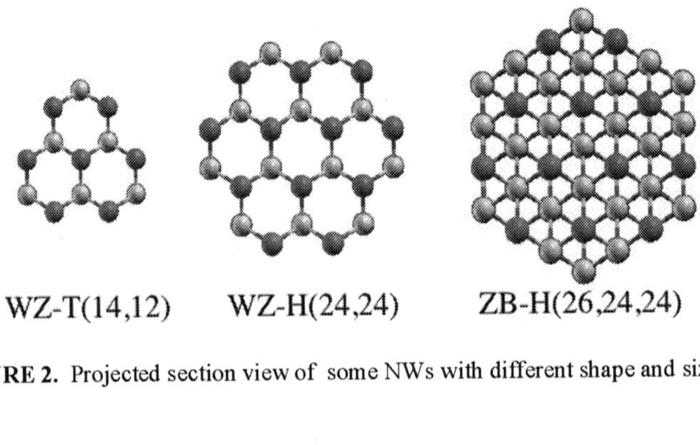

FIGURE 2. Projected section view of some NWs with different shape and size.

RESULTS

Undoped GaAs Nanowires

We have studied the cohesive energy of WZ-H, WZ-T and ZB-H GaAs NWs. We have performed full supercell calculations for NWs with diameter up to about 20 Angstrom, corresponding to the structure WZ-H(54,54). We have calculated the cohesive energy as:

$$E_c = \left\{\sum_i n_i \mu_i - E_{tot}\right\} / \sum_i n_i$$

where E_{tot} is the total energy of the supercell containing the GaAs NW, μ_i is the chemical potential of atoms i (i=Ga or As) and n_i is the number of i atoms.
Similarly to findings reported in Ref. [3] for InP NWs, GaAs NWs are more stable in WZ than in ZB structure for diameters smaller than 20 Angstrom.

The stability is mainly determined by the dangling bonds (DB) at the NWs facets: in WZ NWs 3-fold coordinated atoms are present (we refer to them as DB_1), whereas in ZB NWs also 2-fold coordinated atoms (DB_2) are present.

Following Ref. [3], we can express the NWs cohesive energy in terms DBs contributions:

$$E_c^{NW}(WZ) = E_c^{bulk}(WZ) - \frac{N_{DB}^1}{N_{tot}} E_{DB}^1 \quad, E_c^{NW}(ZB) = E_c^{bulk}(ZB) - \frac{N_{DB}^1}{N_{tot}} E_{DB}^1 - \frac{N_{DB}^2}{N_{tot}} E_{DB}^2$$

where $E_c^{(bulk\ ZB)}$ is the cohesive energy of the bulk ZB (similarly for WZ), $E^{(1)}_{DB}$ and $E^{(2)}_{DB}$ are the energy contributions of 3-fold and 2-fold coordinated atoms respectively and $N^{(1)}_{DB}$ and $N^{(2)}_{DB}$ their number in the NW. Fitting data obtained from full supercell NWs calculations it is possible to estimate $E^{(1)}_{DB}$ and $E^{(2)}_{DB}$, that can be used in the previous Equation to find the cohesive energy for ZB and WZ NWs with larger diameter. We obtain: $E^{(1)}_{DB}(WZ)$=0.54 eV, $E^{(1)}_{DB}(ZB)$=0.48 eV, $E^{(2)}_{DB}(ZB)$=1.17 eV. On the basis of these values, we estimate that undoped WZ NWs are more stable than ZB NWs up to a diameter of at least 50 Angstrom; for larger diameters (between ~50 and 100 Angstrom) the two NW structures become energetically equivalent within the numerical accuracy of the cohesive energy; for even larger diameters the DB contribution reduces with respect to the volume, going towards the limit of the two corresponding bulk structures with ZB slightly favoured with respect to WZ.

We point out that, in spite of the presence of surface DB, GaAs NWs are semiconducting. Surface relaxations (As move outwards, Ga inwards) act in the direction of compensating DBs.

Si doped GaAs Nanowires

It is well known that Si atoms have amphoteric behaviour in bulk GaAs, substituting both Ga and As atoms. However, donors are typically favoured in

Molecular Beam Expitaxy growth conditions. We have calculated the formation energy of neutral Si donors and acceptors in both ZB and WZ bulk GaAs matrices using 32-atoms supercells with one Si atom substituting one Ga or one As respectively. Our results are in agreement with findings in Ref. [4]: the formation energies for Si donors and acceptors with respect to Si chemical potential are:

$$\Delta\Omega_d^{ZB}(\Delta\mu,\mu_{Si}) = 1.46eV + \frac{\Delta\mu}{2} - (\mu_{Si} - \mu_{Si}(bulk))$$

$$\Delta\Omega_a^{ZB}(\Delta\mu,\mu_{Si}) = 1.27eV - \frac{\Delta\mu}{2} - (\mu_{Si} - \mu_{Si}(bulk))$$

$$\Delta\Omega_d^{WZ}(\Delta\mu,\mu_{Si}) = 1.58eV + \frac{\Delta\mu}{2} - (\mu_{Si} - \mu_{Si}(bulk))$$

$$\Delta\Omega_a^{WZ}(\Delta\mu,\mu_{Si}) = 1.39eV - \frac{\Delta\mu}{2} - (\mu_{Si} - \mu_{Si}(bulk))$$

($\Delta\mu$ is the difference between the chemical potentials of Ga and As).

The formation energy of Si dopants in NWs depends on intrinsic characteristics of the host NW (structure, shape and diameter) and on the particular substitutional site, which can vary from a bulk-like position close to the center of the NW to different non-equivalent surface positions (middle of facets, close to edges, fully 4-fold or 3-fold or 2-fold coordinated sites). As a general trend, we found that dopants are more stable on the facets of the NWs. Concerning the relative stability of donors with respect to acceptors, the trend is similar to the bulk case: Si donors are favoured with respect to acceptors. Detailed results will be presented in a future work.

CONCLUSIONS

Undoped GaAs NWs are stabilized for small diameter in WZ rather than in ZB structure. The possibility of bistability for larger diameters is consistent with the experimental observations, depending on the growth mechanism. Similarly to bulk, Si is found to have an amphoteric behaviour also in the NWs, based on energetics only. The experimental observation of p-type behavior mainly of Si-doped NWs could be due to the kinetics of Si incorporation.

ACKNOWLEDGMENTS

Computational resources from CINECA (Italy) have been used, under agreements with CNR-INFM and with the University of Trieste.

REFERENCES

1. M. Piccin, G. Bais, V. Grillo, F. Jabeen, S. De Franceschi, E. Carlino, M. Lazzarino, F. Romanato, L. Businaro, S. Rubini , F. Martelli and A. Franciosi, Physica E **37**, 134 (2007).
2. http://www.quantum-espresso.org and http://www.pwscf.org
3. T. Akiyama, K. Nakamura and T. Ito, Phys. Rev. B **73**, 235308 (2006).
4. J.E. Northrup, Phys. Rev. Lett. **62**, 2487 (1989).

Interaction of carbon dioxide with Ni(110): a combined experimental and theoretical study

X. Ding[a], L. De Rogatis[b,c], E. Vesselli[c,d], A. Baraldi[c,d], G. Comelli[c,d], R. Rosei[c,d], L. Savio[e], L. Vattuone[e,f], M. Rocca[e,g], P. Fornasiero[b], F. Ancilotto[a,h], A. Baldereschi[a,i,l], M. Peressi[a,i]

[a]*Theory@Elettra Group, CNR-INFM DEMOCRITOS National Simulation Center, Trieste*
[b]*Department of Chemical Sciences, University of Trieste, Italy*
[c]*CNR-INFM TASC Laboratory, Area Science Park, Trieste, Italy*
[d]*Department of Physics, University of Trieste, Italy*
[e]*Department of Physics, University of Genova, Italy*
[f]*CNISM, Unita' di Genova, Italy*
[g]*IMEM-CNR, Sezione di Genova, Italy*
[h]*Department of Physics, University of Padova, Italy*
[i]*Department of Theoretical Physics, University of Trieste, Italy*
[l]*Institute of Theoretical Physics, EPFL, Lausanne, Switzerland*

Abstract. We performed a combined experimental and theoretical study of CO_2 adsorption on the Ni(110) surface with the aim of understanding the details of the molecule-surface interaction, thus identifying the adsorption sites and evaluating the relevant energy barriers in the adsorption and in the diffusion processes. In this contribution we mainly focus on the results of numerical simulations, performed within the framework of density functional theory using ab-initio pseudopotentials. We discuss the results obtained together with experimental data.

Keywords: Surface Science; Carbon dioxide; Ni surfaces; Chemisorption; Density Functional Theory; First-principles Pseudopotentials; TPD; XPS; HREELS
PACS: 68.43.-h; 68.43.Bc; 73.20.Hb

INTRODUCTION

CO_2 is a key compound e.g. for methanol synthesis, according to the reaction $CO_2+3H_2 \longleftrightarrow CH_3OH + H_2O$ [1]. Cu-based catalysts are normally used, but it has been pointed out [2] that Cu/Ni catalysts are 60 times more active for CO_2 hydrogenation than traditional pure Cu catalysts. It is of crucial interest to understand the complete reaction path, but also the basic question concerning the CO_2 interaction with pure Ni is still open.

Interestingly, CO_2 chemisorption in Ultra-High Vacuum is observed only on Ni(110) and not on other low Miller index surfaces, where on the contrary it is typically promoted only by pre-adsorbed alkali atoms. The available experimental and theoretical data agree in relating CO_2 chemisorption on Ni(110) with the formation of

CP1018, *Frontiers of Fundamental and Computational Physics*,
edited by B. G. Sidharth, F. Honsell, O. Mansutti, K. Sreenivasan, and A. De Angelis

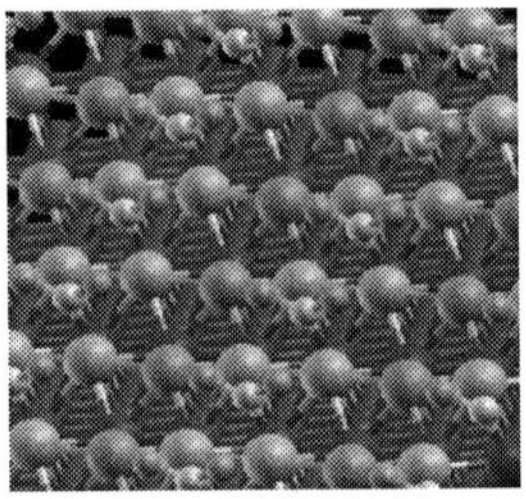

FIGURE 1. Ball-and-stick model of chemisorbed CO_2 on Ni(110) obtained by periodically repeating the supercell used for the simulation in the directions parallel to the surface.

CO_2^-, a charged and bent species [1]. However, it is not yet clear which is the most favorable adsorption geometry or whether there are different possible configurations and how their occupation depends on coverage and temperature. In absence of direct information about the chemisorption configuration of CO_2 on Ni(110), a symmetric C_{2v} configuration with the molecule in a reversed "V" shape (bound to surface with the two O atoms) has been proposed, in analogy with formate (HCOO), which is known to bind to metallic surfaces mainly in a reversed "Y" configuration. Our aim is to clarify this issue by combining experimental and theoretical investigations, the former including Temperature-Programmed Desorption (TPD), X-ray Photoelectron Spectroscopy (XPS) and High Resolution Electron Energy Loss (HREELS) measurements, the latter based on accurate first-principles numerical simulations.

EXPERIMENTAL RESULTS

TPD and XPS indicate the presence of two CO_2 species, physisorbed and chemisorbed. The TPD peaks are at ~100 K and ~220 K, corresponding to desorption energies of ~0.26 eV for physisorption and ~0.60 eV for chemisorption, on the basis of a simple application of the Readhead approximation. No direct information about adsorption sites and geometries can be derived by these experiments only.

THEORETICAL RESULTS

We performed calculations within the framework of Density Functional Theory (DFT) in the local density approximation using first-principles pseudopotentials to describe the valence electrons and plane wave as a basis set to expand the electronic wavefunctions. The Quantum-Espresso package has been used [2]. For a convenient reciprocal space formulation of the problem, periodically repeated supercells with slab geometry are used, therefore describing a fictitious order on surface (Fig. 1) that is not observed experimentally. However, the supercell used (3x2 on surface) is large enough to guarantee that the molecule-molecule interactions are very small.

The results of total energy and force minimization suggest the existence of different inequivalent chemisorption sites, short-bridge (SB) and hollow (H) sites, all characterized by the bent, negatively charged molecule, and a predominant surface-carbon coordination. In the SB site, the molecular plane is perpendicular to the surface and parallel to the [1-10] direction; both a fully symmetric configuration (C_{2v} symmetry) and a slightly asymmetric one (C_s symmetry) are found to be stable. They are however very similar. The most stable configuration turns out to be very similar to the one reported by Wang et al. [4], i.e. with C almost in the hollow site, with a dominant carbon coordination and with Ni and oxygen atoms pointing upwards and the molecular plane inclined with respect to the Ni surface normal (hereafter indicated as *hollow-up* (HU) configuration) (Fig. 2).

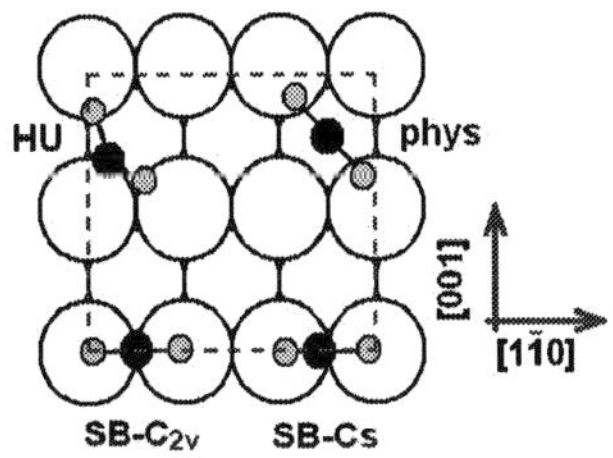

FIGURE 2. Schematic top view of Ni(110) with CO_2 in different possible adsorption sites.

DISCUSSION

SB and HU configurations differ in energy by only 0.07 eV/molecule and are separated by a diffusion barrier SB—>HU of 0.15 eV/molecule, calculated using the Nudged Elastic Band approach [5]. The adsorption energy in the SB site is of 0.26 eV. A further indication coming from numerical simulations is that the chemisorption process is activated, with an activation barrier slightly smaller than 0.1 eV/molecule. These values are compatible with the measured desorption energies and with a simultaneous population of both HU and SB sites at low temperature.

Vibrational frequencies have been calculated for the different chemisorption configurations and for a linear, neutral metastable configuration with the CO_2 molecule parallel to the surface, which is representative of a 'physisorbed' state. A clear correspondence can be established between the calculated values and the main peaks detected by HREELS (Table 1). With the exception of the hindered rotation modes, that are not observable due to dipole scattering rules, all the other modes characterizing the adsorption geometries can be identified; in particular, one peak detected at about 168 meV can be interpreted as a clear fingerprint of the presence of the HU configuration and attributed to its asymmetric stretching mode, which is clearly different from the SB case.

TABLE 1. Vibrational frequencies (in meV) obtained from DFT calculations for different adsorption configurations, and the corresponding (in parenthesis) peaks observed in the HREELS spectra.

Vibrational mode	HU	SB	'phys'
Asymmetric stretching	174 (168)	218 (-)	287 (290)
Symmetric stretching	138 (141)	142 (141)	160 (168)
Bending	90 (90)	81 (81)	79 (81)
Hindered rotation (out of plane)	52 (-)	63 (-)	78 (-)
External stretching	47 (46)	44 (46)	25 (-)
Hindered rotation (in plane)	41 (-)	35 (-)	25 (-)

The transition states reached during the adsorption process both in the SB sites and in the HU sites are characterized by sudden changes of the O-C-O bond angle and atomic charges with respect to the gas phase, indicating that *bending* and *electron transfer* are two important features in the chemisorption process. A charge density plot is shown in Fig. 3 for the molecule in the SB site.

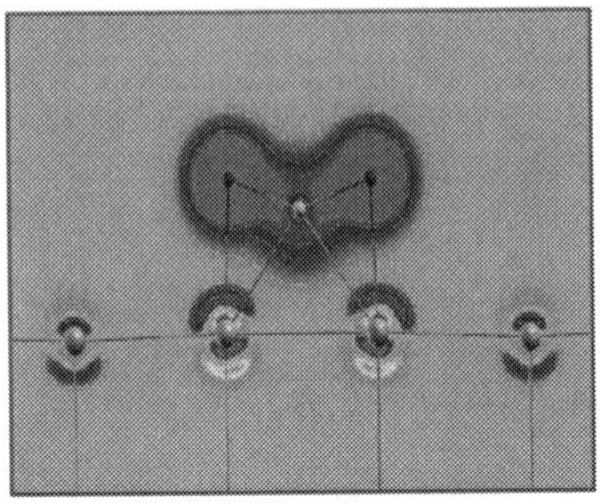

FIGURE 3. Contour plot of the differential charge density distribution of CO_2 adsorbed on SB site on Ni(110). The difference is with respect to the neutral molecule and the neutral Ni slab considered separately but in the same positions of the chemisorbed configuration.

CONCLUSIONS

Experimental and theoretical investigations concerning energetics and vibrational frequencies yield consistent indications about two inequivalent adsorption sites for CO_2 on Ni(110) that can be simultaneously populated at low temperature: short-bridge site with the molecular plane perpendicular to the surface and hollow site with the molecular plane inclined with respect to the surface. In both cases CO_2 is negatively charged and bent. At variance with widely accepted previous literature data, in both sites the molecule has pure carbon or mixed oxygen-carbon coordination with the metal.

ACKNOWLEDGMENTS

Computational resources from CINECA (Italy) have been used, under agreements with CNR-INFM and with the University of Trieste.

REFERENCES

1. H.-J. Freund and M.W. Roberts, Surf. Sci. Rep. **25**, 225 (1996).
2. J. Nerlov and I. Chorkendorff, Catal. Lett. **54**, 171 (1998); J. Catal. **181**, 271 (1999); J. Nerlov, S. Sckerl, J. Wambach, and I. Chorkendorff, Appl. Catal. A **191**, 97 (2000).
3. http://www.quantum-espresso.org and http://www.pwscf.org
4. S.G. Wang, D.B. Cao, Y.W. Li, J.G. Wang, and H.J. Jiao, J. Phys. Chem. B **109**, 18956 (2005).
5. H. Jonsson, G. Mills, and K.W. Jacobsen, in *Classical and Quantum Dynamics in Condensed Phase Simulations*, edited by B.J. Berne, G. Ciccotti and D.F. Coker (World Scientific, 1998).
6. X. Ding, L. De Rogatis, E. Vesselli, A. Baraldi, G. Comelli, R. Rosei, L. Savio, L. Vattuone, M. Rocca, P. Fornasiero, F. Ancilotto, A. Baldereschi, and M. Peressi, Phys. Rev. B **76**, 195425 (2007).

Interaction of a CO molecule with a Pt monoatomic chain: the top geometry

G. Sclauzero*, A. Dal Corso*, A. Smogunov†,* and E. Tosatti†,*

*International School for Advanced Studies (SISSA) and CNR-INFM Democritos National Simulation Center, Via Beirut 2-4, I-34014 Trieste, Italy
†International Centre for Theoretical Physics (ICTP), Strada Costiera 11, I-34014 Trieste, Italy

Abstract. Recent experiments showed that the conductance of Pt nanocontacts and nanowires is measurably reduced by adsorption of CO. We present *DFT* calculations of the electronic structure and ballistic conductance of a Pt monoatomic chain and a CO molecule adsorbed in an on-top position. We find that the main electronic molecule-chain interaction occurs via the 5σ and $2\pi^*$ orbitals of the molecule, involved in a donation/back-donation process similar to that of CO on transition-metal surfaces. The ideal ballistic conductance of the monoatomic chain undergoes a moderate reduction by about 1.0 G_0 (from 4 G_0 to 3.1 G_0) upon adsorption of CO. By repeating all calculations with and without spin-orbit coupling, no substantial spin-orbit induced change emerges either in the chain-molecule interaction mechanism or in the conductance.

Keywords: platinum nanowire, carbon monoxide, ballistic conductance, spin-orbit coupling
PACS: 73.63Rt, 73.23.Ad, 71.15Mb, 71.15Rf

INTRODUCTION

The conductance of metallic nanocontacts produced by scanning tunneling microscopy (*STM*) or by mechanically controllable break-junctions (*MCBJ*) is routinely measured and reported in conductance histograms [1]. The nanocontacts are much shorter in size than the electron mean free path so that the conductance is ballistic and quantized, rather than diffusive and non quantized. Quantization is most readily observed in the form of peaks in conductance histograms obtained by superposing the conductance traces of many break junctions events. In some metals, such as gold and platinum, short tip-suspended monoatomic chains can be formed at the nanocontact [2, 3, 4]; in these cases the main histogram peak with the lowest conductance is due to the atomic chain. This peak is usually found near 1 G_0 ($G_0 = 2e^2/h$) for clean Au nanowires [1, 2] and at about $1.5 - 2.0$ G_0 in Pt, where $5d$ electrons contribute to the conductance [5, 6]. According to Landauer's theory, ballistic conductance is determined by electron transmission and reflection at the contact. The presence of small adsorbates on the nanowire or in the nanocontact may naturally change electron transmission and reflection and alter substantially the conductance. In the case of Pt nanocontacts this effect has been demonstrated experimentally for H_2 [7] and for CO [5, 6]. For instance, in presence of CO the conductance histogram of Pt is modified to exhibit two peaks at about 0.5 G_0 and 1.0 G_0 [5, 6]. Only the lower peak near 0.5 G_0 has been discussed theoretically [8], while that at 1 G_0 has not. At the chemical level, in the experimental work of Kiguchi et al. [6] it was suggested that the relative strength of CO adsorption on Au, Cu, Pt and Ni nanocontacts could be explained by the Blyholder model [9], but the binding between a CO molecule

CP1018, *Frontiers of Fundamental and Computational Physics*,
edited by B. G. Sidharth, F. Honsell, O. Mansutti, K. Sreenivasan, and A. De Angelis

and a Pt nanowire has not been addressed in this respect. Our calculations presently fill this gap. Finally, the effect of spin-orbit (*SO*) coupling on the molecule-chain binding and on the nanocontact ballistic conductance has not been taken into account in previous calculations, even though it is known that *SO* is so important to change significantly the electronic band structure of Pt monoatomic wires [10, 11].

In this work we study the interaction between CO and a Pt nanowire using a very simplified model: we consider a straight, isolated, Pt atomic chain with one molecule of CO adsorbed on-top (see the inset in Fig. 2). Calculations use a periodically repeated supercell containing N_{Pt} atoms of Pt and the molecule with its axis perpendicular to the chain[1]. We perform electronic structure calculations in the framework of density functional theory (*DFT*) [12] within the local density approximation (*LDA*) for the exchange and correlation energy using the *PZ* functional [13] (the optimized distances and the chemisorption energies have been calculated also with the gradient corrected *PBE* functional [14]). The role of *SO* coupling is investigated by comparing the results obtained with scalar-relativistic (*SR*) and fully-relativistic (*FR*) ultrasoft pseudo potentials[2] (*US-PP*) [16], the latter including the *SO* effect [15]. The ballistic conductance has been calculated with the Landauer-Büttiker formula, $G_0 = 2e^2/h\, T(E_F)$, evaluating the transmission T (obtained with the method described in Refs. 17, 18, 11) at the Fermi energy E_{F}. All calculations have been performed with the computer codes contained in the QUANTUM-ESPRESSO package [19].

ELECTRONIC STRUCTURE AND BALLISTIC CONDUCTANCE

We optimized the platinum-carbon distance $d_{\mathrm{Pt-C}}$ and the intermolecular distance $d_{\mathrm{C-O}}$, while keeping the Pt atoms aligned at their theoretical equilibrium distance in the isolated chain (2.34 Å with *LDA*, 2.39 Å with *PBE*). Within *LDA* the optimal geometrical parameters are $d_{\mathrm{Pt-C}} = 1.82$ Å (1.81 Å) and $d_{\mathrm{C-O}} = 1.14$ Å (1.14 Å) in the *SR* (*FR*) case, while the *SR-PBE* calculation gives slightly longer distances: $d_{\mathrm{Pt-C}} = 1.84$ Å and $d_{\mathrm{C-O}} = 1.15$ Å. The adsorbed C-O distance is slightly longer than the equilibrium value calculated for the isolated molecule, which is 1.13 Å with *LDA* and 1.14 Å with *PBE*. The chemisorption energy E_{chem} is calculated as the difference between the energy of the optimized geometry and the sum of the energies of the isolated wire and of the isolated molecule. With *PBE*, which usually gives realistic estimates, we get $E_{\mathrm{chem}} = -1.4$ eV, while *LDA*, which is known to overbind, gives larger values, $E_{\mathrm{chem}} = -1.9$ eV in the *SR* case and $E_{\mathrm{chem}} = -2.0$ eV in the *FR* case. Our results confirm that the interaction between the CO molecule and the Pt chain can be rationalized by the Blyholder model (and its improvements), used to explain the chemical bond between CO and transition metal surfaces [9, 20]. The CO–metal bond is characterized by a balance between donation

---

[1] We verified the convergence of our results with respect to N_{Pt} and to the other parameters of the supercell. In particular the *PDOS* and the transmission have been calculated with $N_{\mathrm{Pt}} = 17$.

[2] Within *LDA*, the Pt pseudopotential is that given in Ref. 15, while for C and O we generated new *US-PP*. With *PBE* we generated new *SR US-PP* for all the atoms (the parameters will be reported elsewhere). We adopt a plane wave basis set with the following cut-off for the kinetic energy of the wavefunctions (charge density): *SR-LDA*, 29 (300) Ry; *FR-LDA*, 32 (300) Ry; *SR-PBE* 32 (320) Ry.

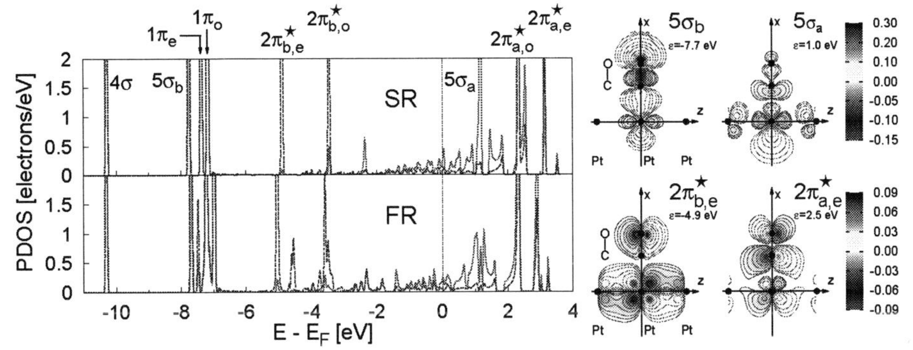

FIGURE 1. Left: sum of the *PDOS* on the atomic orbitals of C (solid lines) and O (dashed lines) in the *SR* and *FR* cases. Right: *SR-KS* eigenstates associated to some of the peaks in the *SR-PDOS*.

of 5σ electrons from the molecule to the metal and back-donation from the Pt d-bands to the empty $2\pi^\star$ molecular orbital, which gets partially filled by electrons. This can be argued by looking at the density of states projected on valence orbitals of C and O (*PDOS*), shown in Fig. 1 for the *LDA* case. The nature of the peaks in the *SR PDOS* has been identified through symmetry arguments or by directly visualizing the Kohn-Sham (*KS*) orbitals. In addition to the sharp *PDOS* peaks due to CO valence orbitals (3σ, not shown since much lower in energy, 4σ, 5σ, 1π below the Fermi energy E_F and $2\pi^\star$ above E_F), we find peaks which are not present in the *PDOS* of the isolated CO and thus are due to the interaction with Pt. In particular, the peak at 1.2 eV corresponds to a new state $5\sigma_a$ (shown in Fig. 1) due to antibonding hybridization between the 5σ orbital and the s and d bands of Pt; it lies *above* E_F and thus gives rise to electron donation. The two sharp peaks at -4.9 eV and -3.5 eV, instead, are due to bonding hybridization between the $2\pi^\star$ orbitals and the d bands of Pt: since these new states lie *below* E_F they are filled with metal electrons by back-donation. We note here that each of the π orbitals is split into an even state ($2\pi^\star_{b,e}$ and $2\pi^\star_{a,e}$, both reported in Fig. 1, and $1\pi_e$) and an odd state ($1\pi_o$, $2\pi^\star_{b,o}$ and $2\pi^\star_{a,o}$) with respect to the mirror plane which contains the chain and the molecule. We finally point out that, as in the case of CO adsorbed on surfaces, the $5\sigma_b$ level has moved below the 1π level, and that there is the same mixing between 4σ and 5σ and between 1π and $2\pi^\star$ (see Ref. 20). Comparing the position of the peaks in the *SR PDOS* with those in the *FR PDOS*, we can argue that this model of the interaction is valid also in presence of *SO* coupling. In fact, although the exact position of the peaks and their intensity may vary, we can recognize in the *FR PDOS* the same pattern of levels identified in the *SR PDOS*. This is due to the fact that the strength of the CO–Pt interaction is much larger than the effects of *SO* coupling on this binding.

Fig. 2 shows the calculated transmission curve of the ideal (junction-free) Pt chain with the adsorbed CO molecule, calculated in the *SR* and *FR* cases within *LDA*. The calculated *SR* (*FR*) conductance is about 3.1 G_0 (3.0 G_0), reduced by about 1.0 G_0 relative to the isolated Pt monoatomic chain, whose theoretical conductance is $G = 4\ G_0$ (see Ref. 11). Below E_F, between -1.6 eV and -0.4 eV we find a larger reduction in transmission with respect to the isolated chain, because the Pt states having orbital

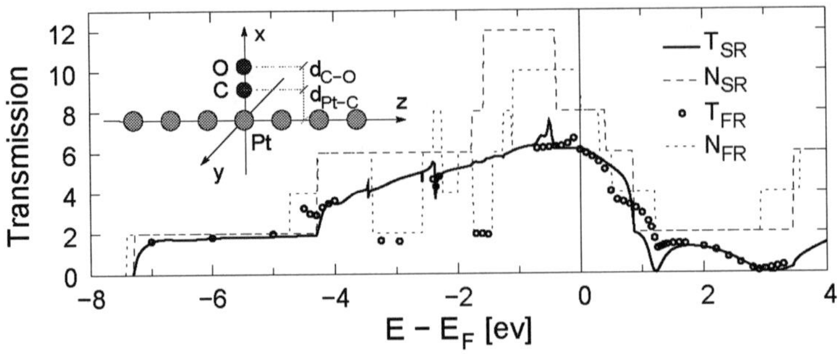

FIGURE 2. Electron transmission as a function of the energy. The *SR* (*FR*) trasmission T_{SR} (T_{FR}) is displayed with solid lines (open circles), while the transmission of an ideal monoatomic chain N_{SR} (N_{FR}), is shown with dashed (short dashed) lines. Inset: geometry and optimized distances.

angular momentum along the chain $m = \pm 2$ are heavily perturbed in the vicinity of CO. Above $E_{\rm F}$, we find a notch in the transmission curve at about 1.2 eV, precisely the energy where the new $5\sigma_a$ state is formed. The *FR* transmission agrees quite well with the *SR* one near $E_{\rm F}$, since the band with total angular momentum along the chain $m_j = \pm 5/2$, which is raised in energy by the *SO* effect, is almost completely blocked by the interaction. In fact this band derives from the $m = \pm 2$ band, which is not well transmitted in the *SR* case. The largest differences between *SR* and *FR* transmission lie in the energy ranges -3.4 eV $\leqslant E \leqslant -2.6$ eV and -1.7 eV $\leqslant E \leqslant -1.4$ eV (where the different hybridization of the *FR* bands depletes the number of available channels), but they are far away from the Fermi level and hence they do not produce, in this system, a measurable effect on the conductance.

REFERENCES

1. N. Agraït, A. L. Yeyati, and J. M. van Ruitenbeek, *Phys. Rep.* **377**, 81 (2003).
2. H. Ohnishi et al., *Nature (London)* **395**, 780 (1998). A. I. Yanson et al., *ibid.* **395**, 783 (1998).
3. R. H. M. Smit et al., *Phys. Rev. Lett.* **87**, 266102 (2001).
4. V. Rodrigues et al., *Phys. Rev. Lett.* **91**, 096801 (2003).
5. C. Untiedt et al., *Phys. Rev. B* **69**, 081401(R) (2004).
6. M. Kiguchi, D. Djukic, and J. M. van Ruitenbeek, *Nanotechnology* **18**, 035205 (2007).
7. R. H. M. Smit et al., *Nature (London)* **419**, 906–909 (2002).
8. M. Strange, K. S. Thygesen, and K. W. Jacobsen, *Phys. Rev. B* **73**, 125424 (2006).
9. G. Blyholder, *J. Phys. Chem.* **68**, 2772 (1964).
10. A. Delin, and E. Tosatti, *Phys. Rev. B* **68**, 144434 (2003).
11. A. Dal Corso, A. Smogunov, and E. Tosatti, *Phys. Rev. B* **74**, 045429 (2006).
12. P. Hohenberg, and W. Kohn, *Phys. Rev.* **136**, 864 (1964).
13. J. Perdew, and A. Zunger, *Phys. Rev. B* **23**, 5048 (1981).
14. J. P. Perdew, K. Burke, and M. Ernzerhof, *Phys. Rev. Lett.* **77**, 3865–3868 (1996).
15. A. Dal Corso, and A. Mosca Conte, *Phys. Rev. B* **71**, 115106 (2005).
16. D. Vanderbilt, *Phys. Rev. B* **41**, 7892(R) (1990).
17. H. J. Choi, and J. Ihm, *Phys. Rev. B* **59**, 2267 (1998).
18. A. Smogunov, A. Dal Corso, and E. Tosatti, *Phys. Rev. B* **70**, 045417 (2004).
19. S. Baroni et al., website: `http://www.quantum-espresso.org`.
20. A. Föhlisch et al., *J. Chem. Phys.* **112**, 1946 (2000).

Monte Carlo Simulation of the Optical Absorption of Hydrogenated Nanocrystalline Silicon Thin Films

Fatiha Besahraoui, Jamal Dine Sib, Yahia Bouizem, and Larbi Chahed

Laboratory of Physics of Thin Films and Materials for Electronics
Department of Physics, University of Oran, P. O. Box 1524, El M'Naouer, Oran, Algeria

Abstract. The optical absorption coefficient measured by Constant Photocurrent Method (CPM) for nanotextured silicon thin films is apparent affected by light scattering produced in these heterogeneous materials. A detailed Monte Carlo simulation of the absorption spectra and the random optical paths traveled by the scattered photons is presented for the case of nano-Si:H layers. The calculated values of apparent absorption coefficient and the mean optical path depend mainly on the variation of the included nanocrystallites fraction, which favors bulk light scattering phenomena. The particular structure of these materials is a key characteristic of efficient thin films solar cells.

Keywords: Monte Carlo simulation; nanocrystalline silicon; apparent optical absorption; Constant Photocurrent Method; bulk light scattering.
PACS: 02.70.Uu; 78.20.Bh; 78.66.Fd; 73.50.Pz.

INTRODUCTION

Hydrogenated nanocrystalline silicon (nano-Si:H) thin films are an heterogeneous mediums [1]. They consist of an ordered silicon nanocrystallites which have a spherical form with size varies from 2 to 4 nm included in amorphous matrix. These nanomaterials constitute an important class with some of their properties distinctly different from either amorphous or single crystals. The important key for the success of nano-Si:H as a photovoltaic (PV) absorbent material is its enhanced optical absorption compared to the monocrystalline silicon. The main reason of this optical behavior is due to its particularly structure which gives place to the bulk and/or surface light scattering phenomena [2]. The CPM measurements [3] of these heterogeneous mediums give access to an apparent absorption coefficient α_{app} affected by scattering effects, different from the true one α_{true} measured for homogeneous mediums. In this study, we will try to calculate $\alpha_{app}(E)$ spectra influenced by the characteristic parameters of a nanotextured silicon film, especially the crystalline/amorphous fraction by introducing a numerical simulation based on Monte Carlo method. This simulation is introduced in order to obtain the optimal parameters of an efficient absorbent thin-film used mainly in photovoltaic applications. The method consists to calculate α_{app} values taking into account the contribution of the

CP1018, *Frontiers of Fundamental and Computational Physics,*
edited by B. G. Sidharth, F. Honsell, O. Mansutti, K. Sreenivasan, and A. De Angelis

absorbed of specular and scattered beams. It based on the probability that the photon from the specular beam will be trapped by the included nanocrystallites and undergoes a several scattering events until its intensity will be attenuate and consequently, it will be absorbed within its traveled optical path.

SIMULATION RESULTS

Results presented in Figure 1(a) show that α_{app} values increase with the nanocrystallites number included in the bulk, especially in infrared region. This enhancement is due to the increase of the mean free path of scattered photon with nanocrystallites percentage [Fig. 1(b)]. This is in agreement with the experiment [4].

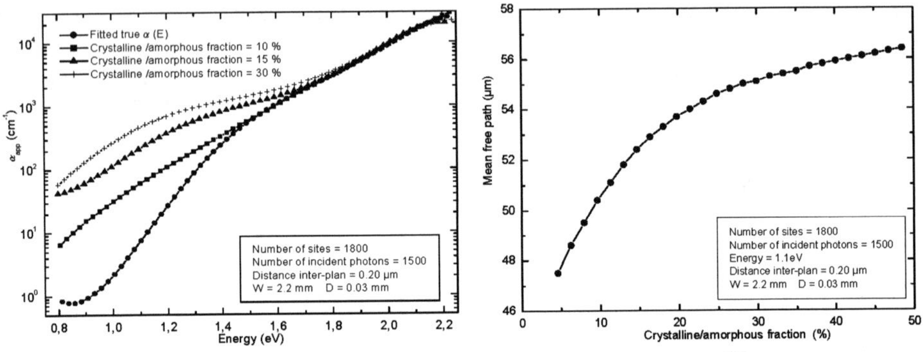

FIGURE 1. Influence of the crystalline/amorphous fraction on: (a) $\alpha_{app}(E)$ spectra calculated for nanocrystalline silicon layer, (b) calculated mean free path of the scattered photon.

CONCLUSION

The results issues from this simulation point to the conclusion that the absorbed scattered light which will contribute to the absorption enhancement in infrared region is mainly affected by the increase of nanocrystallites number. The important presence of nanocrystallites with small size in silicon layer increases the probability that the incident photon from the specular beam will be confined within the film by total internal reflections. It will remain trapped until it is either absorbed or scattered again. However, the optical path of the scattered photon increases also with the increase of the internal reflections events.

REFERENCES

1. P. Roca i Cabarrocas, A. Fontcuberta i Morral and Y. Poissant, *Thin Solid Films* **403-404**, 39-46 (2002).
2. A. Poruba, A. Fejfar, Z. Remeš, J. Špringer, M. Vaněček, J. Kočka, J. Meier, P. Torres and A. Shah, *J. Appl. Phys.* **88**, 148-160 (2000).
3. M. Vaněček, J. Kočka, J. Stuchlik and A. Třiska, *Solid State Communications* **39**, 1199–1202 (1981).
4. F. Diehl, B. Schröder and H.Oechsner, *J. Appl. Phys.* **84**, 3416-3418 (1998).

Principles of Self-Organization in Young Brain - Viewed from Kohonen Model

T. Pallaver*,†, H. Kröger† and M. Parizeau**

*Ecole Supérieure d'Optique, Campus Polytechnique, RD 128, 91127 Palaiseau cedex, France
†Département de Physique, Université Laval, Québec, Québec G1V 0A6, Canada
**Département de Génie Electrique et de Génie Informatique, Université Laval, Québec, Québec G1V 0A6, Canada

Abstract. We propose a variant of the Kohonen model to study principles of self-organization in a model of young brain. We suggest a function to measure aquired knowledge and use it to auto-adapt the topology of neuronal connectivity, yielding substantial organizational improvement relative to the standard model. In the early phase of organization with most intense learning, we observe that neural connectivity is of Small World type, which is very efficient to organize neurons in response to stimuli. In analogy to human brain where pruning of neural connectivity occurs in early life, this feature is found to stabilize neuronal response to stimuli.

Keywords: Young brain, Self-organization, Kohonen model
PACS: 87.10.+e,87.19.La

INTRODUCTION

An important principle of neural organization of the brain are the so called feature maps. This can be (i) sets of feature-sensitive cells, (ii) ordered projections between neuronal layers or (iii) ordered maps of abstract features. An example of the first kind are cells responding to human faces. Examples of the second type is the retinotopic map of the visual field or the somatotopic 'homunculus' in the sensorimotor cortex. Examples of the third type is the color map in the visual area V4, the directional hearing map in the owl midbrain [1] and the target-range map in the bat auditory cortex [2]. More than one response property can be mapped simultaneously within an area, as has been shown in visual, somatosensory and auditory cortex [3, 4, 5, 6]. As no receptive surface exists for such abstract features, the spatial order of representations must be produced by some self-organizing process in brain, which occurs mainly postnatally. It has been proposed that those maps need to smoothly map several response properties onto a two-dimensional cortical surface (dimensional reduction), which has led Kohonen [7] and later Durbin et al. [8], Obermayer et al. [9] and Swindale [10] to propose models of formation of such feature maps. The biological relevance of Kohonen model has been established in a number of cases. Obermayer et al. [9, 11], Goodhill [12], Wolf et al. [13], Swindale and Bauer [14] have shown that the Self-Organized Map (SOM, Kohonen model) is able to produce mappings that bear a detailed resemblance to real visual cortex maps. Swindale et al. [15] analyzed visual cortex maps in cat and found coverage optimization important for cortical map development, lending support to dimension reduction models of Kohonen [7] and Durbin et al. [8]. Swindale [16] used the Kohonen

CP1018, *Frontiers of Fundamental and Computational Physics*,
edited by B. G. Sidharth, F. Honsell, O. Mansutti, K. Sreenivasan, and A. De Angelis

TABLE 1. Standard Kohonen. Neighborhood order $V(t)$ versus total number of connexions $M(t)$.

V(t)	5	4	3	2	1	0
M(t)/V(t)	3.9×10^2	3.6×10^2	3.1×10^2	2.4×10^2	1.8×10^2	0

model to study how many different feature spaces may be represented in cortical maps. SOM have been used to study attentional impairment and familarity preference in autism [17]. By investigation of the primary visual cortex (V1) of ferret Sur et al. [18] found a distorsion in the mapping of the visual scene onto the cortex, in agreement with the predictions of the Kohonen model. Aflalo et al. [19] investigated origins of complex organization of motor cortex and found that the map generated from the Kohonen model contained many features of actual motor cortex in monkey.

STANDARD KOHONEN MODEL

Here we consider self-organization of young brain, based on the Kohonen model [7] of formation and organization of brain maps. This is achieved by reducing (mapping) the high-dimensional manifold of Q stimuli to a low-dimensional (2-D) manifold, represented by $N \ll Q$ neurons, which maintain topological continuity, i.e. proximate stimuli yield proximate neurons. The dynamical rules are as follows. There are stimuli $q = 1,\ldots,Q$, located at $\vec{p}_q$ in a vector space, and there are neurons, $i = 1,\ldots,N$, living on a regular 2-D grid map. To each neuron is assigned a weight vector $\vec{w}_i$ living in the same vector space as the stimuli. At each time t, a stimulus $\vec{p}_q$ is randomly selected, and the neuronal weights (representing synaptic strength) are updated according to a rule derived from unsupervised Hebbian learning, $\vec{w}_i(t+1) = \vec{w}_i(t) + \eta(t)[\vec{p}_q(t) - \vec{w}_i(t)]$, for all $i \in V(t)$ of neuron g. The "winning" neuron g is the one closest to stimulus $\vec{p}_q$. The neighborhood order V defines for each neuron g its neighborhood neurons, located on the regular grid map. For example, $V = 1$ corresponds to the 4 nearest neighbors up to 1 lattice unit away from neuron g, while $V = 3$ includes the 24 next-to-nearest neighbors up to 3 lattice units away. During evolution of organization, it is customary to let the neighborhood order decrease according to the exponential rule, $V(t) = int[V_0 \frac{\exp(-t/\tau)-1/e}{1-1/e}]$ for $0 \leq t/\tau \leq 1$. Here V_0 denotes the initial neighborhood order and τ the learning duration. For the example of $V_0 = 5$, the neighborhood order $V(t)$ versus total number of connections $M(t)$ under temporal evolution is shown in Tab.1. The parameter η denotes the learning rate. Also $\eta(t)$ decreases linearly in time to ensure convergence of neuronal weights. Using this rule, the neuronal map learns the topology of the data set and eventually becomes deformed (see Fig.2). The map topology $V(t)$ as well as the learning rate $\eta(t)$ determine the final absolute error of modelization $E^{ext} = \sum_{q=1}^{Q} \min_{i\in\{1,\ldots,N\}} ||\vec{p}_q - \vec{w}_i||^2$. The error E^{ext} decreases with decreasing neighborhood order $V(t)$.

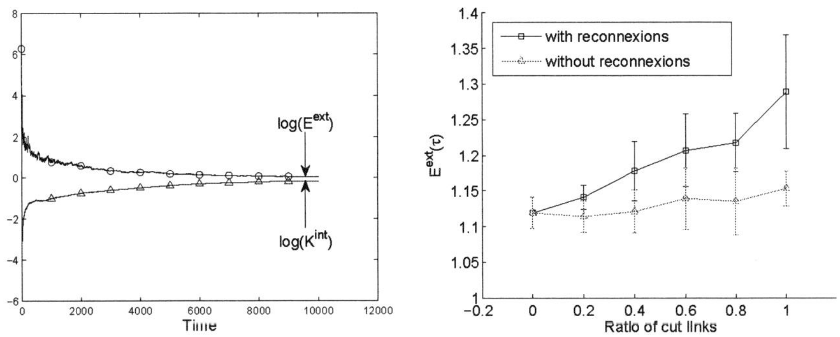

FIGURE 1. Self-adapted model. Left: Correspondence between knowledge K^{int} and error E^{ext} during the learning process. Right: Comparison of error for pruning links with and without reconnections. Links cut from the regular grid. Results correspond to data set 2, replicating the test 10 times, but are almost identical for all data sets.

SELF-ADAPTIVE KOHONEN MODEL

Knowledge function monitors error

Guided by the biological principle of adaptation and plasticity of the brain, we suggest to adapt dynamically $V(t)$ and $\eta(t)$ to the state of the network at time t, and in particular to cumulatively learnt information. Moreover, we suggest to introduce a locally independent learning rate $\eta_i(t)$ for each neuron. For such purpose we construct (inspired by the Growing Neural Gas network [20]) the function of *local attraction* $A_i^{int}(t)$: $A_i^{int}(t) = A_i^{int}(t-1) + \delta_{i,g(t)} \parallel \vec{p}_q(t) - \vec{w}_i(t) \parallel$ for all $i = 1, \ldots, N$, with initial value $A_i^{int}(0) = 0$. This function counts how much the weight vector of a neuron i has moved, being an indicator of how much that neuron has learnt. Low/high attraction means that such neuron has rarely/often been a "winner". It is biologically more plausible to have learning rates individually different for each neuron. Based on local attraction, we suggest a new learning rate of the form

$$\eta_i(t) \propto 1/A_i^{int}(t) \,. \qquad (1)$$

It is chosen inversely proportional to local attraction, because for a neuron not having learnt at all it is advantageous to have a high potential learning speed, while a neuron which has learnt much tends to saturation, i.e. learning speed becomes slow. The individual learning rate, Eq.1, is the first ingredient of the self-adapted Kohonen model. Starting from such learning rate, we define an internal *knowledge function* K^{int}. Intuitively, one expects that learning rates are low and homogeneously distributed among the neurons, when the network has achieved organization, i.e. has learnt much information. This leads to the definition of the knowledge function K^{int}, in terms of an harmonic mean,

$$K^{int}(t) = N\left[\sum_{i=1}^{N}\frac{\overline{A^{int}}(t)}{\overline{A^{int}}(t)+A_i^{int}(t)}\right]^{-1} - 1\,, \qquad (2)$$

with initial value $K^{int}(0) = 0$ and where $\overline{A^{int}}$ denotes the average of A_i^{int} over all neurons $i \in \{1,\ldots,N\}$. The function K^{int} varies in the range $1/N \leq K^{int} \leq 1$, and becomes maximal when all A_i^{int} are equal. Fig.1(left) shows the temporal behavior of function K^{int} and the absolute error E^{ext}. The decreasing error function E^{ext} mirrors the behavior of function K^{int} and hence justifies the interpretation of the latter as knowledge gain. It is important to note that the error function E^{ext} which measures the error (distance from ideal solution) at the end of organization, it is not part of the Kohonen network. However, the knowledge function $K^{int}(t)$, which mirrors the error, is available to the network at any time t by monitoring its history of moving neurons. Because K^{int} is computed from the neural weights, it can be viewed as a measure of memory. We suggest the possibility that mammalian brain can compute such function at any given time. We propose to use the information from $K^{int}(t)$ to dynamically self-adapt the global number of connections $M(t)$ by

$$\delta M(t) \propto -\delta K^{int}(t)\,. \qquad (3)$$

It means the change of topology adapts itself to the rhythm of knowledge gain, which creates a feed-back loop. Eqs.(1,3) guiding local and global adaptation, respectively, define the self-adapting Kohonen model.

Self-organization behavior of adaptive Kohonen model

For a fixed learning duration ($\tau = 10^4$ iterations until stability of organization has been reached), the standard Kohonen model is compared with the self-adapted model, using a 10×10 neural grid map. First, three different 2-D data sets are tested. They are each composed of $Q = 800$ stimuli with different spatial distributions (Fig.2). It shows that the neurons are closer to stimuli in the self-adapted model relative to the standard version. Also a substantial reduction in the final absolute error $E^{ext}(\tau)$ is observed (Tab.2). Ranging between only 19% to 66% of the error of the standard Kohonen model, the self-adapted model is found to be both more efficient and robust. The standard deviations obtained over 10 replications are low ($\approx$ 1%). Second, high dimensional data sets were tested, like the "Pima Indians Diabetes", in $D = 8$ with $Q = 768$ stimuli [21] and the "Hall of Fame", in $D = 15$ with $Q = 1320$ stimuli [22] giving qualitatively similar results (Tab. 2).

Pruning of synaptic connections

It is a biological property that the organization of cortex in mammals occurs during a short time window after birth during which neural connectivity is established. Most remarkably, this is accompanied by genetically controlled neuron cell death and pruning

TABLE 2. Comparison between standard and self-adapted model. Absolute error and standard deviation over 10 replications in parentheses.

Data set	$E_{ext}(\tau)_{standard}$	$E_{ext}(\tau)_{self-adapted}$	$\frac{E_{ext}(\tau)_{self-adapted}}{E_{ext}(\tau)_{standard}}$
1	0.446(0.023)	0.222(0.006)	**0.498(0.013)**
2	1.589(0.029)	1.055(0.014)	**0.664(0.009)**
3	2.302(0.458)	0.442(0.030)	**0.192(0.013)**
Pima	$5.03(0.37)10^5$	$2.33(0.07)10^5$	**0.462(0.014)**
II. Fame	$2.06(0.12)10^5$	$1.156(0.02)10^5$	**0.562(0.009)**

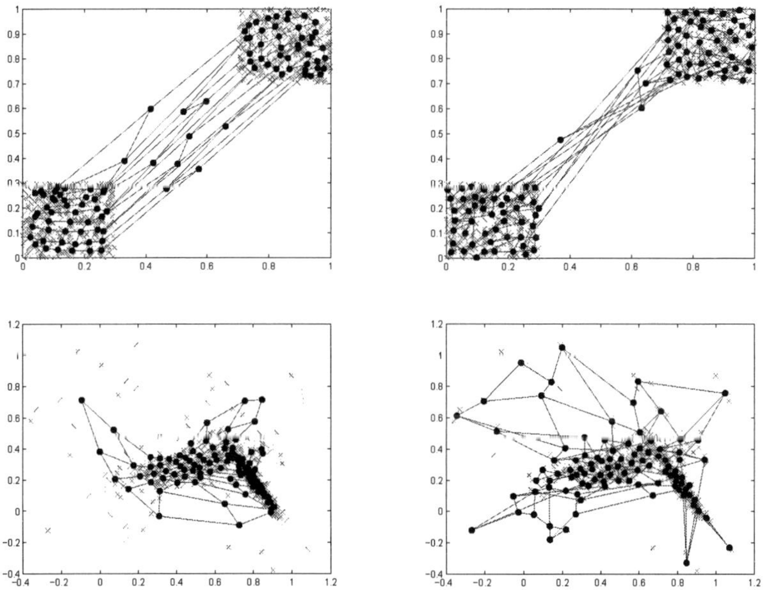

FIGURE 2. Comparison between standard (left column) and self-adapted (right column) map, showing stimuli (light crosses) and neurons (bold dots) for 2-D data sets (top=1, bottom=3). Lines represent the deformed map at the end of organization.

of synaptic connections [23, 24]. The pruning of connections appears to be an advantage because the progressive independence of neurons leads to a better precision in the local placement of neural weights. This has been confirmed by exploring the alternative of allowing for reconnections, which led to a larger error and deterioration of organization (Fig.1 right). This gives a possible explanation for pruning of synaptic connections during organization in mammalian brain during a short postnatal period.

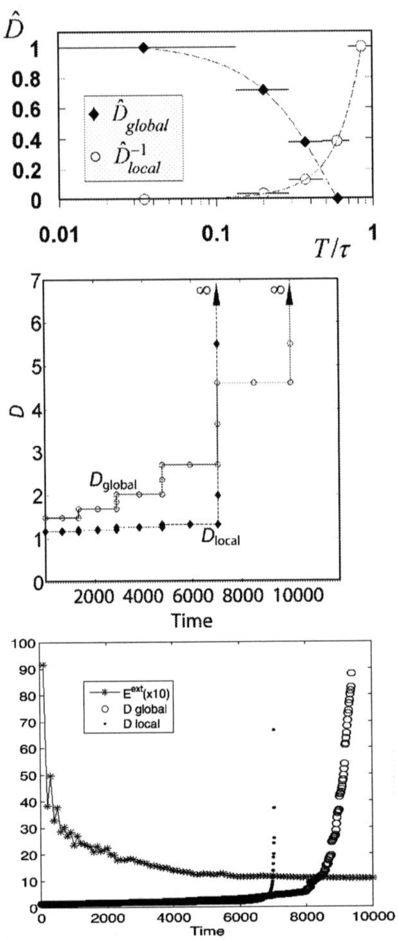

FIGURE 3. Top: Standard Kohonen. Normalized connectivity dimensions $\hat{D}_{local}^{-1} \sim C$ and $\hat{D}_{global} \sim L$ during evolution of organization. Dashed line: guide to eye. Middle: Standard Kohonen. Connectivity dimensions D_{local} and D_{global}. Bottom: Self-adapted model. D_{global}, D_{local} and absolute error E^{ext}.

Role of network connectivity during organization

In brains of humans, functional magnetic resonance imaging (FMRI) has shown evidence for functional networks of correlated brain activity [25], which display *small world network* (SWN) and *scale free* architecture. On the other hand, $1/f^{\alpha}$ frequency scaling has been observed in electric local field potentials [26]. Neural connectivity of SWN architecture has also been found in cat and macaque monkey. SWN neural connectivity has been found in a network of Hodgkin-Huxley neurons to give a fast and coherent system response [27], in an associative memory model to improve restoration of perturbed memory patterns [28] and to reduce learning time and error in a multi-layer learning model [29]. Thus SWN architecture appears to be another organizational principle of brain. In biology, often there is a link between structure and function. In order to measure this quantitatively, we use network connectivity dimensions on global and local scales, D_{global} and D_{local} [30, 31]. Those are defined via global and local efficiency E_{local} and E_{global} in the following way. For a given graph G, $E_{global}(G) = \frac{1}{N(N-1)}\sum_{i\neq j\in G} 1/d_{ij}$, $E_{local} = \frac{1}{N}\sum_{i\in G} E(G_i)$, $D_{global} = 1/E_{global}$ and $D_{local} = 1/E_{local}$, where d_{ij} denotes the shortest path length between vertices i and j and N is the total number of vertices. The local efficiency is given by the average efficiency of subgraphs G_i where the subgraph G_i is formed by the neurons directly connected to neuron i. A small world network (SWN) is characterized by a high local clustering coefficient C and a small average path length L [32]. It has been shown that D_{global} is similar to L, and D^{-1}_{local} is similar to C (although this is not an exact relation). Thus D^{-1}_{local} is high and D_{global} is small in the SWN regime [30, 31]. For standard Kohonen we show in Fig.3(top) the normalized dimensions $\hat{D}^{-1}_{local} = \frac{D^{-1}_{local} - D^{-1}_{local}|_{min}}{D^{-1}_{local}|_{max} - D^{-1}_{local}|_{min}}$ and $\hat{D}_{global} = \frac{D_{global} - D_{global}|_{min}}{D_{global}|_{max} - D_{global}|_{min}}$, scaled to lie in the range $0 \leq \hat{D}^{-1}_{local}, \hat{D}_{global} \leq 1$. One observes that $\hat{D}^{-1}_{local}$ is high and $\hat{D}_{global}$ is low - indicating SWN connectivity i.e. the network is highly connected at all length scales. This persists during most of the organizational phase, i.e. the regime where the error decreases until saturation and learnt knowledge increases (Fig.1 top). Towards the end of organization, the network looses its SWN character (Fig.3 middle). D_{local}, D_{global} both go to infinity, there is refinement of spatial scale of learning and formation of separate islands of neurons representing classes of stimuli. The behavior of the self-adapted model (Fig.3 bottom) is overall similar, however, the end of organization (up-shoot in D_{global}) is reached earlier than in the standard model.

Kohonen model in artificial intelligence

The Kohonen model has been widely used also in artificial intelligence, e.g. for data-mining purposes (using in particular its projection on a low-dimensional map). Recently, optimization algorithms, inspired by biological principles, like evolutionary computing, swarm intelligence, ant algorithms were found to be very sucessful as optimization strategies (example of design of antenna beam holder). Our adaptive Kohonen model can be viewed as another example, where the use of biological principles improves algorithmic performance.

ACKNOWLEDGMENTS

H. Kröger and M. Parizeau have been supported by NSERC Canada.

REFERENCES

1. E.I. Knudsen, M. Konishi, *Science* **200**, 795 (1978).
2. N. Suga, W.E. O'Neill, *Science* **206**, 351 (1979).
3. D.H. Hubel, T.N. Wiesel, *J. Physiol.* **165**, 559 (1963).
4. M. Sur, J.T. Wall, J.H. Kaas, *Science* **212**, 1059 (1981).
5. J.F. Linden, C.E. Schreiner, *Cereb. Cortex* **13**, 83 (2003).
6. R.M. Friedman, L.M. Chen, A.W. Roe, *Proc. Natl. Acad. Sci. USA* **101**, 12724 (2004).
7. T. Kohonen, *Biol. Cybern.* **43**, 59 (1982); **44**, 135 (1982).
8. R. Durbin, G. Mitchison, *Nature* **343**, 644 (1990).
9. K. Obermayer, H. Ritter, K. Schulten, *Proc. Natl. Acad. Sci. USA* **87**, 8345 (1990).
10. N.V. Swindale, *Biol. Cybern.* **65**, 415 (1991).
11. K. Obermayer, G.G. Blasdal, K. Schulten, *Phys. Rev. A* **45**, 7568 (1992).
12. G.J. Goodhill, *Biol. Cybern.* **69**, 109 (1993).
13. F. Wolf, H.U. Bauer, T. Geisel, *Biol. Cybern.* **70**, 525 (1994).
14. N.V. Swindale, H.U. Bauer, *Proc. Roy. Soc. Lond. B* **265**, 827 (1998).
15. N.V. Swindale, et al., *Nature NeuroSci.* **3**, 822 (2000).
16. N.V. Swindale, *Network: Comput. Neur. Syst.* **15**, 217 (2004).
17. L. Gustafsson, A.P. Papliński, *J. Autism Develop. Disord.* **34**, 189 (2004).
18. H. Yu, et al., *Neuron* **47**, 267 (2005).
19. T.N. Aflalo, M.S. Graziano, *J. Neurosci.* **26**, 6288 (2006).
20. B. Fritzke, in: *Advances in Neural Information Processing Systems 7*, G. Tesauro, D.S. Touretzky, T.K. Leen, eds., MIT Press, Cambridge (1995), p. 625.
21. http://www.ics.uci.edu/~mlearn/MLRepository.html .
22. http://lib.stat.cmu.edu/datasets/ .
23. G.M. Shepherd, *Neurobiology*, Oxford (1994).
24. E.R. Kandel, J.H. Schwartz, T.M. Jessel, *Essentials of neuroscience and behavior*, Appleton and Lange (1995).
25. V.M. Eguiluz et al., *Phys. Rev. Lett.* **94**, 018102 (2005).
26. C. Bédard, H. Kröger, A. Destexhe, *Phys. Rev. Lett.* **97**, 118102 (2006).
27. L.F. Lago-Fernandez et al., *Phys. Rev. Lett.* **84**, 2758 (2000).
28. J.W. Bohland, A.A. Minai, *Neurocomputing* **38-40**, 489 (2001) .
29. D. Simard, L. Nadeau, H. Kröger, *Phys. Lett. A* **336**, 8 (2005) .
30. M. Marchiori, V. Latora, *Physica* **A285**, 539 (2000).
31. V. Latora, M. Marchiori *Phys. Rev. Lett.* **87**, 198701 (2001).
32. D.J. Watts, S.H. Strogatz, *Nature* **393**, 440-442 (1998) .

PART VI

PHYSICS EDUCATION

Multimedia in Physics Teaching and Learning

L. Mathelitsch

Institute of Physics, Regional Centre for Didactics of Physics, University of Graz, Austria

Abstract. The huge number of multimedia products that have been developed and are available on internet demands a critical review. A European and an American group have accomplished such a project within the last years. The process of evaluation is reported in this article, recommendations for the best products in different areas of physics are given.

Keywords: physics education, multimedia, evaluation of software
PACS: 01.50.H-, 01.50.hv

INTRODUCTION

Computational physics has developed as an own field of physics. But not only did it become an integral part of physics research, it has been gaining also more and more importance in physics education. Besides its traditional components in physics research, as tool of computation, control of experiments, data acquisition and analysis, other features, like visualization, became of central importance in physics teaching. Simulations and animations, for example, play a special role in physics education stressing the modelling aspects in physics. Furthermore, the different levels of age and knowledge of the learners have to be taken into account.

Thousands of multimedia products on physics teaching, small pieces of software up to full courses, are currently available worldwide. Some are commercial products, but the great majority of them is offered freely on the internet. Some is excellent material, but the percentage of less valuable output is very large. One cannot expect from an individual user, may it be a professor at the university, a high school teacher or a student, to find the time and to invest efforts to effectively search for and evaluate these resources, whenever he/she is looking for a special multimedia support in his/her teaching.

Therefore, a systematic approach is needed: Available material has to be collected and preferably included into a data base; criteria for evaluation have to be established and agreed upon; the material has to undergo an assessment according to these criteria; qualified material has to be recommended and disseminated. Two groups, the network Multimedia in Physics Teaching and Learning (MPTL) in Europe, and the association Multimedia Educational Resource for Learning and Online Teaching (MERLOT) in the United States, accomplished this program in a joint action within the past years.

CP1018, *Frontiers of Fundamental and Computational Physics*,
edited by B. G. Sidharth, F. Honsell, O. Mansutti, K. Sreenivasan, and A. De Angelis

MPTL AND MERLOT

Within the European Physics Education Network (EUPEN, [1]), a group "Multimedia in Physics Teaching and Learning" has been formed about 15 years ago [2]. Under the leadership of Hans-Joerg Jodl from the Technical University of Kaiserlautern several goals have been pursued in order to improve the use of multimedia in physics education: production, presentation and dissemination of products followed by applications and assessments of these products. Annual workshops provided the platform for these actions, the first one took place in Munich in 1996 [3], the last one in Wroclaw in 2007 [4]. The next workshop will be organized as joint meeting with the biannual GIREP conference (Nicosia, August 18 – 22, 2008, [5]).

MERLOT [6] is a consortium of American and Canadian universities with the goal to improve the effectiveness of teaching and learning by increasing the quantity and quality of peer reviewed online learning materials. MERLOT has been performing reviews of multimedia resources in 15 different disciplines since 2000. MERLOT's vision is to be a premiere online community where faculty, staff, and students from around the world share their learning materials and pedagogy.

EVALUATION

Basic criteria for a good quality of multimedia products are interactivity, user friendliness and a high standard of animation. Technical aspects, aspects of content and of teaching should be evaluated. After several rounds of discussion and one year of experience, MPTL set up a rubric form with short questions that are used to judge materials in the overall categories Motivation, Content and Method. A five-level score (-2 to +2) is used to give an overall score for each item [7].

The MERLOT review process is similar, although with a different scheme used for the evaluation. Three general categories Quality of Content, Potential Effectiveness for Teaching and Learning, and Ease of Use are defined by 4 to 5 questions. Reviews provide scores on a 1 to 5 scale for each of the three categories and an overall score. Details of the MERLOT review process can be seen in [8].

Because of the large number of items a two-step evaluation was pursued. Each group performed a preliminary review of the products from their lists in order to select those items suitable for a full evaluation. This makes the review process more efficient because of the otherwise extended time necessary to perform a detailed review. In the preliminary evaluation each resource was accessed by one reviewer who selects those items likely to be reviewed as of good or excellent quality.

The results of the preliminary evaluations by the EPS and the MERLOT group were combined into a single list of materials selected for the final evaluations. This combined list was broken into sub-topics, similar to what is found in standard textbooks. Each of these items was reviewed by one member of each team, EPS as well as MERLOT. These reviews were performed independently to provide a test of the consistency and repeatability of review results.

RECOMMENDATION

The MPTL process started in 2002 by reviewing the field of quantum mechanics followed by optics (2003). Mechanics was the first topic being evaluated together with MERLOT (2004), in the next years came thermodynamics and statistical physics (2005), electricity and magnetism (2006), and condensed matter and nuclear and particle physics (2007). The number of materials depended on the topic, most of the items (more than 300) have been found on mechanics.

There is no space and need to repeat the recommendations on the various topics here, they can be seen on the web-link given in ref. [9]. Despite the different lists of criteria, there was a very good agreement between the results of the two review processes. If there were discrepancies, then only slight ones about the ranking, whether a product was considered as excellent or "only" as very good.

Several general features showed up being more or less independent of the respective topic of physics:

- Most of the sites can be assessed technically without problems.
- Many sites collect material from other sources.
- 80 to 90 % of the material is about standard tasks. These standard examples can be found multiple times.
- Most of the material is of mediocre quality, there are only few very bad sites, but there are not too many really good ones either.
- Many sites are devoted to one, short and typical physical problem; only a few consist of a larger collection of material.
- Only about half of the simulations contain information about the physical background.
- Suggestions on how to implement the material in teaching/learning are very rare. Didactic reflections are missing for almost all products.

EXAMPLES

Two examples may show how multimedia can give an additional dimension to traditional teaching and learning of physics. The first one concerns experiments. Not every experiment can be done in student labs, they can be too dangerous, too large, too expensive, too complicated. In order to give students access also to those experiments, two approaches have been suggested.

In Interactive Screen Experiments [10], an apparatus is set up on some place, and photos, videos are made of different stages of operation. This material is the basis of a multimedia product: a user has a view on the experiment, he/she can vary some parameters and the according response of the apparatus is shown via the prepared videos and photos.

Remote Control Labs [11] are real experiments set up somewhere and kept in operational status for 24 hours a day. A user from outside can log in via internet and he/she has control over several parameters of the experiment. The experiment operates according to the chosen set-up of parameters, the action of the apparatus is transmitted to the screen of the user. Also data can be collected like in a traditional experiment.

The second example is taken from particle physics, it is called Masterclass [12] and organized by the European Particle Physics Outreach Group. Students of age 15 to 18 are invited to a masterclass, usually organized by a local university. There they get lectures on particle physics and explore CERN by virtual tours. Finally, the students have to analyse real data collected by the OPAL or DELPHI detectors. At the end, a video conference is set up for the different groups working on the same day in different universities (usually about five). They compare their results and discuss them with experts at CERN. During such a masterclass many facets of multimedia contribute to an attractive presentation of up-to-date particle physics as well as to an efficient transfer of basic and advanced knowledge on this topic.

CONCLUSION

Only a small fraction of available multimedia products on physics teaching and learning meets the standards of being very good or excellent. A selection was done in a fair and transparent procedure resulting in recommendations for multimedia products in the various fields of physics. Quite a lot of (wo)man-power has been devoted to this task, but, thinking globally, it should pay off and save time and efforts for many colleagues, teachers as well as students.

REFERENCES

1. http://www.eupen.ugent.be/
2. http://www.mptl.eu/
3. http://cripe03.rug.ac.be/MPTL/history.htm
4. http://mptl12.ifd.uni.wroc.pl/
5. http://www.ucy.ac.cy/girep2008/
6. http://www.merlot.org/merlot/index.htm
7. Appendix 3 of http://www.mptl.eu/reports_files/Rep_Recom_Thermodyn_2005.pdf
8. http://taste.merlot.org/evaluationcriteria.html
9. http://www.mptl.eu/reports.htm
10. http://didaktik.physik.fu-berlin.de/home/ibe_beispiele.html
11. http://rcl.physik.uni-kl.de; http://kmi.open.ac.uk/people/martyn/PEARL/PEARL%20Project%20summary.htm; http://www.ises.info/default_a.htm
12. http://www.physicsmasterclasses.org/mc.htm

The SUPERCOMET 2 Project: Teacher Seminar and Teacher Guide

V. Engstrom*, G. Ireson†, H. Latal**, L. Mathelitsch**, M. Michelini‡, W. Peeters§ and G. Rath**

*Simplicatus AS, Norway
†Nottingham Trent University, UK
**Institute of Physics, University of Graz, Austria
‡CIRD, University of Udine, Italy
§Institute of Physics, University of Antwerpen, Belgium

Abstract. The Leonardo da Vinci pilot projects SUPERCOMET (2001-2004) and SUPERCOMET 2 (2004-2007) developed and tested teacher training materials for active, minds-on learning of electromagnetism and superconductivity at the level of upper secondary school. New multimedia materials for the pupils supported the teacher materials, including a teacher seminar in 4 half-day modules detailing the scientific contents, teaching methods, using ICT in physics teaching and learning, online collaboration and further resources for exploring the selected topics. The aim of these projects were to improve the quality of physics teaching on a European level, involving a combined total of 45 partners in 16 countries, and conducting trials at 67 schools with approx. 230 teachers, 280 trainee teachers and 2200 pupils. New follow-up projects develop hands-on materials for carrying out the activities described in the teacher guide and seminars and additional teacher materials involving modelling, simulations and data logging.

Keywords: physics education, physics curriculum, teacher training, teacher guide, teacher seminar, e-learning, hands-on learning, blended learning, teaching methods, teaching materials, electromagnetism, superconductivity, Leonardo da Vinci programme
PACS: 01.30.Os,01.40.ek, 01.40.Fk, 01.40.gb, 01.40.Ha, 01.40.J-, 01.50.ht, 01.65.+g, 01.50.My, 41.20.Gz, 74.20.-z, 74.25.-q, 74.70.-b, 74.72.-h, 84.71.Ba, 84.71.Fk, 84.71.Mn

Background

Data from the ROSE [1] project show a -0,92 correlation coefficient between 15 year old students. attitude towards learning science and the UN Human Development Index of their home countries. For pupils of the "screenagers" generation, learning 100 year old modern physics and 400 year old mechanics seem equally ancient, boring and irrelevant to their lives. In order to motivate pupils for understanding Newton and Einstein we aim to show that physics textbooks are revised and extended every day.

The SUPERCOMET [2] (2001-2004) and SUPERCOMET 2 (2004-2007) pilot projects had a combined total of 45 partners in 16 countries, funded by the Leonardo da Vinci [3] programme of the European Union and by in kind contributions by the partners.

Both projects aimed to develop teacher training materials for active, minds-on learning of electromagnetism as a curricular topic. New multimedia materials for the pupils supported the teacher materials, using superconductivity as a motivational theme connecting the school curriculum to state-of-the-art science and technology [4].

These materials have been translated to more than 10 languages and trialled in 14

CP1018, *Frontiers of Fundamental and Computational Physics*,
edited by B. G. Sidharth, F. Honsell, O. Mansutti, K. Sreenivasan, and A. De Angelis

countries with adaptations to local curricula. More than 65 schools, 230 teachers and 2200 pupils have been involved in the trials, as well as more than 280 trainee teachers in the process of completing a Postgraduate Certificate of Education or similar.

The projects' underlying philosophy of learning is social constructivist and based on the works of Dewey, Piaget, Vygotsky and others. Learning science should be facilitated by providing pupils with an environment where they encounter experiences challenging their preconceptions and allowing them to construct their own understanding of the world, in our case particularly electromagnetic phenomena.

The material of SUPERCOMET 2 goes partially beyond the content of a typical curriculum of a European high school. Also the style of presentation differs from traditional text books. Therefore it makes sense to provide the teacher with background information with regard to the various aspects of SUPERCOMET 2.

Teacher Seminar

The SUPERCOMET 2 project developed high quality teaching materials which include online learning modules [4], a teacher guide [5], teaching methods, as well as a prototype online community called "Physible" [6]. All this is embedded in a structure for training teachers on how to use these high level materials with new teaching methods. A set of support materials for the teacher trainer giving the teacher seminar has been developed, comprising time schedules and practical details, content presentations, videos and descriptions of activities.

During the teacher seminar each of the online modules with animations describing electromagnetic phenomena is studied, often in different ways. For example in this activity called "A/B" used with the module "Electric conduction" during the first half-day module of the Teacher Seminar:

Aims

- Students handle information in group, but independently, and pass that to each other.
- Students apply the content of their new knowledge in a creative task.

Method

Step 1: The aim of the task is explained. The task is rather general and conceptual. Students are divided in groups of 4 or 5. Each student plays a specific role: chairman, secretary, planner, etc. The time is set. The specific roles are explained.

Step 2: Students carry out the first part of the task (A). Handling information (textbook, websites) and sources are essential here. Every student gets a different set of possible sources and a subject. Questions can guide them and stimulate discussions. They also bring in own experiences, knowledge and understanding.

Step 3: Based on the previous phase (A) task B is given: this is more concrete, practical: the knowledge gathered in A is applied in B. The teacher mainly observes and and lets the students work as independently as possible.

Feedback by the teacher is given after the A/B tasks.

Step 4: Each group presents the results of their specific task. Teacher and students give feedback and suggestions.

Evaluation of the product (how A is applied in B) and the attitude (during B) is possible. These methods link to research on different approaches towards effective learning and aim at improving motivation, engagement, understanding, and life long learning both for teachers and their students (future teachers and scientists). During the teacher seminars teachers attend demonstration experiments and participate in some experiments themselves. The teachers also get videos of fascinating experiments dealing with the Meissner effect and pinning, supporting hands-on experiments if liquid nitrogen is available. Materials for carrying out such experiments is currently being developed in the MOSEM [6] Project.

Teacher Guide
There exists already a teacher guide developed within the former SUPERCOMET project [5]. The book contained information in four different languages, English, Norwegian, Italian and Slovenian. Since SUPERCOMET 2 covers fifteen partners with more than ten different languages, it has been decided to produce the teacher guide, in first instance, as electronic version and only in English. It is up to the individual partners to translate the guide in the respective language, if necessary. The format is printable, so it is very easy to produce booklets, as it exists now in English language [8]. In the following we will comment on the different chapters of the teacher guide.

Introduction
An overview is given on the aims of SUPERCOMET: Superconductivity is introduced as an exciting way to teach many concepts of physics. Suggestions are given how to navigate through the different chapters of the modules.

ICT in science teaching
Different ways how to use information and computer technology in physics teaching are presented and discussed with examples taken from the SUPERCOMET material.

Physics of superconductivity
Experience with seminars and in-service training has shown that many teachers do not have enough information on superconductivity to feel safe in their teaching. But it is not easy to find background information on the appropriate level. Therefore one of the editors (Heimo Latal) provided for an introduction to the topic of superconductivity.

The e-modules
Necessary prerequisites and learning objectives are stated for each of the nine modules. In order to check the gained knowledge test questions are given which the teacher use or adapt for his or her system of examination. Suggestions for activities round up this information on the respective modules.

Examples of activities
Further examples for activities are given in this separate chapter. Detailed lesson plans are suggested, in case a teacher wants to follow such a guide. A variety of interesting, stimulating worksheets have been created at the university of Udine. The same group has also developed experiments how to measure accurately a magnetic field.

Experiments - teacher seminar
Because of the novelty of the material, teacher seminars are as important as a teacher guide. Wim Peeters (Univ. of Antwerp, Belgium) proposes efficient ways to organize such seminars where experiments play a crucial role, as mentioned above. Marisa Michelini and her group in Udine provide for hands-on experiments, whereas Bernadette Schorn from the university of Munich describes how one can "cook" a superconductor in a school or university lab. Finally, Gren Ireson (Univ. of Loughborough, UK) gives

details how the transition temperature of a superconductor can be measured applying the four-point method.

Evaluation
Evaluation is an important but not always easy business in teaching. Questionnaires are provided for the evaluation of the attitudes of students and teachers and also to perform classroom observations. Two case studies are presented, one from the university of Udine and one from the university of Murcia in Spain.

Further Resources
References on books and websites might help interested teachers to explore the topic in more detail.

Teachers are not always curious about teacher guides. We hope that this multifaceted approach to a non-traditional topic will find its interested audience.

MOSEM project
Running from 2007 to 2009, the MOSEM [7] [3] multilateral transfer of innovation project with 28 partners in 7 countries is financed by the Leonardo da Vinci programme as part of the Lifelong Learning Programme of the European Union. MOSEM will follow up the results of the two previous projects by developing specific kits of experimental materials matching the multimedia materials and teacher training materials, aimed at complementing the existing collections of such materials at each school. Combining hands-on investigative experiments with critical questioning, interactive animations, videos and theoretical explanations allows for what we call a blended approach to minds-on learning. Following the MOSEM project, the partners have applied for funding for a MOSEM2 project to develop further teaching materials, methods and training involving modelling, simulations and data logging.

REFERENCES

1. Relevance of Science Education. http://www.ils.uio.no/english/rose
2. Superconductivity Multimedia Educational Tool. Website for the SUPERCOMET project family. http://supercomet.eu
3. Leonardo da Vinci programme. http://ec.europa.eu/education/programmes/llp/leonardo/index_en.html
4. SUPERCOMET Online Learning Modules. http://online.supercomet.no
5. A. Earle, J. Frost, V. Engstrom, M. Cepic, G. Planinsic, G. Ireson, S. Ciaparelli, SUPERCOMET Teacher Guide, Simplicatus AS, 2004. ISBN 82-8130-045-0
6. Physible Online Community Prototype. http://physible.eu
7. Minds-On experimental equipment kits in Superconductivity and Electromagnetism for the continuing vocational training of upper secondary school physics teachers. Website for the MOSEM project results (to be launched in 2008). http://www.mosem.eu
8. V. Engstrom, H. Latal, L. Mathelitsch, G. Rath (Eds.), SUPERCOMET 2 Teacher Guide, Simplicatus AS, 2007. ISBN 978-82-8130-055-2

Superconductivity in Italian Secondary Schools: the experimentation carried out by Udine University with Supercomet2 (SC2) materials

Marisa Michelini[a], Lorenzo Santi[a], Rossana Viola[a] and Federico Corni[b]

[a] *Research Unit in Physics Education, University of Udine, Via delle Scienze n208, 33100 Udine, Italy*
[b] *University of Modena and Reggio Emilia, Italy*

Through a highly interactive tools on CD, which comprises animation and films of demonstrative experiments, and uses modern pedagogical methods, such as collaborative learning and problem solving, the Supercomet 2 Project (SUPERCOnductivity Multimedia Educational Tool phase 2 of the European Union) aims to introduce superconductivity to European high school curriculums. During the first year of the project the following were produced: a CD-ROM with didactic material, a teacher's guide to clarify characteristics and roles of support material (texts, worksheets and computer presentations) and preview didactic courses. During the second year of the project It was: translated the material into the languages of countries participating in the project, experimented the application in high school classes in various parts of Europe, tested the updating course and the teacher's guide with a group of reference teachers. At the end of the first period of experimentation and dissemination were revised and new material was integrated in order to produce a final version at the end of the project. Further integrated proposals have been included with the revised material. In this paper the main characteristics realized during the project will be described.

Keywords: Supercomet Project, superconductivity, science
PACS: 40, 70, 80

The **methodologies** used in the experimentations are the following: A) Use in Info-Lab the multimedia material of SC2; B) Combined Info-Lab with multimedia SC2 tools and Experimental Lab activities (mainly qualitative); C) Experimental Lab based course (qualitative and quantitative experiments).

There are 3 different **way to carried out school experimentations**:

Type A) Research experimentation with setting up of didactic tools for analyzing learning processes and didactic innovation produced during class activities. 6 different monitoring tools were used to analyze the experimentation. This was carried out in 2 contexts with 2 very different modalities: A1) As apprenticeship of a perspective teacher in collaboration with an expert teacher of a secondary school and the Research Unit in Physics Education of the University of Udine using PhReserch TOOLS; A2) Didactic Research conducted by an expert teacher, following a protocol of evaluation of didactic innovation individuated in previous researches.

Type B) Action Research (AR) by a web network of schools (Drago Web). Drago Project is a web online collaboration between 3 teachers (expert in using ICT) from 3 schools (from Sicily and Calabria which are areas in the south of the country with a strong need for innovation and the support required by this). The 3 levels of study: that

CP1018, *Frontiers of Fundamental and Computational Physics*,
edited by B. G. Sidharth, F. Honsell, O. Mansutti, K. Sreenivasan, and A. De Angelis

of students; the teachers who lead and follow these with blended modalities; the researchers, who in different environments carried out the task of analyzing the work of students and the activity of the teachers. The project received an award by national projects of support to scientific vocations (PLS – Physics). The prize-giving was held at a conference where 500 students participated and exhibited the experimentation Supercomet (around 60 students).

Type C) Testing of SC2 material produced using the usual style in teaching activities by teachers involved: in not-traditional secondary school at 14-16 years old students (3) and in traditional classes of the last years of secondary school (15) following the suggested path foreseen by the project. Meetings for comparison and discussion during the experimentation.

Three main **didactic paths** were adopted into the experimentation. They were proposed by teachers from Pordenone and Udine after teachers' course based on Action Research (AR) style. The path are both for experimental low level of secondary school and for traditional classes, using the Supercomet (SC2) materials.

P1_Magnetic field path. "Faraday's way to the magnetic properties of the superconductor, or rather the path of field lines" is characterized by the following steps: A) Analysis of the electric field lines due to a dipole and of the field lines in the case of a magnetic dipole (mapping and characterization in both cases). B) Analysis of the situations of suspension associated with field lines and situations of equilibrium (stable and not stable): experiences of ring-shaped magnets and not ring-shaped magnets. C) Slowing down of the magnet and the superconductor thanks to induced currents; analogy with the Fountain Effect; D) Possibility to surround the superconductor with field lines: experience of magnetic levitation; introduction of the hypothesis of diamagnetism; possibility of correction of the position of instable equilibrium.

P2. Resistivity behavior. " The master way of the properties of the superconductor, or the course of resistivity" follows these steps: A) Analysis of temperature dependence of resistivity, with particular attention the below 0 degree Celsius temperatures. B) Falling of resistivity and electrical transport properties of superconductors. Effects of superconduction. C) Analysis of electrical circuits with elements of virtual superconductors.

P3. Energy approach implemented in second classes of PNI course, put in the curriculum after the study of energy transformations and electric field. The proposal contains a new path in that it departs from the conservation of mechanical energy and its transformation into other forms and in particular into an electrical form; it looks at electric field and electrical conduction (module 4 of CD-Rom) and then analyzes the magnetic behavior of currents (Module 1 and 2), electromagnetic induction (Module 3), superconductivity (Module 5). It is constituted by a combined use of laboratory experience, modules of the project Supercomet and interactive applets. It was favored group work for the laboratory experiences, some of which were conducted in qualitative form, others quantitative, guided by the worksheets. The strategy of inter-group discussions with the aim of sharing conclusions is implemented as the way to discuss experimental results.

Electric Conduction in Solids: a Pedagogical Approach Supported by Laboratory Measurements and Computer Modelling Environments

A. Bonura, M. C. Capizzo, C. Fazio and I. Guastella

UoP_PERG (University of Palermo Physics Education Research Group)
Dipartimento di Fisica e Tecnologie Relative, Università di Palermo, Italia

Abstract. In this paper we present a pedagogic approach aimed at modeling electric conduction in semiconductors, built by using NetLogo, a programmable modeling environment for building and exploring multi-agent systems. 'Virtual experiments' are implemented to confront predictions of different microscopic models with real measurements of electric properties of matter, such as resistivity. The relations between these electric properties and other physical variables, like temperature, are, then, analyzed.

Keywords: Conduction, semiconductors, NetLogo, simulation, emergent phenomena.
PACS: 01.40.-d, 01.40.Fk, 01.40.J-, 01.50.H-

INTRODUCTION

Modelling matter, in particular in the solid state, has become since long time part of the curricula at college level and it is now coming to high schools, given the need to introduce students to the understanding of technological devices. Modelling matter can be also considered a good way to give students a meaningful understanding of physics as a modelling endeavour and of the quantum behaviour at microscopic level that reflects on macroscopic, measurable properties of matter.

We present here some pedagogic procedures aimed at relating measurements of electric properties of semiconductors with "virtual experiments" built by using the well known programmable modelling environment NetLogo. The reported activities have been experimented during the laboratory courses of the Italian Graduate School for Physics Teacher Education (S.S.I.S.), in the context of a broader approach to electric conduction in conductors and semiconductors, discussed in more details in [1] and [2]. Models of electric conduction in semiconductors have been implemented and the different model predictions about the resistivity vs. temperature dependence and some related parameters have been compared to experimental results, in order to make prospective teachers reflect about actions and interactions of individual microscopic objects resulting in emergent, measurable system properties.

CP1018, *Frontiers of Fundamental and Computational Physics*,
edited by B. G. Sidharth, F. Honsell, O. Mansutti, K. Sreenivasan, and A. De Angelis

EXPERIMENTAL RESULTS

Resistance vs. temperature measurements have been performed by using a commercial electronic board allowing to take data of the electric current circulating in a sample when this is subject to a given voltage, as a function of the sample temperature. The resistance data at low temperature values are taken by suspending the sample in a thermally insulated container and letting it cool under the action of liquid nitrogen vapours placed in the container.

Figure 1 reports the natural logarithm of the ratio R/R_o between the resistance, R, of a slightly doped semiconducting sample (metal oxide) and the sample resistance at T_o= 273 K , as a function of $(1/T - 1/T_o)$, in the T range going from 220 K to about 350 K.

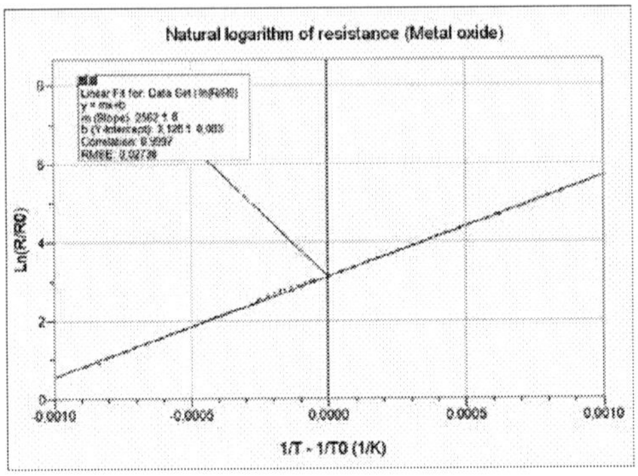

FIGURE 1. Natural logarithm of resistance in a doped semiconductor (metal oxide) as a function of the reciprocal of temperature. The line best fitting data and the related parameters are shown

A linear fit is applied to figure 1 data and results give a clear exponential relation between the semiconductor resistance and $(1/T - 1/T_o)$. By using these data it is also possible to give an estimate of the value of a relevant parameter of semiconductors, the energy gap value. From the known resistance vs. temperature formula

$$R = R_o e^{\frac{\Delta E}{k}\left(\frac{1}{T}-\frac{1}{T_o}\right)}$$

where ΔE is the energy gap and k is the Boltzmann constant, it is possible to find the value $\Delta E = 0.221 \pm 0.001\, eV$, a reasonable one for doped semiconductors. These results are in good agreement with other measurements [3] aimed at helping students to better understand the electrical behaviour of conductors and semiconductors.

MODELLING

Microscopic models describing and explaining the experimental results have been implemented by using models implemented in NetLogo 3.0.2 and NetLogo 3D preview 1 [4]. The main pedagogical purposes of such simulations can be summarized as follows: 1) to emphasize the fundamental concepts related to electron and hole motion and electron/hole-crystal lattice interactions as key concepts to understand the electric transport phenomena and the effects of temperature variation; 2) to supply a deep understanding of microscopic conduction mechanisms and their modifications due to the temperature, emerging from different and more refined microscopic models of semiconductors.

Some of the implemented models share some characteristics with well known historical models and will be briefly described in the following. In order to model the electric conduction in semiconductors it is necessary to study the charge carrier concentration, n, as well as carrier mobility, μ, as a function of temperature. The semiconductor behaviour is described by substituting the traditional statistical approach (requiring a deep mathematical background) with microscopic models, based on a Monte Carlo method, in which simple rules applied to microscopic particles and quasi-particles determine the macroscopic properties.

Different models have been considered. The first two are aimed at modelling the band structure of the semiconductor and its internal dynamics. The third one is aimed at reproducing the transport properties, like the mobility.

Two level or 'Einstein' model. In this simple model, the band structure of the semiconductor (conduction and valence band) is described by two well-defined energy levels characterized by the population numbers N_2 and N_1, which represent the electron number for volume unit occupying the conduction and the valence band.

Multi-level model. In this model a structure to bands is given by dividing them in sublevels. So, every sublevel in the valence band, as well as in the conduction band, is characterized by an energy value, a number of electrons and a number of holes. Unlike the 'Einstein' model, a sublevel is also typified by a number of available states. The transition probability depends on the number of electrons, holes and photons. More details can be found in [2].

Carrier concentration graphs as a function of reciprocal of temperature in different kind of semiconductors, obtained by applying the Multi-level model, are obtained and compared to actual experimental data [2].

Transport model. In order to analyze the transport properties in an intrinsic semiconductor, we assume the free carrier model, in which: i) carriers with an energy close to a band minimum behave as free carriers possessing only kinetic energy. The effective mass concept is used in order to take into account the effect of the periodic potential of the crystalline solid; ii) the mobility of the carriers depends on the mean free path. This is function of the carrier energy and it is strictly related to the scattering mechanisms; iii) carrier interact with lattice vibrations, that can be modelled as phonons.

The motion of a carrier drifting in an intrinsic semiconductor due to an applied electric field, E, is analysed by taking into account collisions with lattice waves including the absorption or emission of either acoustical or optical phonons. Since the density of phonons in a solid increases with temperature, the scattering rate due to these mechanisms will increase with temperature, while the mobility will decrease. The scattering probability is proportional to the carrier energy and to the number of phonons.

This model allows to directly obtain the curve of mobility of electrons and holes as a function of temperature in Si, as shown in [2]. Figure 2 reports a graph of resistivity vs. temperature built by using results obtained by merging the Transport model with the Multi-level one, a result in qualitative accordance to experimental data reported in figure 1. More details are reported in [2].

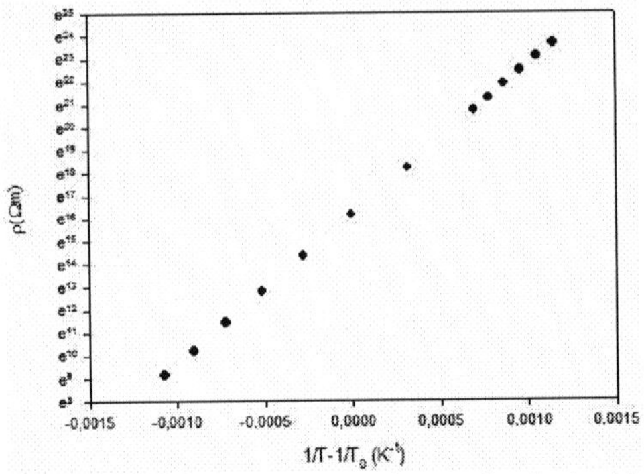

Figure 2. Graph of resistivity vs. temperature built by using results obtained by merging the Transport model with the Multi-level one.

DISCUSSION AND CONCLUSION

Our models are based on a few fundamental rules of atomic interactions that exhibit emergent behaviours reproducing important aspects of the properties of materials, as electric conductivity. There are several features of our pedagogical approach that may contribute to student learning. In our simulations, the same underlying computational model appears in different contexts and gradual modifications of the model are outlined. This provides several advantages: 1) students gain familiarity with different representations and ways in which electrons and holes behave and interact with nuclei/phonons; 2) the model parameters can be set by the student, in the same way in which they interact and control the experimental parameters; 3) the macroscopic properties of the model emerge from the details of the atomic-scale interactions in just the way that the corresponding properties emerge in real atoms.

A preliminary analysis of prospective teachers worksheets allows us to infer that laboratory activities, supported by modelling, can improve students' learning about actions and interactions of individual objects resulting in emergent system properties. Prospective teachers showed a better understanding of the difference between simply describing a situation in terms of equations and interpreting it on the basis of mechanisms of functioning. They showed a renewed attitude to search for interpretative models, even involving microscopic interactions, explaining why a phenomenon develops in a given way or some specific experimental results are obtained.

REFERENCES

1. A. Bonura, M.C. Capizzo and C. Fazio, "A Pedagogical Approach To Electric Conduction In Solids". Proceedings of XXI GIREP International Conference, Amsterdam, 20–25 August 2006
2. M.C. Capizzo, R.M. Sperandeo-Mineo and M. Zarcone, "Modelling electrical conduction in semiconductors: a pedagogical approach". *European Journal of Physics* (in press).
3. A. Sconza, G. Torzo, G. Delfitto, M. Michelini, 1996. "Teaching Electrical transport Properties in Solids: an Experimental Approach", in *Teaching the Science of Condensed Matter and New Materials*, GIREP-ICPE Book, Forum, p.122
4. http://ccl.northwestern.edu/netlogo

Disciplinary knots and learning problems in electromagnetism

Alberto Stefanel[a]

[a]*Research Unit in Physics Education, University of Udine, via delle Scienze 208, 33100 Udine.*

Abstract. The concept of field acquire the ontological status of a physical entity when we consider dynamic situations. In this context rise up typical disciplinary knots that are source of learning problems for students, connected to: the sources of time depending fields, the electrodynamic of moving conductors, the nature of electric and magnetic fields as components of the electromagnetic field. A discussion of these principal knots on electromagnetism, and an analysis of the related learning problems provide one of the tree vertex of the triangle founding a research aimed to design a teaching/learning path on electromagnetic field.

Keywords: Electromagnetism, teaching, learning.
PACS: 01.40.G; 03.50.De

INTRODUCTION

The concept of field is fundamental in physics to describe interactions. It is useful in the static case. Yet, it is of capital importance in dynamical situations, because we do not provide any other concept to describe the phenomenology. Only when we consider time depending cases, the field concept necessarily acquires the ontological status of a physical entity, a perturbation propagating with a finite velocity into the space-time across vacuum or matter. In this context the knots, generically connected to the concept of field as superposition [1], sources of fields [2-4], fields representation and systems [4-7], field lines and trajectory [8], are partially related to new learning problems. Moreover, typical disciplinary knots rise up. These knots are new sources of learning problems for students connected, for instance, to the sources of time depending fields [9-11], the electric and magnetic fields as changeable components of a unitary physical entity (the electromagnetic field), the electromagnetic induction [13-15].

A discussion of these principal knots, and an analysis of the related learning problems provide the basis of one of the tree vertexes of the triangle founding a research aimed at designing a teaching/learning path on electromagnetic field [16].

LEARNING PROBLEMS OF STATIC-DYNAMIC BORDERLINE

Before considering some specific examples of the dynamic case, it is important here to point out that well known learning problems which emerge from the analysis

CP1018, *Frontiers of Fundamental and Computational Physics*,
edited by B. G. Sidharth, F. Honsell, O. Mansutti, K. Sreenivasan, and A. De Angelis

of the static case, are connected in several ways also to aspects of the dynamics of fields. The knot related to the missed recognition of interaction reciprocity, in the case of two segments of wire carrying a current, is intimately connected to the knot of impossibility of instantaneous transmission in interactions, or to the fact that interaction between wire elements carrying current is not explicitly in accord to the third law of dynamics. [9]

The knot of field ontology (i.e. gravitational, magnetic, electric) acquires a new perspective. The recognition of magnetic fields generated by wires carrying a current is in close relation with the learning problems bound to: the recognition of fields sources, the geometry of fields, the representation of fields themselves. This, however, opens more problematic issues. The following questions synthesize the problem: Is field material? Is it a state of space? And if it is a material entity, how come that in the case of an electromagnet this entity may be created or destroyed by means of an electric current? Or how can a material entity that has properties be created or destroyed by the motion or the acceleration of something else (a charge carrier)? Which exact properties has this entity? In the same way, the nature of perturbation propagating at a finite velocity in the space emerges in dynamics: as soon as the current passes through a wire, is there a field in all points of the space? [11]

If we move from the static case, where electrostatics and magneto-statics are different phenomenological areas, to the dynamic case (where the electric and magnetic field are components of the electromagnetic field) we amplify the panorama of the recalled problems about the concept of field. Nevertheless, we may also identify new and significant roads both from a disciplinary point of view and in constructing field concepts.

THE KNOT OF EM INDUCTION

In a clearly electromagnetic context, we highlighted difficulties in interpreting e.m. induction. In this section we consider e.m. induction on the phenomenological point of view. In literature we can find many information about the lack of a correct application of e.m. induction laws. We can consider in detail the results obtained in two examples of the CSEM assessment test [2] and summarize the outcome of two other researches.

The CSEM assessment test provides a quite complete example of items. In Figure 1a question 29 of the CSEM test is presented. The question focuses on evaluating if students recognize the different ways in which one can obtain the em induction. In particular it aims at exploring if students associate the induction to the relative motion of the inductor and of the circuit where the induction is, or if they recognize the role of the flux, i.e. of the several parameters which come together to generate it. Only answer C stresses all the aspects related to the fact that the induced current is produced by a flux variation. Overall, 72% of students in the sample considered to validate CSEM, chose answers that used the idea that "motion" from either the loop or the magnet is necessary to create an induced current (Answers A-B-D). Just few students see the collapsing loop as changing the magnetic flux (answer C, 22-23 %) or the rotating loops as not changing the magnetic flux (Answer A) .

In the CSEM test also the following question (item 31 reproduced in Figure 1b) is proposed, regarding a moving conductor in a magnetic field. The authors report this process still as an em induction phenomenon. Answers D and E (from 30% to 43%) could indicate that students think that there is an induced ''emf'' that causes charges to move to the top (or bottom) of the metal bar.

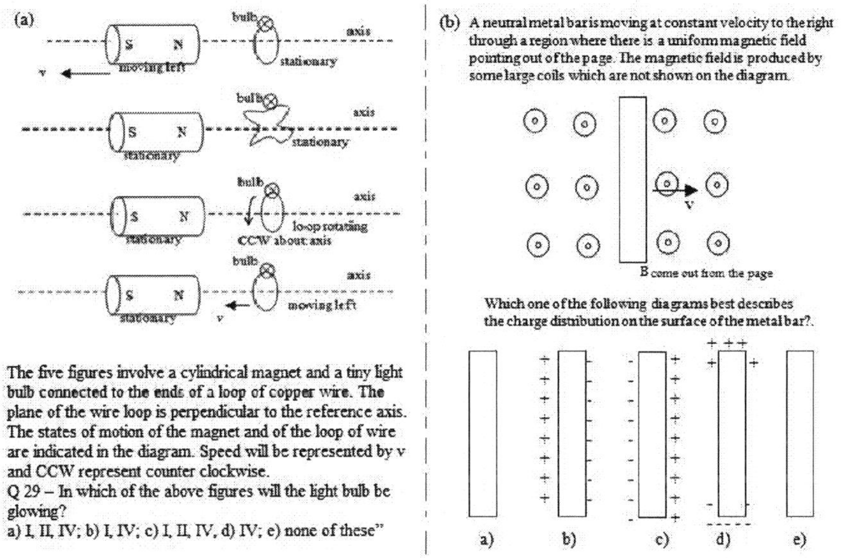

FIGURE 1. (a) Item 29 and (b) item 31 of the test assessment CSEM.

Answers B and C (49%-40%) indicate those students not distinguishing electrical and magnetic effects. Answer A (18%) may indicate that students do not recognize Lorentz force or that there are no moving charges inside the conductor.

In other researches led by Bagno Sheva-Eylon [10] emerged problems in the application of dB/dt. Students have frequently difficulty in determining the direction (versus) of the induced magnetic field. The major source of difficulty has to do with fuzzy encoding (typology a in Fig. 1b), probably caused by an incorrect interpretation of textbook sentences like ''The induced current resists its cause'', where for example, opposes the change is interpreted as being in the opposite direction (versus).

In their research, Thong, Gunstone interviewed 15 second-year undergraduate students [13]. They proposed the situation in which a coil with 1-2 bulbs is placed near to the middle of a long solenoid with the coil plane perpendicular to the solenoid axis [13]. They found three alternative conceptions 1) induced current varies proportionately with the current in the solenoid; 2) there must be contact between the magnetic flux and the external coil for any emf to be induced in the coil; 3) Coulomb or electrostatic potential difference is present in an induced electric field. The researchers suggested there were five areas of deficiency in students' knowledge structures: 1. Often the student's knowledge structure does not include central relationships (e.g. Maxwell's equations) in any form, neither mathematical nor qualitative. 2. There is an over-emphasis on subsidiary information (i.e. Ohm's law).

3. They have difficulty retrieving even the partial information that they store. 4. Most students seem to represent the relationships only in mathematical form and do not have access to more qualitative representations. 5. Students hold many inaccurate ideas in electromagnetism and erroneously interpret the central relationships [10].

THE CONCEPTUAL LEVEL

From the presented examples it is clear how the attention of researchers was mostly concentrated on gathering information about the way students face specific phenomenological situations. The outcome effect is to simply reproduce the phenomenology of students ideas and conceptions. It is therefore important here to recosider the problem from a conceptual point of view, instead of adding more examples. This is essential because it provides also a different way of reading the shown learning difficulties.

Questions 29 and 31 of the CSEM test analyzed above, propose two situations which can be reinterpreted according to tree different interpretative schemes:

1. These constitute two phenomenologies thus require two interpretations (Faraday –Neumann – Lenz law; Loretz force).

2. They concern just one phenomenology, the em induction phenomenology, and they are kinematically equivalent, even if in classic e.m. we need two interpretations.

3. They regard a single phenomenology, the kinematical equivalence is given by the relativity principle, thus there is just one interpretation, the special relativity one. The apparent different phenomenology of the first scheme derives only from the change of the reference system, which implies the change in the components of the E-M tensor, i.e. in each of the Maxwell equations and also in the complete Lorentz force. What is really surprising is the fact that the symmetry suggested by the intuition leads to the correct high level interpretation. [15]

CONCLUSIONS

The concept of field, useful in the case of static interactions, is the basic concept to describe dynamical situations. It emerges as a perturbation propagating with a finite velocity into the space-time across vacuum or matter. In this context typical disciplinary knots rise up. They are source of learning problems for students as: the sources of time depending fields, the electrodynamic of moving conductors, the nature of electric and magnetic fields as interchangeable components of the same unitary physical entity i.e. the em field.

From several researches it emerged that students have difficulties in recognizing that time-depending electric and magnetic fields are themselves sources of magnetic and electric fields changing in time.

Moreover students often reach a partial understanding of em induction, due to a partial knowledge of the different situations producing induced currents or, from an opposite point of view, related to a wrong use of the Lenz law or circuits laws.

If we analyze the conceptual framework we can recognize three different ways to interpret the diverse situations concerning the em induction:

a) when we have a circuit, a flux variation occurs, therefore the theoretical reference is the Faraday-Neumann-Lenz law; when we consider a moving conductor, there is not a flux variation, and we interpret the phenomenology as a consequence of the Lorentz force.
b) even though the two previous situations are kinematically equivalent, in classical em we do not have the instruments to give a compact and unique interpretation.
c) in the framework of special relativity the role of the reference system emerges and we can re-combine the two separated phenomenologies, (but the students do not have the mathematical instruments for treating the em field in the covariance form. From this point of view these learning problems are connected to the basic problem related to the birth of Einstein special relativity theory.

Two main indications or priority research problems for a instructional path on em induction emerged:
- How can the Faraday-Neumann-Lenz law be addressed? (exploring several situations underlying the role of the flux variation and of the diverse quantity involved (intensity, surface, rotation in B constant, relative motion...)?
- How can the electrodynamic of the moving system be treated? How can this point in a coherent way be addressed?

Using a weak relativistic approach, as Galili suggested, it would be possible to provide students with a vision closer to the relativistic invariance one. Formal difficulties of this approach undermine its potentialities. On the other hand it is clear that the choice of bounding the induction law to the relative motion is the basis of the several difficulties emerged from researches. At last, an approach keeping separated the two phenomenologies has the limit to drive students away from the recognition of the symmetry in a situation.

REFERENCES

1. S. Rainson and L. Viennot, *Int. J. Sci. Educ.* 14(4), 475-487(1992).
2. D. P. Maloney, T. L. O'Kuma, C.J. Hieggelke and A. Van Heuvelen, *Phys. Educ. Res., Am. J. Phys.* Suppl. 69 (7), pp S12-S23(2001).
3. J. Guisasola, J. M. Almudi and M. Ceberio, *Enseñanza de las Ciencias*, 21 (2) 281-293(2003).
4. J. Guisasola, J. M. Almudi and M. Ceberio, *Science Education*, 8 (3) 443-464 (2004).
5. M. Faraday, *Experimental Researches in Electricity*, London: *Taylor and Francis* (1955), Vol I, II, III.
6. J. C. Maxwell, *A Treatise on Electricity and Magnetism*, Dover Publications, INC (1954).
7. A. T. Borges and J. K. Gilbert, *Int. Journal of Sci. Ed.* 20 (3) 361-378 (1998).
8. S. Tornkwist, K. A. Pettersson, and G. Transtromer, *Am. J. Phys.* 61(4) 335-338(1993).
9. J. Guisasola, J. M. Almudi and M. Ceberio, Students ideas about source of magnetic field, II int. Esera Conf., pp. 89-91(1999).
10. E. Bagno and B. S. Eylon, Am. J. Phys. 65 (8) 726-736 (1997).
11. IPN Kiel, *Die elektromagnetische Induktion* (IPN Curriculum Physik) Stuttgard, (1985).
12. E. R. Salverberg and T. De Jong, M. G. Ferguson-Hassler, *Am. J. Phys.* 39 (10) 928-951 (2002).
13. W. M. Thong and R. Gunstone, *Res. Sci. Educ.* 38, 31-44 (2008).
14. P. C. Peters, *Am. J. Phys.* 52(3) 208–211(1984).
15. I. Galili and D. Kaplan, *Am. J. Phys.* 65 (7) 657-668 (1997).
16. R. Duit, ESERA Summer School, Braga, 2006, at http://www.naturfagsenteret.no/esera/summerschool2006.html

Surface charges and fields in stationary conductors with steady currents

Julio Akashi Hernandes* and Andre Koch Torres Assis†

*Departamento de Ciências Exatas e Tecnológicas – DCET,
Universidade Estadual de Santa Cruz – UESC,
45662-000 Ilhéus, Bahia, Brazil
†Instituto de Fíica 'Gleb Wataghin'
Universidade Estadual de Campinas – Unicamp
13083-970 Campinas, São Paulo, Brazil

Abstract. Is there a force between an external stationary charge and a resistive stationary conductor carrying a steady current? The answer to this question is positive. In this work we present the main results of this interaction. We present experiments published in the literature which measured these effects. We also show the analytical solution for the most common situations, namely: straight wires, strips and toroidal conductors. This force is due to charges spread along the surface of the current carrying conductor. This distribution of surface charges is maintained by the battery, and keep the current flowing along the conductor. This had been pointed out by Kirchhoff and Weber. These surface charges keep the potential gradient along the resistive circuit. They also create an electric field inside and outside the conductor. With this approach we show that there are no fundamental differences between electrostatics and current-carrying conductors.

Keywords: Electric potential, electric field outside steady currents, surface charges
PACS: 41.20.Cv

INTRODUCTION

Consider a stationary resistive wire, in the form of a closed loop circuit, connected to a battery which generates a voltage V between its terminals. If the wire carries a steady current I, will the resistive wire exert a force on a stationary charge q located nearby? Do some components of this force depend on the voltage generated by the battery? These questions can be rephrased in terms of electric fields: does the wire generate an electric field in the surrounding space, depending on the battery voltage? Is the wire carrying a steady current charged along its surface, or inside it?

These fundamental questions have been answered by many physicists as "no," and this opinion has been held for a long time. We show that conducting wires with steady currents are not neutral along its surface, and there is in fact an electric field outside the conductor, proportional to the voltage of the battery.

In this work we show theoretically the existence of a force upon the stationary external charge exerted by a resistive wire connected to a battery and carrying a steady current when there is no motion between the test charge and the wire. We also compare the theoretical calculations with the experimental results which proved the existence of this force. Detailed results can be found in the book [1], which can be freely downloaded in PDF format.

CP1018, *Frontiers of Fundamental and Computational Physics*,
edited by B. G. Sidharth, F. Honsell, O. Mansutti, K. Sreenivasan, and A. De Angelis

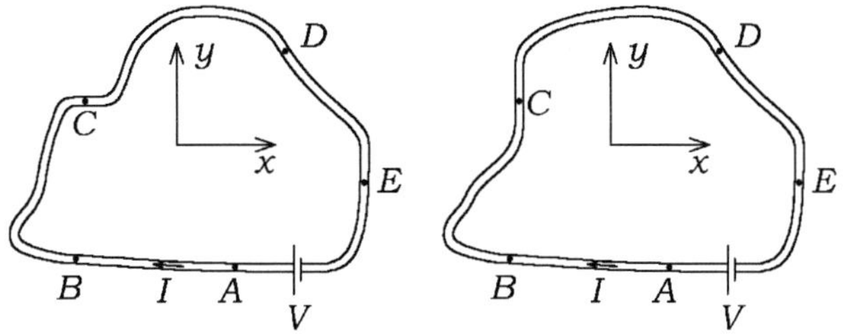

FIGURE 1. A wire is bent around point C, changing the direction of the electric field.

ELECTRIC FIELD INSIDE AND OUTSIDE THE WIRE

Consider a resistive wire of finite conductivity connected to a battery V carrying a steady current I, as in the left side of Figure 1. Where are the charges which generate the electric field at each point along the wire located? It might be thought that this electric field is due to the battery, but this is not the complete answer. We can see that the battery does not generate the electric field at all points along the wire looking at the right side of Figure 1. At a specific point C we bend the wire, changing the direction of the electric field. However, at the other points there is no change in the electric field. The battery can not be responsible for such local change. This was Weber's and Kirchhoff's idea [2].

What creates the electric field inside the wire are surface charges, distributed as a continuous gradient density along the length of the wire, more positive towards the positive terminal of the battery, and increasingly negative towards the negative terminal.

CYLINDRIC WIRE

Consider a long straight wire of circular cross-section, of length ℓ and radius $a \ll \ell$. The uniform longitudinal electric field inside the wire keeps the steady current flowing, and is generated by surface charges along the conductor. By supposing a linear distribution [3], we can calculate the electric potential and the electric field. The mathematical treatment is detailed in [4]. Figure 2 shows the electric field lines along a longitudinal cross-section of the wire.

The solution is valid in the case of a solid cylinder or a hollow cylindrical shell. The potential is linear along the cylindrical surface, and obeys Laplace's equation in the space surrounding the conductor. Similar results can be obtained for a conducting strip (of length ℓ and width $a \ll \ell$) [5].

TOROIDAL CONDUCTOR

Consider a resistive toroidal conductor with a steady current I in the azimuthal direction, flowing along the circular loop. There is a battery located in $\varphi = \pm\pi$, as indicated in Figure 3. This case is detailed in [6].

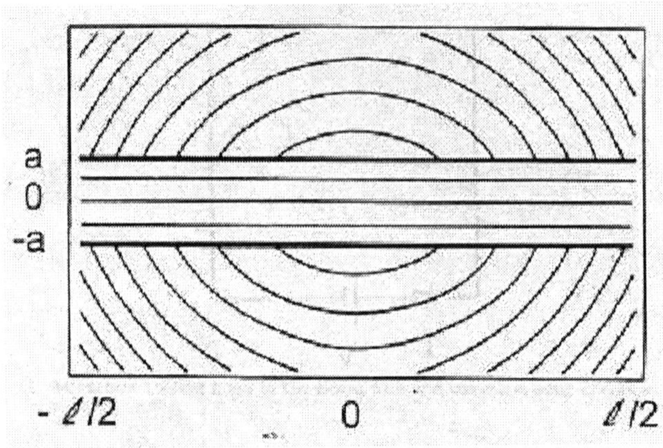

FIGURE 2. Electric field lines along a longitudinal cross-section of the wire. The wire surface is represented by the two horizontal thick lines.

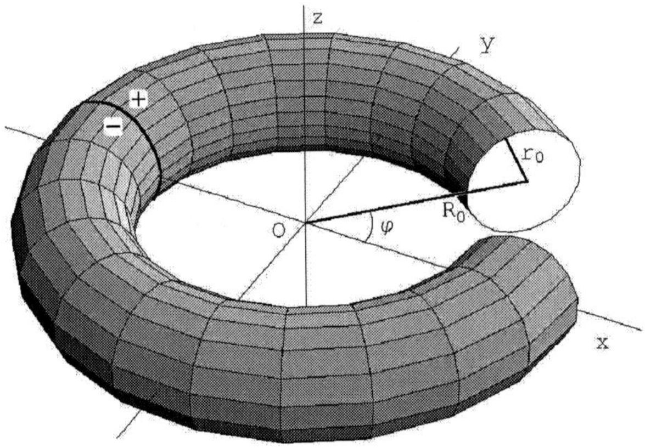

FIGURE 3. Toroidal conductor with a steady azimuthal current. The battery is indicated by the signs "+" and "-."

We solve Laplace's equation in toroidal coordinates, using the azimuthally linear potential along the conductor surface as a boundary condition. The equipotential lines are shown in Figure 4. The surface charges along the conductor, in this case, is not linear as in the case of a straight conductor.

Experiments performed by Jefimenko mapped the electric field lines [7]. He utilised a transparent conducting ink to make a two-dimensional printed circuit on glass plates of 10 × 12 inches. After the power supply (of about 10^4 V) was turned on, he spread some fine grass seeds over the plate and conducting system. The seeds lined up in the direction of the electric field over and outside the conductors.

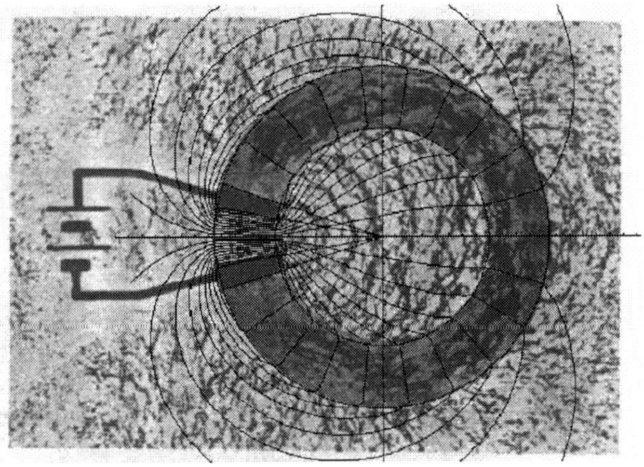

FIGURE 4. Equipotential lines in the plane of the toroid. This image was superimposed on an experimental plate by Jefimenko.

CONCLUSION

There is a non-null electric field outside stationary conductors carrying steady currents. This electric field is proportional to the battery *emf* and is generated by surface charges along the conductor. This approach shows a very important connection between electrostatics and magnetostatics (stationary currents). A related topic is to consider in detail the behavior of surface charges and the corresponding external field in the transition from steady-currents to low and high frequency circuits with alternating currents [8]. Another is a possible connection between the external electric field around a resistive cylindrical conductor carrying a steady current and the Aharonov-Bohm effect, as discussed in [9].

For further discussion, with many more examples, exact mathematical expressions and references, see the book [1].

REFERENCES

1. A. K. T. Assis, and J. A. Hernandes, *The Electric Force of a Current: Weber and the Surface Charges of Resistive Conductors Carrying Steady Currents*, Apeiron, Montreal, 2007, iSBN: 978-0-9732911-5-5. Available in PDF at: http://www.ifi.unicamp.br/~assis/.
2. G. Kirchhoff, *Philosophical Magazine* **37**, 463–468 (1850).
3. B. R. Russell, *American Journal of Physics* **36**, 527–529 (1968).
4. A. K. T. Assis, W. A. Rodrigues Jr., and A. J. Mania, *Foundations of Physics* **29**, 729–753 (1999).
5. J. A. Hernandes, and A. K. T. Assis, *American Journal of Physics* **71**, 938–942 (2003).
6. J. A. Hernandes, and A. K. T. Assis, Electric potential for a resistive toroidal conductor carrying a steady azimuthal current (2003), physical Review E, Vol. 68, 046611.
7. O. Jefimenko, *American Journal of Physics* **30**, 19–21 (1962).
8. J. D. Jackson, *American Journal of Physics* **64**, 855–870 (1996).
9. H. Torres S., and A. K. T. Assis, *Revista de la Facultad de Ingenieria de la Universidad de Tarapaca (Chile)* **9**, 29–34 (2001).

Perspective Primary Teachers Explores Magnetic Interactions As A Base To Understand Lorentz Force

Marisa Michelini[a] and Rossana Viola[a]

[a] *Research Unit in Physics Education, University of Udine, Via delle Scienze n208, 33100 Udine, Italy*

Lorentz's Force represents a conceptual gap as for the analysis of the simple effects related to the properties characterizing the electric and magnetic nature of phenomena. Searching for the definition of a teaching/learning path according to MRE model [5] a research dedicated to primary school prospective teachers has been carried out. This learning path is structured and based on experimental practice [4] according to PCK guide lines and main results are here presented on the side of teachers training.

Keywords: Electromagnetism, perspective primary teachers.
PACS: 40, 70, 80

Electromagnetic interaction have an important role in the interpretation of electric and magnetic phenomenology, but different students' difficulties are identified: students adopt different strategies to solve problems and so there is a lack of a unitary interpretative framework of the electromagnetic phenomenology because the problems are solved utilizing the laws in a fragmentary way [7] and these conceptual knots are revealed by research at primary level and at university level too on the organization of contents, on the analysis of the ideas about the nature of force and magnetic field, on the principle of superposition [9]. Students don't related phenomena, concepts and laws [8], adopt different strategies to solve problems and so there is a lack of a unitary interpretative framework of the electromagnetic phenomenology.

Research literature on teachers formation underlines that a great part of learning difficulties of students are in common to the teachers and then become professional problems in approaches and strategies. In prospective primary science teachers' courses, one of the principal problems is to build the didactic competence in science education combining pedagogical and content knowledge to guarantee them as prerequisite [6].

In the prospective of a vertical curriculum we have to look at learning in primary school. In this perspective is carried out an empirical research aimed to Teaching/Learning proposals for prospective primary teachers. In the theoretic frame of the Model of Educational Reconstruction [5] this preliminary study focuses on magnetic and electromagnetic interactions to build the bases, in particular for Lorentz Force, which represents a conceptual gap as for the analysis of the simple effects related to the properties characterizing the electric and magnetic nature of those phenomena. In the framework of CLOE Labs [1] this research focuses on the identification of the magnetic effects of an electric current as a first step towards the study of electromagnetic interactions.

CP1018, *Frontiers of Fundamental and Computational Physics*,
edited by B. G. Sidharth, F. Honsell, O. Mansutti, K. Sreenivasan, and A. De Angelis

Starting from the explorations of magnetic phenomena and magnetic effects of an electric current, the first aim is to explore by a compass the characteristics of the magnetic field produced by one or more magnets and to identify the equivalence between the field lines produced by a magnet and by a solenoid. Than the identification of Lorentz Force is explored as a new aspect as regards the superposition of field. The **sample** consisted of 105 students of the 2nd year of University Course for prospective primary teachers formation (21 years old) before the university courses of physics and after the visit of GEI exhibit and in particular of the didactic proposals on magnetic and electromagnetic phenomena.

The main **research questions**, related to some situations are shown in Table1:

TABLE 1. The main research questions, related to some situations.

Situation (S)	Research Questions (R)
Students have some objects of different materials: 1. Are there any magnets? 2. How do you recognize a magnet? 3. Describe the behaviour of a magnet with objects of different materials (magnet, wood, ferromagnetic material, needle) 4. Investigate the reciprocity of the attraction	R1. a) Which are the conceptual referents to recognize the magnetic behaviour? b) Which are the characteristics of the description of an electromagnetic interactions? c) Which are the way of description of them?
5. Exploration with needles of the space around a magnet.	R3. a) Do students recognize the tridimensionality of the field lines? b) Are the reasoning in terms of field activated? c) Are the use of an explorer and of the field lines an effective angle of attack for the reasoning in terms of field? d) Which angles of attach are effective for the reasoning in terms of field in the conceptual reconstruction of the phenomenology?
6. Several magnets: analysis of the field lines.	R3. e) Is the principle of superposition utilized in the description with the field lines?
7. Needle and magnet: exploration of the interactions. 8. Interactions between magnets when the rotation is not permitted.	R4. a) Which is the role of project tasks to reach explicative and interpretative needs and to induce explorative behaviour to overcome description in a common language? b) Do students recognize that there are two different aspects, the orientation and the repulsion?
9. Students have a current currying wire: how to explore magnetic effects? 10. A current carrying coil and a solenoid: prevision about the field lines.	R2. Is the magnetic field recognized in the case in which there isn't a permanent magnet? Is the representation of field with lines effective to allow a prevision about the field of a coil and a solenoid? Do students recognize the analogy between the field lines of a magnet and a solenoid?
11. A current carrying wire in a magnetic field (Lorentz Force)	R3. d)

The work carried out consists of two sessions of three ours each of cognitive laboratory Conceptual Lab for Operative Exploration (CLOE) in which prospective teachers worked in small group of 6-7. Prospective teachers worked to an explorative hands-on path with stimuli sheet [2,3,4] based on SPPEA cycles (Situation, Prevision, Planning, Experiment, Analysis). The tackled knots in the sequence: A) exploration of interactions between a magnet and: A1) an other magnet , A2) ferromagnetic objects, A3) objects done by other materials, A4) a compass. B) reciprocity of the action being not in contact. C) the compass is a magnet. D) magnet ordinary suspension. E) the space around a magnet: magnetic field and their representation through field lines. F) superposition of fields. G) magnetic effects of a current: field lines of a current, a coil, a solenoid. H) interaction between a magnet and a solenoid I) a current into a magnetic field: Lorentz Force. Data analysis was done with the following **methodology**: classifications of answers in a priori classes, statistical analysis of standard answers, qualitative case study for special interpretative models, comparison with quantitative and qualitative data.

The 105 students recognizing there are magnets in the box with different objects explain this identification with the attraction (67/105 students), or the sticking (4) to ather objects (65). Few students refers to a specific explorer (the needle of the compass) as indicator related to the attraction (16) or to the rotation (8). The reciprocity of the interactions is recognized in 3 cases only and the subject of the action is often (88) the magnet. The main conceptual referent for the descriptions of magnetic interactions is the attraction. Interaction is described in terms of behaviour of systems and rarely in terms of physics quantities/entities (force, field, torque). The categories of behaviour: A) 2 types of behaviour (magnets); B) Only attraction (ferromagnetic materials); C) No interactions (the other materials); D) Compasses. We see that the compass favours the recognition of the interaction being not in contact (102). Planning tasks are useful to overcome the simple description of a behaviour and to focus on causes. Exploring the space around a magnet, 50 students say that there is a magnetic field but only 24 of them justify it with an experimental procedure using, for example, iron dust, but they don't know the nature of these objects, It seems that they are reproducing a procedure without understanding it; only 6 students speak about the position of metal objects as indicator of field. Exploring the superposition principle drawing field lines of two magnets, only in 8 cases there is a qualitative idea of superposition, in 86 cases they refers to vectorial sum. When we arrive to the exploration of magnetic effects of a current, the preliminary part (aimed to the recognition of the necessity of an explorer and of the compass as explorer) is effective: 100/105 suggest a compass to study existence and characteristics of a magnetic field around a currying current wire. After the observation of the lines of a straight current carrying wire and of a coil, 83/105 students foresee that magnetic field lines path of a solenoid is similar to a magnet. Doing the experiment, 94 students draw field lines similar to a magnet even if 10 students don't write any comment, 36 students talk about a "generic" similarity, 48 recognize the equivalence of the systems, 25 recognize the only equivalence of the field lines. Giving a prevision about the behaviour of a current carrying wire in a magnetic field, the answers are divided in two groups: a) Students who look at the interaction; b) Students who look at what changes in the space around. Doing and interpreting the phenomenon, 25 identify it with the birth of a force as a new element, 9 thinks at a composition of actions between the force by the magnet and the force by the wire as the consequence of two magnetic field, other students speaks about forces and energy, 3 write about interaction between magnet and electric field, some students think the force as generated by magnet or wire, by both, 5 thinks that field lines of the magnet and of the wire repelling or there is an interaction.

In relation to each research question (see Table 1), we can conclude:

R1. It is necessary to build the specificity of these different types of interactions for prospective teachers to give clarity to students that in this way do not remain on a descriptive phenomenological level. The identification of the behaviour of systems is local and contingent and often without explanations. It raise up that it is the need on one side to build the subject prerequisites and, on the other side, to make teachers responsible in their concepts' organization to be able to reframe the main elements of an interpretation of phenomena.

R2. The familiarity with magnetic interactions produces effective analogy to the identifications of systems of currents with same behaviours. The need to learn the specificity of magnetic interaction before of that electromagnetic rises up from 3 elements: 1. the need to recognize the characteristics of magnetic interactions (rotation and attraction); 2. the awareness of the need of an explorer of space properties; 3. the recognition of reciprocity of interactions. Introductive work is effective for the right use of explorers and students immediately recognize a magnetic field around a current.

R3. The representation in terms of field lines is an useful angle of attach to overcome local and animistic interpretations of magnetic interactions, to build reasoning in terms of field and to conceptual reconstruction of phenomenology and is effective when has a role in superposition of field and interactions between magnet and current.

R4. We find evidence that planning tasks turn out effective to overcome the common diffuse attitude to be reductive and reproductive. Project tasks active explicative needs and make students awareness of processes not analyzed before. Through the responsibility of explanation, students overcome descriptions of a local type and in common language: they start to use specific terms of physics entities, even if sometimes in a not so correct way, instead of common sense ideas. It is necessary to recall often theoretic frame to avoid that teachers precipitate in local vision of single phenomenon or underestimate the exploration of simple phenomena and give a sequence of non related descriptions and that, often, are only their previsions.

REFERENCES

1. Corni F, Michelini M, Formal and informal aspects in planning educational paths for primary school on simple machines in an open distance-learning context, in Girep book of selected contributions, Ljubljana (SLO), 2006 [ISBN 961-6619-00-4]
2. Michelini M (2003), New approach in physics education for primary school teachers: experimenting innovative approach in Udine University, Inquiries into European Higher Education in Physics, H Ferdinande, E, Valcke, T Formesyn eds., European Physics Education Network (EUPEN), vol.7, 2003, p.180
3. Corni F, Michelini M, Stefanel, A (2003) Strategies in formative intervention modules for physics education of primary school teachers: a coordinated research in Reggio Emilia and Udine, in Quality Development in Teacher Education and Training, sel papers of the GIREP international seminar, Udine 2003
4. R. Martongelli, M Michelini, L Santi, A Stefanel, Educational Proposals using New Technologies and Telematic Net for Physics, in Physics Teacher Education Beyond 2000 (Phyteb2000), R.Pinto, S. Surinach Eds., Girep book - Selected contributions of the Phyteb2000 International Conference, Elsevier, 2001, p.615
5. Duit, R., Science Education Research – An indispensable prerequisite for Improving Instructional Practice, ESERA Summer School, Braga, July 2006 at http://www.naturfagsenteret.no/esera/summerschool2006.html
6. Shulman, L.S. (1987). Knowledge and teaching: Foundations of the new reform. Harvard Educational Review, 57(1), 1-21.
7. Greca, I., Moreira, M. (1998). Modelos mentales y aprendizaje de Fisica en Electricidad y Magnetismo. Ensenanza de las ciencias, 16(2), 289-303.
8. Bagno, E., Eylon, B.S. From problem solving to a knowledge structure: An example from the domain of electromagnetism. American Journal of Physics -- August 1997 -- Volume 65, Issue 8, pp. 726-736
9. C. Furió , J. Guisasola , J. M. Almudí , M. Ceberio , Learning the electric field concept as oriented research activity, Science Education, Vol. 87, Issue 5 , Pages 640 - 662

Virtually Exploring A Pillar Of Experimental Physics: The Hertz Experiment

A. Bonanno, P. Sapia, M. Camarca, A. Oliva [a]

[a]*Physics Education Research Group – Physics Department – University of Calabria. Ponte Bucci, cubo 31/C – I87036 Rende (CS) - ITALY*

Abstract. In the present work we report on the implementation and early assessment of a multimedia learning object, developed using the Java programming language, which also integrates in a creative way some internet freely available educational resources, intended to support the teaching/learning process of the historical Hertz experiment.

Keywords: physics education; learning object; java; Hertz experiment; electromagnetic waves.
PACS: 01.30.la; 01.40.Fk; 01.40.gb; 01.50.ht

INTRODUCTION

The remarkable series of experiments conducted by Heinrich Hertz between 1885 and 1889 had an exceptional influence on subsequent development of science and technology [1-4]. In these experiments, in fact, by generating and detecting electromagnetic waves, he succeeded in demonstrating that the predictions done by J. C. Maxwell some decades before were correct. In particular, he used an oscillator made of polished brass knobs, connected to an induction coil and separated by a tiny gap over which sparks could leap. If Maxwell's predictions were correct, electromagnetic waves would be transmitted during each series of sparks. To confirm this, Hertz made a simple receiver of looped wire. At the ends of the loop were small knobs separated by a tiny gap. The receiver was placed away from the (possible!) transmitter. If electromagnetic waves were radiating from the oscillator sparks, they would induce a current in the loop that would send sparks across its gap. This occurred when Hertz turned on the oscillator, producing the first transmission and reception of electromagnetic waves. Among other things, with its experiments Hertz found out how to make the electric and magnetic fields detach themselves from wires and go free as Maxwell's waves, so allowing all subsequent well known technological applications.

Nevertheless, notwithstanding their exceptional importance, it is fairly rare to find these experiments treated in educational context. This is also due to the difficulty in reproducing them in a significant and quantitative way, by only using simple school

CP1018, *Frontiers of Fundamental and Computational Physics*,
edited by B. G. Sidharth, F. Honsell, O. Mansutti, K. Sreenivasan, and A. De Angelis

laboratory equipment. As we know, very few are the available examples denying this picture [5].

In this connection, we have developed and classroom-tested a multimedia learning object (MLO) aimed to interactively illustrate the main features of Hertz-like experiments, by allowing learners to virtually manipulating the main characteristics of them. The proposed virtual learning environment is intended as the first step of an approach to the teaching/learning (T/L) process which integrates multimedia technologies as an illustrative support of real experiments made up by using low cost, easily found, materials.

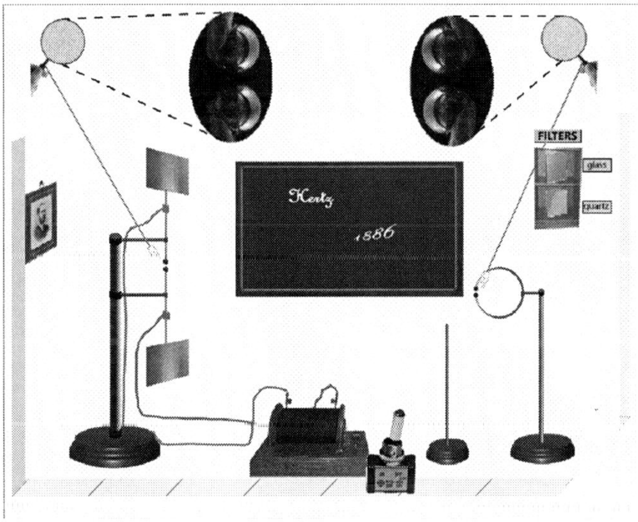

FIGURE 1. Virtual interactive setup of Hertz experiment. By clicking on the switch, sparks glow on the transmitter's spark-gap (left), then causing the receiver loop (right) to spark too. Magnifying glasses show the detail of respective spark-gaps. To strengthen the perception of causal relation between sparking events, that one occurring on the receiver side is shown exaggeratedly retarded (tenths of a second!) with respect to that one occurring in the transmitter.

MLO IMPLEMENTATION AND ASSESSMENT

The Multimedia Learning Object

The MLO consists of a Java developed simulation of the Hertz' experiment interactively illustrating how it works. The virtual experimental setup presents (Fig. 1) an interactive artist's concept of the original apparatus [2] in which the learner can act on a switch to trigger a spark in the transmitter spark-gap, subsequently observing sparks glowing in the receiver one. The simulation is embedded in a web-like environment which conducts the student through a learning path in which some Hertz-related educational resources, freely accessible on-line, are proposed. As, for example, a simulation of the Lodge experiment [6], which in 1894 first demonstrated the principle of tunable radio receiver

The classroom assessment

The MLO was employed in the 2006/2007 scholastic year in 12 last year classes belonging to 5 Italian secondary schools focusing on sciences (*licei scientifici*), for a total number of 213 students. Classes have been divided in two groups, numerically nearly equivalent, one of which (*control-group*) studied Hertz' experiment theme by using a traditional "book, paper and chalk" approach; the other one (*MLO-group*) employed instead the MLO. A pre-test and a post-test were conducted on both groups, and results compared. The mean normalized score achieved by the *MLO-group* ($\overline{S}_{MLO} = (63 \pm 5)\%$) was clearly higher than that gained by the *control* one ($\overline{S}_{ctrl} = (45 \pm 4)\%$), unambiguously suggesting a greater efficacy of the MLO-based T/L process.

Hereafter an example of a test item is given:

Which characteristic of the Hertz experiment would result unaccountable without supposing the electromagnetic waves existence?

a) *The sparking in the circuit directly connected to the Ruhmkorff coil.*
b) *The sparking in the right side loop.*
c) *BOTH sparking were unaccountable.*
d) *None of the previous answers is correct.*

CONCLUSIONS

We have reported on the development and classroom testing of a multimedia learning object aimed to interactively illustrate the historical 1888 Hertz' experiments on production and detection of electromagnetic waves. The MLO was implemented by using the Java programming language to create an interactive simulation of proposed experiments that was embedded in a web-like explorable multimedia, also integrating freely available web resources. The classroom test, comparatively conducted on a sample of 213 high school students split into two groups (one of which was a *control group*), clearly showed a greater efficiency of the MLO-based teaching/learning process with respect to the traditional one. The assessed MLO is the first step of an integrated T/L approach to the proposed argument which employs multimedia technologies as an illustrative support for real experiments made up by using low cost, easily found, materials. Results from the performed assessment will be helpful in designing and developing the integrated real/virtual learning object.

REFERENCES

1. H. Hertz, *Electric Waves*, London: MacMillan, 1893. This book contains Hertz' papers on the propagation of electromagnetic waves. It contains a preface by Kelvin and a long introduction by Hertz himself about his experimental and theoretical researches.
2. An excellent on-line resource dedicated to H. Hertz: http://chem.ch.huji.ac.il/history/hertz.htm (retrieved February 12 2008).
3. J. Z. Buchwald, *The Creation of Scientific Effects: Heinrich Hertz and Electric Waves*, Chicago: University of Chicago Press, 1994.
4. I. Adawi, *Am. J. Phys.* **57**, 125–127 (1989).
5. D. Faccio and M. Clerici, *Am. J. Phys.* **74**, 992–994 (2006).
6. An interactive simulation of the Lodge experiment: http://www.magnet.fsu.edu/education/tutorials/java/lodgeexperiment/index.html (retrieved February 12 2008).

Understanding the Physics of changing mass phenomena

A.L. Ellermeijer

AMSTEL Institute, Faculty of Science, Universiteit van Amsterdam, 1098 Amsterdam, The Netherlands

Abstract. Changing mass phenomena, like a falling chain or a bungee jumper, might give surprising results, even for experienced physicists. They have resulted in hot discussions in journals, in which for instance Physics professors claim the impossibility of an acceleration larger then g in case of a bungee jumper. These phenomena are also interesting as topics for challenging student projects, and used as such by Dutch high school students. I will take these phenomena as the context in which I like to demonstrate the possibilities of ICT in the learning process of physics. Especially dynamical modeling enables us to describe these phenomena in an elegant way and with knowledge of high school mathematics. Furthermore tools for video-analysis and data from measurements with sensors allow us to study the phenomena in experiments. This example demonstrates the level of implementation of ICT in Physics Education in The Netherlands [1].

A high school student project

When finishing high school, Dutch students have to make a 'profile master piece' (student research project) in their last years. Profiles are groups of subjects organized according to a well defined field of social interest. There are four profiles in the Netherlands: Culture & Society, Economy & Society, Nature & Health and Nature & Technology. In the Nature profiles, usually teams of two students collaborate in creating their piece of work as an independent experimental research in a topic of their own choosing.

In 2003, Niek Dubelaar and Remco Brantjes of the Bonhoeffer College (Castricum, The Netherlands) teamed up to investigate the physics of bungee jumping, triggered by their own interest and articles on the Internet. More specifically, they were excited by the alleged 'greater-than-g-acceleration' of a bungee jumper in some of the published papers on the subject.

They formulated the open research question: *how large is the acceleration at a bungee jump and to what degree is this acceleration influenced by the relative mass of the rope and the jumper?* They used video measurements on a scale model and computer modeling as tools of investigation. Soon they realized that the mass relation rope/jumper was too low to see an outstanding result and they repeated their experiment with a larger mass ratio. They also constructed a computer model based on the theory of Kagan and Kott [2], who derived an expression for the acceleration from

CP1018, *Frontiers of Fundamental and Computational Physics*,
edited by B. G. Sidharth, F. Honsell, O. Mansutti, K. Sreenivasan, and A. De Angelis

energy considerations. Measurements were not very accurate, but good enough to see the effect clearly and to be able to compare the results with the computer model. Their conclusion: it is really true, a > g!

They submitted their work to a national contest in student research projects organized by the University of Amsterdam, and were awarded the second price. Later, they published their results in the Dutch Physics Journal [3], triggering quite a number of reactions. Comments on the subject continued at least a year on the internet. It seemed that the whole Dutch physics community was all of a sudden playing with chains, ropes, elastics etc. There were even complaints on the level of physics teaching in the Netherlands, arguing that obviously(!) a=g and that this project proofed that the level of Physics education in The Netherlands has decreased in the last decades.

Two theoreticians from the Technical University Eindhoven [4] agreed with the findings of the students, and explained that our physics intuition is so easily fooled as we are all raised with the Galilei's paradigm of the motion of constant masses, according to which every acceleration must be produced by a force. Naturally, a rocket is an important counterexample to this line of thought.

In summary, it is nice to see that even juniors like high school students, supported by ICT tools like video measurements to do the experiment and computer modeling to solve the differential equations, are able to evoke such a heated discussion in the physics community with all the misunderstandings and new creativity (like an article on the lasso paradox) that go with it.

ICT tools used for analyzing and understanding the changing mass phenomena

The research project 'Greater then g acceleration of the bungee jumping' serves as an interesting example of using of ICT in the process of learning physics. For such student investigations often the Coach 6 software environment is used.

Coach 6

Coach 6 is a versatile learning and authoring environment for Science, Mathematics and Technology Education. It integrates tools for:

Measurement - on- and off-line with interfaces, data loggers and sensors

Control of processes and devices

Data Video – measurements on digital video clips (capture of own video clips included) and images

Modeling (System Dynamics approach)

Processing and Analysis of data.

It also offers authoring facilities to create multimedia activities which includes instructions, models, videos, images, and links to Internet and can be customized to be used by students starting at primary up to undergraduate level (age 10 to 20). Coach is often used in student investigations and research projects.

Activities of changing mass phenomena

The phenomenon of changing mass during the first phase of the bungee jumping can be studied by using different ICT tools such as video analysis, data logging experiment, and modelling.

Video analysis

Imagine the following experiment. Two identical blocks of wood are fixed in a clamp. One of the blocks is fixed to the chain. The whole construction hangs about 6 meters above the ground. The swivel screw holding the two blocks should be slowly un-tighten, making the two blocks starting to fall at exactly the same moment. Which block hits the ground first?

The block fixed to the chain reaches the ground before the free falling block does. By making a video of this experiment student can use the recorded material to observe the (slow) motion of the two blocks and to measure the acceleration of the falling blocks.

Data-logging experiment

A data-logging experiment gives an opportunity to measure acceleration of the falling block in a different way. An acceleration sensor mounted to the wooden block hanging on the chain is used to collect data during the fall. The measured acceleration is displayed on graphs and tables.

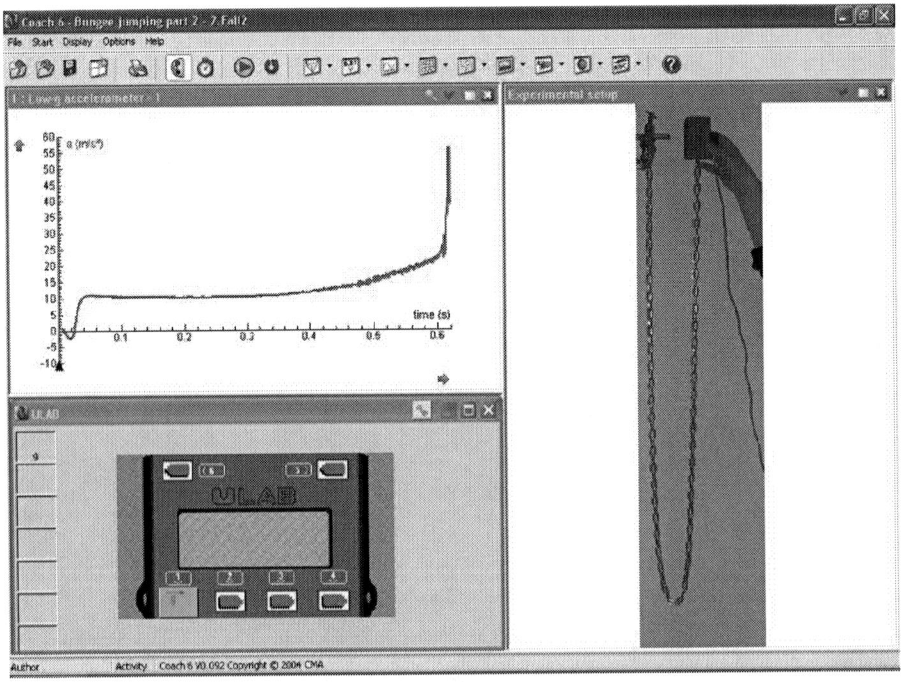

FIGURE 1. Coach 6 Data-logging Activity.

Modeling

The Modeling environment allows creating a (graphical) model which theoretically describes the phenomenon. The model applies the second Newton's Law for an object which mass is changing in time.

$$F = \frac{dp}{dt} = m*(\frac{dv}{dt}) + (\frac{dm}{dt})*v + = m*a + (\frac{dm}{dt})*v$$

Such model can be built with knowledge of high school physics and mathematics. The result of the model can be compared to the experimental data collected in the data logging experiment (or a video measurement).

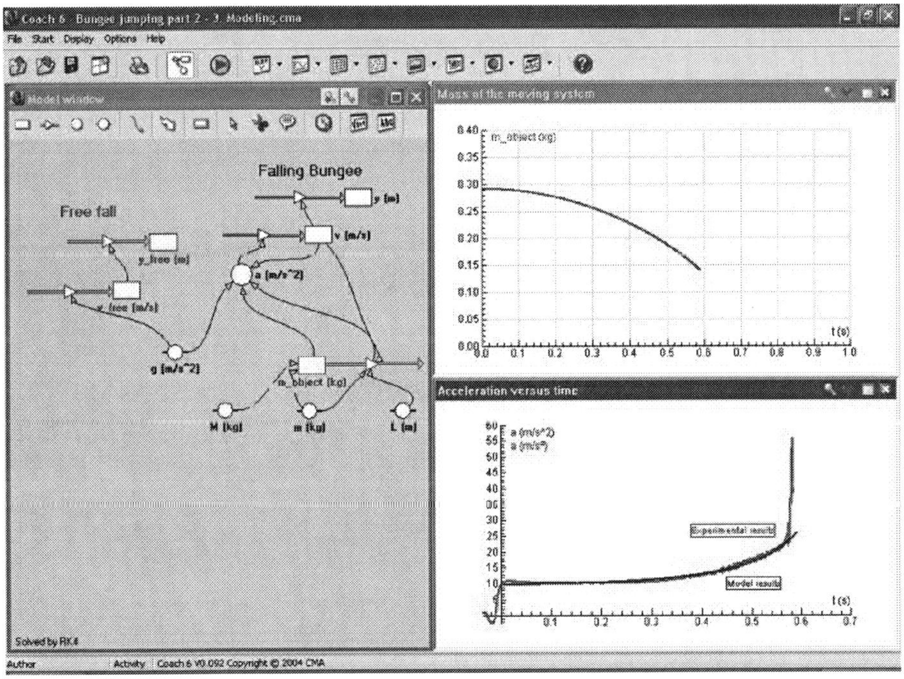

FIGURE 2. Coach 6 Modeling Activity. The experimental results are compared with the theoretical results.

Conclusions

The project of 2 high school students raised a discussion amongst professionals in Physics. This project nicely demonstrates ICT can empower students to enable advanced research activities. In fact the use of a computer has turned out to be very popular and effective during these projects. It facilitates them to perform projects that reach the level of authenticity and relevance.

In The Netherlands after 2 decades of stimulation of the integration of ICT, now students are well-prepared to use the computer for data-acquisition, data-processing, modelling and even measurements based on digital video they themselves have captured. During the physics lessons in grade 7 – 12 (age 13 – 18) they have used these techniques and become familiar. It has turned out to be very effective and important during this period the formal curriculum and recently even the mode of examinations has changed accordingly.

We can conclude, that after 2 decades of implementation activities, the use and integration of ICT in the teaching and learning process in physics education in The Netherlands has reached a rather good level. In our opinion crucial factors have been:

Consistent approach for many years

Concerted action: innovation, together with facilities, teacher training AND change of curriculum and examination

Integration of ICT is reflected in national examination programs, textbooks and in questions in national exams. Teachers are rather familiar through initial training and inservice training, and schools rather well equipped with computers with data-logging facilities (as an average 6 computers per physics department). The concluding step in this process is the actual testing of the students' ICT-abilities during the national examination by using a computer for answering the questions. Since 2002 pilot projects are performed with this respect.

REFERENCES

1. Ellermeijer, A.L., Mulder, C., *The approach to ICT in Science Education in Holland*, In: Barton R. (ed.) *Learning and Teaching Secondary Science with ICT*, Open University Press, England
2. Kagan, D.T. and A. Kott, 'The Greater-Than-g-Acceleration of a Bungee Jumper', *The Physics Teacher* **34**/6 (1996), p 368-373.
3. Dubbelaar, N. and R. Brantjes, Nederlands Tijdschrift voor Natuurkunde (NTvN) **69**/10 (2003), 316.
4. Pasveer, F. and W. de Muynck, NTvN **69**/12 (2003), 394
5. Mioduszewska E., Ellermeijer A.L., (2000). *Authoring environment for multimedia lessons*, In: Pinto R., Surimach S. (ed.) *Physics Teacher Education Beyond 2000*, proceedings of the Girep conference 2000, 689-690, Elsevier.

Master IDIFO for In-Service Teacher Training in Modern Physics

Marisa Michelini[a] and Lorenzo Santi[b]

[a]*Physics Department, University of Udine, via delle Scienze 208, 33100 Udine, Italy*
[b]*Interdepartmental Center for Education Research, via delle Scienze 208, 33100 Udine, Italy*

Abstract. Within the context of a national project aimed to promote actions against disaffection for scientific studies in Italy, a Master in Didactic Innovation in Physics and Orientation was designed, as a result of researches carried out in this field by PERG of 9 Universities of Italy and aimed at the in-service training of teachers on modern physics.

Keywords: Master, teacher training, modern physics.
PACS: *01.40.jh, 03.*

Italy's main response to the fall in motivation with regard to scientific studies has been the PLS Project (Progetto Lauree Scientifiche-Scientific Degree Project), promoted by the Science Faculties of Italian Universities and organised in three areas: mathematics, physics and chemistry. In the field of PLS, the Physics Education Research Groups (PERG) of the Universities of Italy worked together to design and implement a Master in Didactic Innovation in Physics and Orientation (IDIFO), aimed at the in-service training of teachers on the themes of modern physics, as a result of research carried out in this field. The Master project was presented by the University of Udine, as an initiative promoted and supported by the university centers in **FIGURE 1.**

Director
Michelini M., University of Udine
Scientific Committee (*) and UniversityRepresentatives (°):
*Bochicchio M. (LE), *Gagliardi M. e °Levrini O. (BO), *°Giliberti M. A. (MI), *°Guidoni P. (NA), *Giordano E. (MI-Bicocca), *°De Ambrosis A. (PV), *Michelini M. e °Santi L. (UD-TS), *Mineo Sperandeo R. M. e °Fazio C. (PA), *°Rinaudo G. (TO), *°Tarsitani C. (Roma1), *Ottaviani G. & Corni F. (MO-Re), °Stella R. (BA), °Corni Federico (BZ), °Oss S. (TN)
Advisory Board
Marisa Michelini, director of Master
Lorenzo Santi, Faculty of Science MMFFNN
Giancarlo Fava, SSIS of Udine
Gian Luigi Michelutti, Dept. of Physics, Univ. Udine
Carlo Tarsitani, PRIN-Fis21
Cristina Moschetta, course student.

FIGURE 1.. The 9 gray centers offer the Master IDIFO course, while those in white indicate the 5 universities cooperating with the project, the latter are also involved in PRIN research with the former. Project site http://www.fisica.uniud.it/URDF/laurea/index.htm. E-learning platform: http://idifo.fisica.uniud.it

The IDIFO Masters Degree was launched at the University of Udine, with a course length of two years from March 2006 to June 2008 with a total of 600 hours of didactic activity both on campus and by email correspondence. For study by correspondence a specific 'platform' was developed, in collaboration with the Research Unit in Artificial Intelligence by the University of Udine. The financing obtained was for the most part

CP1018, *Frontiers of Fundamental and Computational Physics,*
edited by B. G. Sidharth, F. Honsell, O. Mansutti, K. Sreenivasan, and A. De Angelis

earmarked for 20 scholarships. The selection for admission to the Master was based upon qualifications and exams which were carried out in a single session at 8 universities selected by the Advisory Board. Thirty-one of the 48 applicants were enrolled and 20 scholarships were awarded.

The training objectives of the Master are:

a) training of teachers on the themes of modern physics (i.e. quantum, relativist, statistical physics, and the physics of matter) highlighted through didactic research in the international context carried out by the 9 PERGs proposing the present project [1];

b) increasing the competence of teachers in the operative aspects of the learning/teaching strategies proposed on modern physics (experiments for schools laboratory on crucial interpretative problems of physics of the twentieth century, techniques in analysing the physics of matter, computer modelling, computer simulation of phenomena to interpret processes)

c) preparation of and experimentation with materials for training teachers and developing innovative activities for students (adopting 'blended' modality)

e) perfection of proposals for training orientation, based upon problem-solving for orientation (PSO).

Much time has been dedicated to the individual and group discussion of teaching/learning proposals, the analysis and assessment of questions brought to light by research into science education. Each university offering the course has made a contribution of specific teaching proposals which constitutes a reference for those in the area, for whom they are responsible. The online tutor (Patrizia Magnoler) will perform the role of facilitator, while each lecturer will act as a subject tutor.

Attendance is obligatory for at least 70 hours of the time allotted for training activities of the Master. Assessment will be made on the base of 4 assignments from the teaching Modules (A-Quantum mechanics; B: Relativity; C: physics of matter; D: particle physics and astrophysics; E: formative guidance), and the final thesis (6 cfu) will consist of a written analysis of the experiments carried out with secondary school students. Each of the 4 assignments must contain a didactic experiment activity with other training teachers, following the Module themes. The thesis will be discussed before a Commission appointed by the Master Advisory Board. Completion of the final paper must occur within the third year.

By July 2007 a significant part (70%) of the teaching had terminated and results had been evaluated. Around twenty students had already carried out at least some of the experimental project work. The workload required of the students is very high: much higher than for a standard Master degree. However, the course students have proven to be of a high cultural and professional standard, determined to become competent professionals in their fields. Some were not able to sustain the workload alongside scholastic studies and had to give up(3),others requested extensions and reassessment(6). A Corso di Perfezionamento (Advanced Course) was established, which partially finances the Master to offer those in difficulty the chance of completing a shorter training programme and we have received several requests for this some time ago

REFERENCES

1. Michelini M, *Quality Development in the Teacher Education and Training*, Girep book of selected papers, RT1, Forum, Udine, 2004, p. 127-132 [ISBN: 88-8420-225-6]

The first Italian doctorate (PhD Course) in Physics Education Research

Marisa Michelini[a] and Lorenzo Santi[a]

[a]*Physics Department, University of Udine, via delle Scienze 208, 33100 Udine, Italy*

Abstract. The first PhD Italian course in Physics Education Research in Udine aims to qualify young researchers and teachers coming from all the Italian groups of research in the field. It becomes a context for developing research projects carried out following parallel research lines on: Teaching/Learning paths for didactic innovation, cognitive research, ICT for strategies to overcome conceptual knots in physics; E-learning for personalization; d) Computer on-line experiments and modelling; e) Teacher formation and training; f) Informal learning in science.

Keywords: PhD Course; Physics Education Research.
PACS: 01.40.Di; 01.40.Fk

Italian Universities are not very motivated in education research, but researches in physics and mathematics didactic[1] are carried out at international level since 1974[2] in official financed research projects[3]. There is a great need of research in this field both for teaching/learning paths able to overcome conceptual knots and realize didactic innovation for teachers' formation[4]. Finding qualified human resources is one of the main problems: for many years the little number of young researchers in physics education obtain the doctorate in foreign countries. In academic year 2002-03 the Physics Education Research (PER) line was founded in Udine University in the frame of Physics and Mathematics Doctorate (PMD), adopting procedures, criteria for selecting students, conducting courses and final examinations rules of the existent PMD. This first PhD course in Physics Education in Italy is considered a common resource for Italian groups of research in the field and it becomes a particular fertile and power context for the discussions between our well prepared young co-operators, with different research experience and for developing cooperation among different research groups (Lecce, Modena, Napoli, Padova, Palermo, Trieste and Udine). Three thesis defenses are already done, with international jury. One of the PhD thesis has been prepared in "co-tutela" with Paris VII and now we are involved in similar joint thesis with Universities of Ostrava and Yahunde.

The research project, which presently involves nine PhD students, concerns the following topics: a) Teaching/Learning paths for didactic innovation, for example on

---

[1] We do not distinguish here between researches in physics and mathematics education and in didactic of phys and math, but we consider the last formulation more appropriate with our work of research for teaching/learning the subject.
[2] Researches start many years before, but an official recognition occurs in this years.
[3] Researches selected and supported mainly by Italian Center of National Researches (CNR) and Ministerium (Projects of National Relevance (PRIN)). European Researches Projects are carried out only in the last 8 years.
[4] After 1994 law for university teacher training, it starts in Italian universities in 2000 for the initial teacher training of primary (Faculty of Education) and secondary school (Specialization School).

CP1018, *Frontiers of Fundamental and Computational Physics*,
edited by B. G. Sidharth, F. Honsell, O. Mansutti, K. Sreenivasan, and A. De Angelis

quantum physics in secondary school; b) ICT as strategy to overcome conceptual knots in physics at different levels; c) E-learning for personalization; d) Computer on-line experiments and modelling to create deep path in exploring and interpreting phenomena; e) Teacher formation and training; f) Informal learning in scientific field; h) learning processes and paths.

Developing formal thinking and role of conceptual and operative personal involvement in exploration of phenomena are basic common problems in the different research approaches. Following Millar [1], we think that the role of research does not consist only in discovering what works. Some of the most valuable research studies have been those that made people aware of problems in current practices. Research can inform practice in a range of ways that stop short of providing clear and definite answer: by providing the kind of insights that enable us to see the familiar in a new way, by sharpening thinking, directing attention to important issues, clarifying problems, challenging established views, encouraging debate, stimulating curiosity [2]. For this reason each PhD research includes empirical studies on learning path, with particular attention to the mental models and their evolution. In addition each research project involves more than one research line listed.

The empirical research in the theoretical frame of Model of Educational Reconstruction (MER) [3] implies different phases in the research process: A) Analysis of content structure (science subject matter) includes two processes which are closely linked, clarification of subject matter and the analysis of educational significance; B) Research on teaching and learning comprises empirical studies on various features of the particular learning setting, research on students' perspectives including their pre-instructional conceptions, role of instructional methods, experiments and other instructional tools are also available; C) Development and evaluation of instruction concerns the design of instructional materials, learning activities, and teaching and learning sequences. In the following, an overview on the nine topics of the researches carried out: 1) Developing theoretical thinking through Quantum Mechanics in Secondary School and pre service teacher training; 2) An interpretative model to evaluate the thermal conductivity of the filament in a bulb lamp; 3) Learning problems related to the concept of field; 4) Light polarisation properties explored by computer on-line sensors in an informal learning environment; 5) Children's naive ideas/reasoning about some logic circuits explored in an informal learning environment; 6) Perspective primary teachers' functional models of electric and logic circuits; 7) Role play as a strategy to discuss spontaneous interpreting models of electric properties of matter: an informal education model; 8) Perspective primary teachers and electromagnetic phenomena: a teaching/learning research; 9) Educational ICT tools and modeling activity to improve wave physics understanding

REFERENCES

1. Millar R., "What can we reasonably expect of res. in sci. educ.?", in *Science Education Research in the knowledge based society*, edited by Psillos D. et al., Dordrecht: Kluwer 2003, pp.3-8.
2. Shulman L.S., *Harvard Educational Review*, 57(1), (1987). 1-21.
3. Duit R., *Science Education Research – An Indispensable Prerequisite for Improving Instructional Practice*, ESERA Summer School, Braga 2006.

Secondary school teachers discussing the pedagogical and cultural aspects in teaching-learning quantum physics

Marisa Michelini[a] and Alberto Stefanel[a]

[a]*Research Unit in Physics Education, University of Udine, via delle Scienze 208, 33100 Udine*

Abstract. The instructional path and its tutorials of a teaching-learning proposal for quantum mechanics based on a Dirac approach and focused on building the theoretical thinking were used in a Master blended module for in-service teacher training. The proposal was discussed in a web laboratory and in a workshop proposed in-presence, warranting the personal involvement of forming teachers in concept analysis. We document enhancement of the competencies in order to discuss the crucial point of quantum theory and design instructional path centered on them.

Keywords: Quantum Mechanics, Teacher training.
PACS: 01.40.ek; 01.40.jh; 03.65.-w

Quantum Mechanics (QM) constitutes the cultural reference paradigm for the description of the microscopic world, therefore its methodological elements can contribute to the construction of theoretical thinking and the formation of all citizens.

In literature, there is no common point of view for educational treatment of QM; different formulations and interpretations of QM has been used as starting points for different proposals. Three main approaches are followed: historical; construction of the formalism using analogy with classical situations; conceptual [1,2].

The 3rd one follows the vector approach, proposed by Dirac and adopted by many other authors. This approach is the reference for proposals in which the conceptual aspects of the theory on which one constructs the formalism and recognizes its role in QM are pointed out. Following this conceptual approach we developed a teaching-learning proposal for QM focused on building the theoretical thinking. It is based on simple experiments with polaroids and birefringent crystals in order to discuss the superposition principle and the quantum state concept [3,4]. Our proposal feasibility was tested in different secondary schools [5], and in pre-service secondary school teachers training [6].

The developed teaching-learning path and their tutorials were used in the Master IDIFO, coordinated by the University of Udine and realized with the cooperation of 15 universities for in-service teacher training on relativity and quantum physics. The proposal was discussed in a web distance laboratory and in a in-presence workshop. The formative module on Dicac approach to QM included: a) a on-line course on basic aspects of the theory (3 fc); b) an on-line course focused on the discussion of our approach and its tutorials (3 fc); c) an on-line didactical laboratory, in which the teachers produce a proposal based on the Udine-Dirac approach (1 fc); d) an in-

CP1018, *Frontiers of Fundamental and Computational Physics*,
edited by B. G. Sidharth, F. Honsell, O. Mansutti, K. Sreenivasan, and A. De Angelis

presence discussion about the knots of the proposal simulating an interaction with students (September 2006 -1/2 fc); e) an in presence discussion addressed to conceptual and pedagogical main topics of QM (July 2007 - 1/2 fc).

In the following we consider the items of the final question and we present the corresponding answers given by the 24 in-service teachers attending the IDIFO-Master at the beginning of the module.

1. Elements characterizing/identifying the quantum behaviour. Pre (Pre-Test answers): wave-particle dualism; Heisenberg relations; black body radiation; Frank-Hertz/photoelectric effects. Post (final Questionnaire): Linearity and superposition; Uncertainty and intrinsic probabilistic behaviour of measurement processes; Incompatibility and Uncertainty principle.

2. Possibility to make previsions on measurement results. Pre: Only in a probabilistic way [without any other explanation]. Post: Discussion using simple examples of the quantum probabilistic prediction, according to the superposition principle, stressing the role of the scalar product; sure results preparing the system in an eigenstate; correlation of results with entangled systems (un-separability).

3. Knowing the property of a system implies the existence of this property? Pre (in the Web discussion): a quantum state is like a state of a macroscopic system with not well perceived properties; or: we can attribute a property to a system only after a measurement. Post:(All); Discussion on diverse levels, according to different interpretation (orthodox framework; alternative bohemian scheme).

4. Base concept that must be included in a learning/teaching proposal on QM. Pre: Black body; photoelectric effect; uncertainty relations. Post: conceptual aspects; superposition principle, state, incompatibility(all); entanglement; different meaning of the superposition principle in QM and in classical physics; intrinsic uncertainty; Applications of QM proposed only after conceptual aspects have been treated.

The documented changes show enhancement of the forming teachers competencies on nodal concepts of QM. The crucial aspect of change was the personal involvement in the analysis of the concept during the module. At the educational level, teacers evidenced a radical change: in the pre-test the teachers consider prevalently the old physics of quanta as topics for an educational proposal; at the end of the formative path all of them propose to analyze the conceptual aspects that found the theory with students. Their attitude changed because they experienced the feasability of the experiments, of the conceptual path, of the simple mathematical adopted formalism.

REFERENCES

1. Phys. Educ. 35, 6 (2000)
2. Am. J. Phys. 70, 3 (2002)
2. G. C. Ghirardi, R. Grassi and M. Michelini, "A Fundamental Concept in Quantum Theory: The Superposition Principle", in *Thinking Phys. for Teach.*, edited by C. Bernardini et al., Aster: Plenum P. C., 1995, pp. 329-334.
3. M. Michelini, R. Ragazzon, L. Santi and A. Stefanel, *Phys. Educ.* 35 (6) 406-410 (2000).
5. M. Michelini, R. Ragazzon, L. Santi and A. Stefanel, "Implementing a formative module on quantum physics for pre-service teacher training", in *Quality Development in the Teacher Education and Training*, edited by M. Michelini, Udine: Forum, 2004, pp. 429-435.
6. M. Michelini, R. Ragazzon, L. Santi and A. Stefanel, *Il Nuovo Cimento* 27 C (5) 555-567 (2004).

A round table as tool for getting in touch Science and School

Cristina Disint[a], Marisa Michelini[a,b], Alessandra Mossenta[b], Lorenzo Santi[b], Alberto Stefanel[b], Rossana Viola[b], Fulvia Vogric[a]

[a]*Center for Guidance and Tutoring (CORT), University of Udine, Via Palladio, 2 - 33100 Udine*
[b]*Research Unit in Physics Education, University of Udine, via delle Scienze 208, 33100 Udine.*

Abstract. The FFP9 symposium and the materials produced within the European project SUPERCOMET 2, for the development of teaching proposals on superconductivity in the secondary school, are linked to give an opportunity for formative guidance of secondary schools students in the scientific field. A path was organized to prepare students to meet scientists who won important prizes, like the Nobel one, during a Round Table. The activated process constitutes an effective action for guidance on scientific curricola.

Keywords: Guidance, science, superconductivity.
PACS: 01.40.ek; 74.20.-z; 74.25.-q

The OECD's PISA surveys have evidenced the problem of scientific illiteracy of the citizens, enhancing the problem of reducing enrollment in scientific study [1-3]. An emergency in motivating young people in science has therefore been declared, leading to the promotion of initiatives in all the World [4]. International researches in physics education show that high levels of interest arise when there is a personal involvement of students and when cultural challenges make them protagonists in the process of solving interesting problems [5,6]. A significant example is the 1st National Summer School organized by Udine University in July 2007 on topics of quantum physics and relativity. The number of applying students was 9 times bigger than the number of available positions. The selected students were intensively involved at "high level" in problem solving activities during 10 hours per day in a week [7]. The topics related to the last century physics are of explicit interest for students. If we aim at avoiding the fact that physics is perceived as a not-usable discipline and far from the real world and its evolution, we can not ignore this interest.

On this idea, an European project for the development of teaching-learning proposals on superconductivity in the secondary school (SUPERCOMET 2 project - SC2 hereafter), has been recently concluded. More than 15 countries participated to the project [8,9]. We carried out it in Italy, contributing with a curricular research approach in vertical perspective, building on phenomena and experimental activities the conceptual bases of electromagnetism and superconductivity [10-13].

The SC2 project linked with the FFP9 event gave the opportunity for the University of Udine, founded for popular will, to accomplish one of its most important mission: develop constant and continuous relationship with their surrounding areas, in particular by means of a strong cooperation with schools. Therefore the FFP9

CP1018, *Frontiers of Fundamental and Computational Physics*,
edited by B. G. Sidharth, F. Honsell, O. Mansutti, K. Sreenivasan, and A. De Angelis

symposium and the SC2 materials are looked as opportunities for formative guidance in the scientific field: we organized a path to prepare the FFP9 Round Table (RT), where secondary students and eminent scientist (as Nobel prize winners A. Leggett and D. Osheroff) meet together, discussing with secondary teachers and students some relevant aspects of the formative guidance for scientific curricula, as well as superconductivity topics.

The human resources involved in this process were: physics education researchers and PhD students, experts of the Center for Guidance and Tutoring (CORT) of the University of Udine, teachers of four selected schools of three city of the Friuli Venezia Giulia region and their students: 70 of 3 classes from two scientific Secondary Schools of Udine, 70 of 3 classes from a scientific Secondary School of Pordenone, 15 of 1 class from a Technological School of Cervignano.

Blended a***ctivities was offered to schools***, in presence and in a web collaborative environment, at two levels. At level I, researchers of the Physics Education Research Group (PERG), teachers, CORT experts individuated strategies and materials for students. At level II, teachers and their students are involved in discussions activities on: Issues for scientific guidance (1. science for citizen and as culture for everyone. 2. Scientists! Born or formed? 3. How could we study science to become scientists? 4. What is the contribution of research to education? 5. how the study of scientific disciplines contributes to the knowledge of the natures of those disciplines?); Questions about Superconductivity (after reading an introductory simple paper - 1. Phenomenology characterizing the superconductivity state. 2. How superconductivity phenomena modify the classical electrical current model? 3. Ideas of classical physics radically changed by the phenomena of superconductivity. 4. Interpretations of superconductivity).

The steps of the path to prepare the FFP9 Round Table are below presented:

A) Collection of more than 30 popularization papers on superconductivity and SC2 materials

B) Papers analysis by PERG researchers and recognition of their inadequacy to offer a clear idea of the processes involved in the phenomenology everywhere described.

C) Selection of a paper prepared for physics education (SC2).

D) Three meetings of the CORT-PERG group with teachers to individuate student needs and strategy for the discussion process.

E) Organization of a protocol work-activity including the following operative phases: E1. Project presentation and student discussion (about: scientific guidance aspects; superconductivity issues); E2 Collection of questions emerged during the in-class activity (5-15/12/07); E3 Web-forum discussions on: guidance to science, superconductivity (5-15/12/07).

F) Lectures (in Udine and Pordenone): “Superconductivity: experimental results and models” (17-18/12/07).

G) Collection by teachers of the questions proposed by students.

H) Data analysis by PERG of student web discussions and questions.

I) Grouping student questions in clusters, according with relevance and subject criteria: Guidance, Scientific culture of citizens, Superconductivity (Informative topics; Applicative topics; Knowledge topics) (18/12/07-06/01/08).

J) Round Table between secondary school student and scientists (09/01/08).

The students contributions were divided in the following way: 29 questions about general topics (13 on "Guidance"; 11 on "science for Citizens"; 3 on scientist life curiosity); 16 questions about Superconductivity topics (11 knowledge questions; 3 informative questions; 2 questions about SC applications). 2/3 of the questions were posed by single student, 1/3 were elaborations of groups of students. Here we report only some examples chosen between the more frequent questions posed by students on different topics. ***General questions:*** Can science be useful in everyday life? Are all the behaviors of the world knowable or is the man not able to know it? Qualities to become a researcher (curiosity; aptitude towards science studies). Contribution of science, physics, for the everyday life of the citizens. Future limit of science knowledge. Is the scientific knowledge so specific that citizen can not create their scientific culture by themselves?

Superconductivity questions: Application of superconductivity in everyday life, industry. Peculiar characteristics of I and II types of superconductors. Theoretical interpretation of superconductivity (role of polarization of lattice, role of phonons; role of impurity in high-T superconductors). Could superconductivity be used to solve the environmental problem and the abuse of energy? Will a unified theory comprehensive of all superconductors types be possible?

The reported students questions document the richness of the contributions, giving an indication of the guidance value of the activated process.

REFERENCES

1. PISA - Knowledge and Skills for Life. First Results from the OECD Programme for International Student Assessment (PISA) (2000) Ed.: OECD: Paris; (2001), at http://www.pisa.oecd.org/knowledge/home/intro.htm.
2. M. Euler, "Imagery and formal thinking: approaches to insight and understanding in physics education, in *Developing Formal Thinking"*, edited by M. Michelini & M. Cobal, Udine: Girep, Forum, 2001, pp.13-26.
3. M. Euler, "The role of experiments in the teaching and learning of physics". in *Research on Physics Education*, edited by E. F. Redish & M. Vicentini, Amsterdam: IOS, 2004, pp.175-221.
4. http://www.wyp2005.org/
5. R. Duit, Bibliography – STCSE, Students' and Teachers' Conceptions and Science Education, at http://www.ipn.uni-kiel.de/aktuell/stcse/stcse.html
6. M. Michelini ed., *Quality Development in the Teacher Education and Training*, Udine: Girep, Forum, 2004.
7. http://agenda.fisica.uniud.it/difa/conferenceDisplay.py?confId=4
8. http://supercomet.no/gb/supercomet_2
9. V. Engstrøm, M. Michelini,F. Corni,G. Karwasz, in *Modelling in Physics and Physics Education, edited by T Ellermajer,* Amsterdam: Girep -AMSTEL Institute, 2006, *pp.40*
10. F. *Bradamante, M. Michelini,* Field Lines Representation and Maxwell's Tubes of Flux to Build the Model of Field, in Selected Contrib. of *Modelling in Physics and Physics Education, edited by T. Ellermajer, E. van den Berg, Amsterdam: Girep,Amstel, 2007, in press.*
11. J. Guisasola, M. Michelini, A. Mossenta, A. Testa, I. Testa and R. Viola, Teaching Electromagnetism: Issues And Changes, in Selected Contrib. of Girep Conference : Frontiers of Physics Education, edited by R. Jurdana-Sepic, M. Zuvic-Butorac eds., Rijeka: Girep, Rijeka University, 2007
12. R. Viola and M. Michelini, Prospective Primary Teachers And Electromagnetic Phenomena: A Teaching/Learning Research, in Selected Contrib. *Girep Conference: Frontiers of Physics Education*, edited by R. Jurdana-Sepic, M. Zuvic-Butorac, Rijeka: Girep, Rijeka University, 2007.
13. R. Viola, L. Santi. M. Michelini, F. Corni, The secondary school experimentation of supercomet materials in Italy, in Selected Contrib. *Girep Conference: Frontiers of Physics Education*, edited by R. Jurdana-Sepic, M. Zuvic-Butorac, Rijeka: Girep, Rijeka University, 2007.

AUTHOR INDEX

A

B

C

D

E

F

G

H

I

J

K